Electrical Safety and the Law

Electrical Safety and the Law describes the hazards and risks from the use of electricity, explaining with the help of case studies and accident statistics the types of accidents that occur and how they can be prevented by the use of safe installations, equipment and working practices. It describes British legislation on the safety of electrical systems and electrotechnical machinery control systems, much of which stems from European Directives and which will therefore be affected by the UK's decision to leave the EU (Brexit), and the main standards and guidance that can be used to secure compliance with the law. There are detailed descriptions covering the risks and preventive measures associated with electrical installations, construction sites, work near underground cables and overhead power lines, electrical equipment and installations in explosive atmospheres, electrical testing and electrotechnical control systems. Duty holders' responsibilities for designing, installing, and maintaining safe systems are explained, as well as their responsibilities for employing competent staff.

The fifth edition has been substantially updated to take account of considerable changes to the law, standards and guidance; it has been expanded to include:

- a new chapter on the Corporate Manslaughter and Corporate Homicide Act;
- a new chapter describing landlords' legal responsibilities for electrical safety in private rented properties and social housing;
- a new chapter on the Electricity Safety Quality and Continuity Regulations;
- new information on offences, penalties, sentencing guidelines, and relevant case law;
- a description of the main requirements of BS 7671:2008 and other principal standards, many of which have been amended in recent years;
- new cases studies to illustrate the hazards and risks;
- information on changes to GB's health and safety system.

John M. Madden CEng FIET is a former HM Principal Specialist Inspector (Electrical Engineering) with the Health & Safety Executive and now an electrical safety consultant. He has substantial experience of accident investigation and acting as an expert witness in criminal trials, civil proceedings, Fatal Accident Inquiries, and inquests.

Electrical Safety and the Law

John M. Madden

LONDON AND NEW YORK

Fifth edition published 2017
by Routledge
2 Park Square, Milton Park, Abingdon, Oxon, OX14 4RN

and by Routledge
711 Third Avenue, New York, NY 10017

Routledge is an imprint of the Taylor & Francis Group, an informa business

First edition published by BSP Professional 1990

Fifth edition published by Routledge 2017

British Library Cataloguing-in-Publication Data
A catalogue record for this book is available from the British Library

Library of Congress Cataloging-in-Publication Data
Names: Madden, John M., author. | Oldham Smith, K. (Ken). Electrical safety and the law.
Title: Electrical safety and the law / John M. Madden.
Description: Abingdon, Oxon ; New York, NY : Routledge, 2017.
Identifiers: LCCN 2016035200 | ISBN 9781138670495 (hardback) | ISBN 9781138670501 (pbk.) | ISBN 9781315617626 (ebook)
Subjects: LCSH: Electric utilities—Safety regulations—Great Britain. | Electrical engineering—Safety regulations—Great Britain. | Electric lines—Safety regulations—Great Britain.
Classification: LCC KD2545 .O43 2017 | DDC 343.4109/29—dc23
LC record available at https://lccn.loc.gov/2016035200

ISBN: 978-1-138-67049-5 (hbk)
ISBN: 978-1-138-67050-1 (pbk)
ISBN: 978-1-315-61762-6 (ebk)

Typeset in Bembo
by Apex CoVantage, LLC

For my three wonderful grandsons, Bruce, Ewan and Nicolás, the highly energetic live wires in my life. Also for my wife, Linda, in thanks for her continuing patience and support, and my daughters Lorna and Kirsty.

Contents

Figures

Tables

Preface

It is quite remarkable how much has changed in the fourteen years since the fourth edition of this book was published. In the engineering field, the closure of most of our fossil-fuelled power stations and the move towards 'renewable' sources of energy; the introduction of 'smart' technologies into distribution networks to cater for increasingly distributed and intermittent forms of generation; the increasing use of programmable and networked systems in safety-related control and electrical protection applications, coupled with the ever-increasing and concerning cybersecurity threats; the rise of the robot; and the exponential uptake in the use of digital devices such as smartphones and tablet computers are just some examples of the pace and scope of change.

Although the Health and Safety at Work Act has stood the test of both time and innumerable government reviews, there have been significant changes in the law governing health and safety at work as well as important developments in case law in which the courts have clarified matters such as foreseeability. Perhaps the most noteworthy legislative change has been the introduction of the Corporate Manslaughter and Corporate Homicide Act, aimed at ensuring that businesses whose gross negligence in discharging their duty of care causes work-related deaths are properly held to account. Another controversial development has been the exemption of hundreds of thousands of self-employed people from the scope of health and safety legislation, thereby reducing their burden of compliance. In addition, the European Union has remained active in producing directives and regulations with health and safety requirements that have been transposed into UK law, but the UK's decision on 23rd June 2016 to leave the EU (so-called 'Brexit') is likely to have a profound effect on the future of this legislation.

The regulatory bodies have not escaped unscathed the changes in the political and economic landscape. The Health and Safety Executive has seen its budget tightly squeezed and has been forced to introduce a fees-for-service policy that has changed its relationship with those it regulates. It has also pulled back from traditional preventive inspections of all but higher risk premises, and relaxed its incident investigation criteria. However, it will have to work hard to effect changes brought on by the momentous Brexit decision. Local authorities are providing fewer resources to the enforcement of health and safety in those premises they regulate, such as shops, offices, cafes and restaurants.

There have been wholesale changes in the published standards and guidance addressing the safety of electrical and control systems. The headline change was the publication in 2008 of a new edition of BS 7671, but there have been many changes to other standards and the introduction of new ones. The HSE has not been idle with its guidance material, being required to make considerable changes as a result of major reviews carried out by

Lord Young and Professor Löfstedt; yet more changes will be required in response to the Brexit decision.

Societal changes have led to a dramatic increase in the number of households occupying rented accommodation rather than being owner-occupiers. This has focussed attention on the quality and safety of properties in the rented sector, not least the safety of the electrical installations.

The scale of the changes means that it is timely to produce the fifth edition of this book to explain them and to introduce new chapters on topics such as the corporate manslaughter and homicide legislation and electrical safety in the domestic rental sector.

The first three editions of the book were authored by Ken Oldham Smith, with the first edition published back in 1990. It is my sad duty to report that Ken recently passed away. His book made a significant contribution to the improvements seen in electrical safety and I am proud to have the opportunity to continue his work with the publication of this 5th edition.

John M. Madden
June 2016

Acknowledgements

This book contains public sector information licensed under the Open Government Licence v3.0. The text of the Electricity at Work Regulations 1989 reproduced in Chapters 6 and 19, Figure 2.15 in Chapter 2 and Figure 14.4 in Chapter 14 are subject to Crown copyright and are reproduced in accordance with the Open Government Licence v3.0 as described on The National Archives website www.nationalarchives.gov.uk.

The book makes extensive use of information contained in British Standards, although no content or extracts of British Standards are directly reproduced. Copyright in British Standards is owned by BSI. Copies of British Standards may be purchased through the BSI website www.shop.bsigroup.com.

My grateful thanks go to Peter Blakley of Blakley Electrics Ltd and Chris Ross of J & K Ross Ltd for permission to print images of their products. Also, my grateful thanks go to Gordon Mackenzie and Mushtaq Gulnar of Scottish Power for their help with photographs of electrical equipment, and to Bill Cuthbert of Scottish Power for his very welcome support and input at the proposal stage for this 5th edition.

Abbreviations

ACB	Air Circuit Breaker
ACOP	Approved Code of Practice
AFDD	Arc Fault Detection Device
ALARP	As Low As Reasonably Practicable
AOPD	Active Optoelectronic Protective Device
ASTA	Association of Short Circuit Testing Authorities
BASEC	British Approvals Service for Electrical Cables
BASEEFA	British Approvals Service for Electrical Equipment in Flammable Atmospheres
BIS	(Department of) Business Innovation and Skills
BREXIT	Britain's Exit (from the EU)
BSEN	British Standard Euronorm
BSI	British Standards Institution
CASS	Conformity Assessment of Safety-related Systems
CAT	Cable Avoidance Tool
CB	Circuit Breaker
CBM	Condition-Based Maintenance
CDM	Construction (Design and Management)
CEGB	Central Electricity Generating Board
CEN	European Committee for Standardisation
CENELEC	European Committee for Electrotechnical Standardisation
CI	Customer Interruptions
CIBSE	Chartered Institution of Building Services Engineers
CITB	Construction Industry Training Board
CML	Customer Minutes Lost
CNE	Combined Neutral and Earth
CPC	Circuit Protective Conductor
CSCS	Construction Skills Certification Scheme
CTE	Centre-Tapped-to-Earth
DECC	Department of Energy and Climate Change
DNO	Distribution Network Operator
DSEAR	Dangerous Substances and Explosive Atmospheres Regulations
DTI	Department of Trade and Industry
EAWR	Electricity at Work Regulations
EC	European Community

ECA	Electrical Contractors' Association
EEBADS	Earthed Equipotential Bonding and Automatic Disconnection of Supply
EECS	Electrical Equipment Certification Service
EEMUA	Electrical Equipment Manufacturers and Users Association
EHV	Extra-High Voltage
EICR	Electrical Installation Condition Report
ELV	Extra-Low Voltage
EMC	Electromagnetic Compatibility
EMF	Electro-Motive Force
EN	Euronorm
ENA	Energy Networks Association
EOA	Extension Outlet Assembly
EPL	Equipment Protection Level
EPS	Equipment for Use in Potentially Explosive Atmospheres Regulations
ESC	Electrotechnical Certification Scheme
ESPE	Electrosensitive Protective Equipment
ESQCR	Electricity Safety Quality and Continuity Regulations
ETSI	European Telecommunications Standards Institute
EU	European Union
FELV	Functional Extra-Low Voltage
FFI	Fees For Intervention
HBC	High Breaking Capacity
HHSRS	Housing Health and Safety Rating System
HMO	Houses in Multiple Occupation
HSE	Health and Safety Executive
HSWA	Health and Safety at Work etc Act
HV	High Voltage
IDNO	Independent Distribution Network Operator
IEC	International Electrotechnical Commission
IEE	Institution of Electrical Engineers
IET	Institution of Engineering and Technology
IP	Ingress Protection
ISA	Incoming Supply Assembly
ISDA	Intake Supply and Distribution Assembly
LEL	Lower Explosive Limit
LS	Limit Switch
LSF	Low Smoke and Fume
LSHF	Low Smoke Halogen Free
LSZH	Low Smoke Zero Halogen
LV	Low Voltage
MAG	Metal Active Gas
MCB	Miniature Circuit Breaker
MCCB	Moulded Case Circuit Breaker
MDA	Main Distribution Assembly
MHRA	Medicines and Healthcare products Regulatory Agency
MIG	Metal Inert Gas
MIMS	Mineral-Insulated, Metal-Sheathed

MVA	Million Volt Amperes
NAPIT	National Association of Professional Inspectors and Testers
NFPA	National Fire Protection Association
NICEIC	National Inspection Council for Electrical Installation Contractors
NJUG	National Joint Utilities Group
NLF	New Legislative Framework
OCB	Oil Filled Circuit Breaker
OFCOM	UK communications regulator
OFGEM	Office of the Gas and Electricity Markets
PE	Protective Earthing
PELV	Protective Extra-Low Voltage
PL	Performance Level
PLC	Programmable Logic Controller
PME	Protective Multiple Earthing
PPE	Personal Protective Equipment
PRHP	Private Rented Housing Panel
PUWER	Provision and Use of Work Equipment Regulations
RCBO	Residual Current and Overcurrent Circuit Breaker
RCCB	Residual Current Circuit Breaker
RCD	Residual Current Device
RCM	Reliability Centred Maintenance
REC	Regional Electricity Company
RIDDOR	Reporting of Injuries Diseases and Dangerous Occurrences Regulations
RMS	Root Mean Square
RMU	Ring Main Unit
SCADA	Supervisory Control and Data Acquisition
SELECT	Electrical Contractors' Association of Scotland
SELV	Safety Extra-Low Voltage
SF_6	Sulphur HexaFlouride
SI	Statutory Instrument
SIL	Safety Integrity Level
SNE	Separate Neutral and Earth
SOA	Socket-Outlet Assembly
TA	Transformer Assembly
TO	Transmission Operator
TIG	Tungsten Inert Gas
TRS	Tough Rubber Sheathed
UEL	Upper Explosive Limit
VIR	Vulcanised India Rubber (sometimes abbreviated VRI)

1

The hazards and risks from electricity

Introduction

Electricity is such an important part of our everyday lives that it is difficult to countenance how we would manage to live without it. However, despite its obvious benefits as a source of energy and power, electricity is also a hazard in so far as it is capable of causing injury, including fatal injury.

Electricity is just one of the many types of hazards that we experience in our daily lives, other examples are chemicals, noise, machinery, falling from height, stress, impact with moving vehicles, using a keyboard and so on. Although these hazards do not routinely cause us harm, they have the potential to do so. We therefore need to be concerned not just with the existence of a hazard but also the likelihood, or probability, of it causing personal injury or ill health. The probability of a particular hazard causing harm is the concept of risk, which relates the severity of the harm to the probability of it occurring. The purpose of the legislation and protective measures described in this book is to reduce the risk of injury or death from electricity to a level that we find acceptable, the legal term for which in the UK is 'so far as is reasonably practicable', a concept which is explained further in Chapter 4.

As a starting point in considering electrical risks, it is helpful to have a good understanding of the types of injury that electricity can cause. This chapter explains the injuries so that later discussions on the legislative framework and precautionary techniques can be put into context.

Electricity can cause the following types of injuries:

- electric shock;
- burns caused by electric current passing through the body;
- burns caused by electric arcing, commonly known as 'flashover' or 'arc flash';
- the effects of fire that has an electrical origin;
- the effects of an explosion that has an electrical source of ignition; and
- the physiological effects of electromagnetic radiation.

The chapter is not concerned with other types of injuries, such as crushing and shearing injuries, that may occur as a consequence of, for example, a machine operating aberrantly because of a fault in its electrical control system. However, the topic of the safety integrity of electrotechnical machinery control systems is considered in Chapter 17.

Although each of the injury mechanisms is considered separately, in reality they often occur together. For example, current passing through the body causes an electric shock, but current passing through tissues also causes the tissues to heat up, often to the point at which burn injuries occur. So shock and burn injuries are frequently, but by no means always, experienced together.

Electric shock

If electric current of sufficient magnitude and duration passes through the body, it may disrupt the nervous system, causing the painful sensation of electric shock. The effects of such a shock can range from being mildly unpleasant to being fatal, with death most commonly resulting from cardiac arrest or ventricular fibrillation of the heart, a condition in which the heart's ventricular muscles quiver in a way that prevents blood being pumped around the body. Incidents are most frequently caused by making simultaneous contact with a live conductor and a conductor that is earthed, although some incidents occur when contact is made with two conductors energised at different potentials, neither of which is at earth potential.

Shock injuries are almost always associated with alternating current (a.c.), while direct current (d.c.) electric shock injuries are rare. This is partly because a.c. systems dominate in the workplace and in the home, but also because the excitation effects of direct current on the nervous system are less severe than those of the equivalent magnitude of alternating current. For this reason, only the effects of current at the most common power frequencies of 50 Hz and 60 Hz will be considered in detail, where Hz or Hertz is the SI unit of frequency measured in cycles per second. Broadly, 50 Hz is the power frequency in Europe, Africa, Russia, China, Australia and New Zealand, whereas 60 Hz is used on the North and South American continents, and in countries such as South Korea, the Philippines, Taiwan and Saudi Arabia.

The seriousness of the physiological effects of current passing through the body is directly related to the current's magnitude and duration and to the path that the current takes through the body. The magnitude of the current is related to the voltage across the body (the touch voltage) and to the body's impedance according to Ohm's law, which simply states that the current, I amperes, flowing through an impedance of X Ω (ohms) is derived by dividing the voltage, E Volts, across the impedance by the value of that impedance such that $I = E/X$ amperes. Impedance, rather than pure resistance, must be used because the body contains capacitive reactance at power frequencies.

There has been some research carried out over the years to characterise the effects of current flowing through the body. Most of this has been carried out on animals and human cadavers, although some has been carried out on a very limited number of human volunteers. The small number of volunteers is, perhaps, quite understandable. Some general conclusions can be drawn from the published information, much of which is very usefully summarised in an International Electrotechnical Commission (IEC) standard published in the UK by the British Standards Institution (BSI) as British Standard DD IEC/TS 60479–1:2005, *Effects of current on human beings and livestock. General aspects.*

Body impedance

At power frequencies (50 Hz and 60 Hz), the body's impedance comprises resistance and capacitive reactance. The capacitance is concentrated in the skin, whilst the resistance is associated with both the skin and the internal path through the body; however, the largest contribution of resistance is the skin barrier. The instantaneous value of impedance for any individual depends on a wide range of factors, not least of which is the applied voltage. In general terms, the higher the voltage, the lower will be the total body impedance, with the asymptotic low value being in the order of 750 Ω for 50% of the population for touch voltages greater than 1000V and for a current path from hand to feet. The impedance falls as the voltage increases, because at higher voltages the skin barrier breaks down. At 230 V, which is the voltage at which most electric shock accidents occur, or 120 V in countries such as the United States and Canada, the impedance ranges between 1000 Ω and 2500 Ω for most of the population, again for a hand-to-feet path.

The path that the current takes through the body has a significant effect on the impedance. For example, the impedance for a hand-to-chest path will be in the order of 50% of the impedance for a hand-to-foot path. If the impedance is low, the current will be equivalently high for the same touch voltage.

Other factors affect the impedance. For example, the impedance will be lowered if a sharp object forming part of the circuit, such as a strand of wire, punctures the skin. It will also be lowered by increasing the area of contact, and by wetting the skin surface. Some of the experiments with volunteers used cylindrical metal electrodes, about 100 mm long by 80 mm diameter, held in each hand. It was found that wetting the hands with tap water lowered the impedance a little on the value measured when the hands were dry, but wetting with a 3% salt solution halved the impedance value. Other factors such as age, gender and state of health also affect the impedance.

Effects

The physiological effects of current passing through the body from hand to feet are summarised in Table 1.1, derived and adapted from information in IEC/TS 60479–1. The figures are a rough yardstick and apply to an average person in good health and for a sustained shock exceeding a duration in the order of 1 second. For a shock duration less than

Table 1.1 The effect of passing alternating current (50/60 Hz) through the body from hand to feet

Current in milliAmps (mA)	*Physiological effect*
0.5–2	Threshold of perception
2–10	Painful sensation, increasing with current. Muscular contraction may occur, leading to being 'thrown off'.
10–25	Threshold of 'let go', meaning that gripped electrodes cannot be released. Cramplike muscular contractions. May have difficulty breathing leading to danger of asphyxiation from respiratory muscular contraction.
25–80	Severe muscular contraction, sometimes severe enough to cause bone dislocation and fracture. Increased likelihood of respiratory failure. Increased blood pressure. Increasing likelihood of ventricular fibrillation (uncoordinated contractions of the heart muscles so that it ceases to pump effectively) as current exceeds about 40 mA. Possible cardiac arrest.
Over 80	Burns at point of contact and in internal tissues. Increasing probability of death from ventricular fibrillation, cardiac arrest or other consequential injuries.

1 second, greater currents can be tolerated without such adverse reactions, particularly concerning the probability of ventricular fibrillation.

The most common electric shock effect when a conductor energised at mains voltages is touched is for the consequential muscular contraction to result in the individual being 'thrown off' the conductor. Whereas a painful lesson may have been learned, there are rarely any significant physiological effects unless the physical reaction leads to injuries such as impact injuries or a fall from a height should, for example, the individual be working on a ladder at the instant the shock is experienced. However, if the individual were to be gripping a conductor when it is energised, and if the current flowing though the hand/arm exceeds the threshold of let-go of about 10 mA, rather than being thrown off the conductor, it is possible for the hand to grip onto the conductor and for the muscular reaction to prevent release. The current may increase in magnitude as the palm of the hand becomes moist from sweat. If the current flows across the heart muscle and exceeds about 40 mA, there is the potential for ventricular fibrillation to occur.

The heart is vulnerable to ventricular fibrillation when shock current flows during the first part of the 'T' wave in the cardiac cycle, which is approximately 10 to 20% of the cycle. The shock stimulus produces extra systoles (heart beats), and if it embraces the vulnerable period, and particularly if it persists over several cardiac cycles, the risk of ventricular fibrillation is increased. This explains why the probability of ventricular fibrillation is greater for longer durations of current flow. For example, it has been found from experiments with dogs, and allowing for the difference in weight, that humans should be able to tolerate 650 mA for 10 to 80 milliseconds and 500 mA for 80 to 100 milliseconds without much risk of ventricular fibrillation. However for longer durations, above 1 second, these current magnitudes cannot be tolerated, and the effects tend to be as described in Table 1.1.

It is this type of experimental data, together with data derived from accidents, that explains why residual current devices (RCDs) with a residual operating current ($I_{\Delta N}$) of 30 mA provided for personal protection are rated to trip within 40 milliseconds for a residual current of 150 mA ($5 \times I_{\Delta N}$). RCDs do not restrict the level of current flowing, but they do restrict the duration of the current to a nominally safe level. This topic is covered further in Chapter 3.

The overall relationship between current, duration and probability of ventricular fibrillation is illustrated in Figure 1.1, which is derived and adapted from information presented in DD IEC/TS 60479–1:2005, to which the interested reader who may require more detailed information is referred.

It should also be noted that the probability of ventricular fibrillation depends upon the path that the current takes through the body. For example, a shock from hand to feet is more dangerous than a shock of the same magnitude from hand to hand. DD IEC/TS 60479–1:2005 defines a heart-current factor that can be used to determine the different effects for different paths through the body. The factor is defined as the ratio of the electric field strength in the heart, for a given current path, to the electric field strength in the heart for a current of equal magnitude flowing from the left hand to both feet.

Refer to Figure 1.2, and assume that the current flowing in path BB' from left hand to both feet is 100 mA and its duration creates a 25% probability of ventricular fibrillation in a particular individual. The heart-current factor for path AA', hand to hand, is 0.4, meaning that the current would need to be 250 mA to create the same probability of ventricular fibrillation. Similarly, the heart-current factor for path CC', between the feet, is 0.04, meaning that 2500 mA would need to flow through that path to create the same probability of ventricular fibrillation.

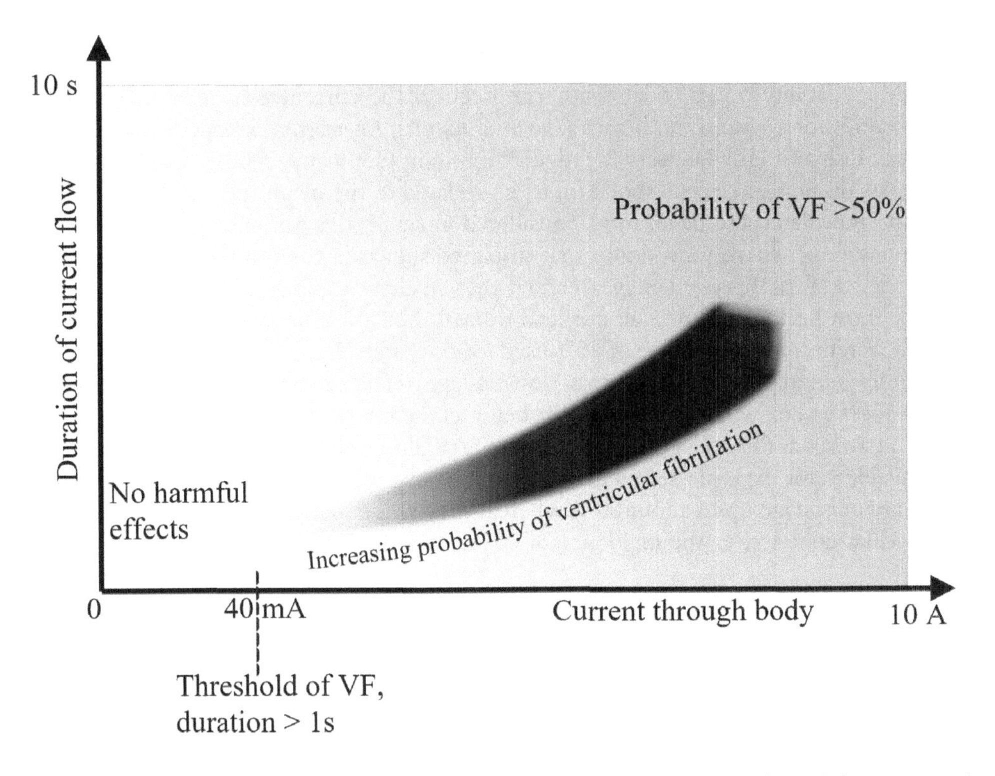

Figure 1.1 Summary diagram of relationship between current magnitude and duration and probability of ventricular fibrillation (VF)

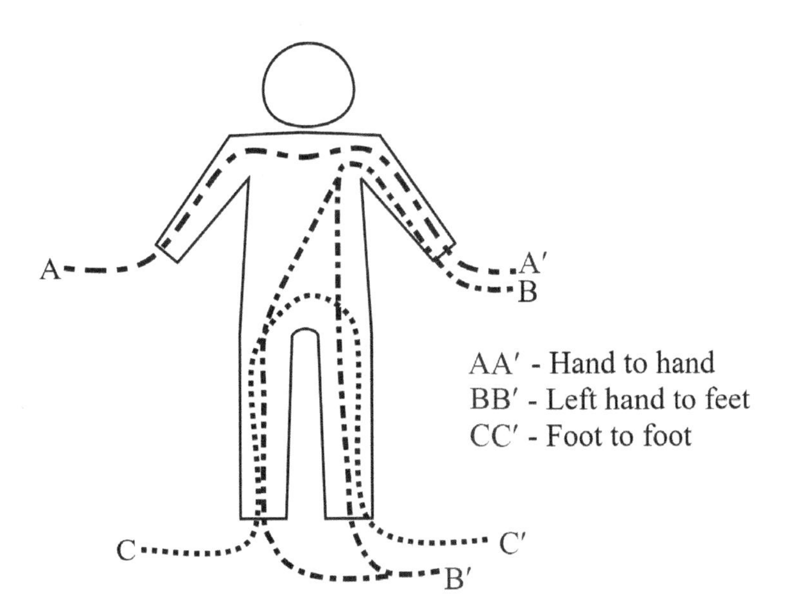

Figure 1.2 The effect of the heart-current factor

Although there is a tendency to concentrate on cardiac effects of electric shock, current flow can also lead to respiratory paralysis. Immobilisation of the respiratory muscles, with the possibility of asphyxial death, can occur if the current is in the order of 18 to 30 milliamps for a limb to limb path (*Electrical Injuries; Engineering, medical and legal aspects*; Nabours, Fish and Hill; Lawyers & Judges Publishing Company; 2000).

The seriousness of electric shock injuries is related to the magnitude of the shock current, but the voltage also needs to be considered since it is this parameter that most people will be aware of. There is no doubt that sustained shocks at the normal mains voltage of 230 V can be fatal. It is generally accepted that, in dry conditions, 50 V a.c. is a voltage that for most healthy people will not lead to fatal injuries, although a margin of safety is needed, as reflected in the value of 25 V used for the Safety Extra-Low Voltage (SELV) systems described in Chapter 3; the term 'low voltage' refers to a voltage lower than 1000 V a.c. or 1500 V d.c., with higher voltages being classed 'high voltage'. The risk of serious injury increases as the voltage increases above 50 V, although it is not a step change at that level. Welders, for example, frequently touch their live welding electrodes which have an open circuit voltage up to about 80 V; whereas they will experience minor shocks leading to muscular contraction, the incidence of serious electric shock injuries at that voltage is very low.

Unusual waveforms

The discussion so far has concentrated on a.c. sine or cosine waveforms at the very common power frequencies of 50 Hz and 60 Hz and at the public supply voltage of 230 V single phase. This is reasonable because by far the large majority of electrical accidents occur on systems energised at those voltages. However, not all electrical systems operate at power frequencies, so some assessment of the risk of injury needs to be made for them.

A good example is electric fence energisers, which are commonly used to power electric fences for stock control and, increasingly, for security. These energisers typically transmit pulses onto the connected fence wire with peak voltages in the order of 10,000 V, with a duration of 1 millisecond, and a pulse repetition frequency of 1 Hz. The idea is that any animal touching the fence will experience an electric shock of such magnitude that it will be deterred from touching the fence again and will therefore remain in its designated field. Of course, it would be highly undesirable for the animal to be killed, and it would also be unacceptable if there were to be an appreciable risk of electrical injury to any humans who may inadvertently touch one of the electrified fence wires. Waveforms such as this cannot be assigned the common safety-related limitation of current and duration. The approach that has been adopted by standards makers is to consider the amount of energy needed to cause fibrillation. Thus, in the case of fence energisers, a safe limit of 5 Joules per pulse has been set, together with limitations on pulse width and pulse repetition frequency. This is the amount of energy delivered into a 500 Ω load, where that figure of resistance represents the low-end value of resistance for most people for the high voltages at which fence energisers operate.

Guidance on how to assess the risk from non-sinusoidal waveforms is published in Part 2 of DD IEC/TS 60479.

Treatment

When somebody experiences an electric shock, it is important that the person rendering first aid should first remove the cause by switching off or otherwise breaking contact between the casualty and the live conductor. Care must be taken to ensure that the rescuer does not make contact with the live parts, including the casualty's skin, and thereby become a victim as well.

If the casualty is suffering ventricular fibrillation, the only effective way to restore normal heart rhythm is by the use of a defibrillator. In that respect, the increasing availability of automatic external defibrillators in some places of work, and in public places such as shopping centres and railway stations, is a welcome development. Unfortunately, in most accident scenarios, a defibrillator is not immediately available. The first aider should therefore summon the emergency services and carry out cardio-pulmonary resuscitation until either the casualty recovers or professional assistance arrives.

It used to be the case that an electric shock treatment placard had to be exhibited in most industrial premises, but this is no longer a legal requirement. However, it is good practice to exhibit such a placard in substations and in places where live work, such as fault rectification and testing, is being carried out.

Contact burns

Current flowing through the body can cause burn injuries at the points of contact and deep-seated burns in the muscle tissues.

The extent of contact injuries is determined by the current density at the point of contact; the higher the current density, the more severe will be the injuries. Nabours et al. report that the estimated minimum current necessary for first degree burns is 100 mA for a period between 1 second and 9 seconds. They observe that, at these currents and durations, cardiac effects will occur before most people experience significant burns.

IEC/TS 60479–1 describes the effects in terms of current density, measured in mA/mm^2. For example, it advises that for current densities between 20 mA/mm^2 and 50 mA/mm^2, a brownish colour develops extending into the skin and blistering may occur for current flow lasting several tens of seconds.

At low voltages, including 230 V, it is uncommon to see significant burn injuries, although that is not to say it does not happen, as exemplified by the case of an electrician who, when working live, had 230 V applied between the hand and his elbow on one arm, leading to the loss through burning of all the tissues in his forearm. It is not too surprising that he should have suffered such severe burn injuries; the resistance of his forearm would have been in the order of a few hundred ohms, leading to a current of 1 A or so flowing. As he became progressively more burned, the resistance would have reduced due to the skin barrier breaking down and carbonising, leading to an increase in the current and the severity of the heating.

The most usual indication that current has flown in low voltage incidents is small white blister-like marks on the skin, which often indicate the entry and exit points for the flow of current. At higher voltages, especially in incidents where somebody has come into contact with an overhead high voltage power line, the burning can be severe and can be the main cause of death.

A quite common occurrence is for there to be no significant burn marks at the point of entry of the current, but very significant burning or charring of the skin where the current exits the body, or vice versa; this relates to the current densities at the entry and exit points. A plumber who inadvertently touched a live 240 V a.c. terminal on a temperature controller with a finger of his left hand was also simultaneously leaning against some earthed copper pipework. He could not let go when the shock current flowed, and he was electrocuted. There was severe charring on his finger where the current density had been high but no indications of current flow on his arm where he had been leaning against the pipework and where the current density would have been much lower.

Arc burns

Arc burns are commonly associated with the failure of insulation in electrical equipment, leading to an arc developing in the air between adjacent conductors. The arc is characterised by a bright light, loud noise and extremely high temperatures. In high energy arc events, the arc is sufficiently hot to vaporise metal such as copper or aluminium with which it is in contact, and the hot gases or plasma which explosively evolve from the affected area can cause severe burn injuries to anyone in the vicinity. The event is usually called a 'flashover' or 'arc flash'.

A widespread misconception is that an arc will jump through air over quite long distances even at low voltages such as mains voltage. The distance over which an arc will 'jump' between conductors is determined by the voltage and the dielectric strength of the medium, typically air. Dielectric strength is the electric field strength, measured in volts/metre, that a material can withstand without breaking down. The dielectric strength of dry air is about 3 million volts per metre (3 V/m), meaning that the air between two conductors separated by 1 metre will break down and conduct electricity if 1 million volts or more is applied across them. Equally, the air between two conductors separated by 1 mm will break down if 3000 volts or more is applied between them. So it can be seen that quite high voltages are required to create arcs in dry air.

Arcs are most commonly generated during the separation of two live conductors that have been in contact with each other, such as when the contacts in a switch separate from each other as it is switched off or a fuse is withdrawn from a circuit that is on load and passing current. An arc is drawn in the small gap between the conductors as they separate, particularly if the load is inductive as this creates an effect known as 'back-EMF', where EMF is the abbreviation for Electro-Motive Force, which momentarily increases the voltage across the gap. Since the air is now ionised, its dielectric strength reduces dramatically and it becomes an effective conductor. As the switch's conductors continue to separate, the arc extends in length and will continue to extend until it can no longer be supported and it self-extinguishes. Switchgear such as a circuit breaker is often constructed with internal arc quenching features designed to ensure that the arc is extinguished as the contacts open, especially under fault conditions when high levels of current have to be interrupted safely.

Some flashover events occur when metal objects such as screwdrivers and spanners short a phase conductor to earth, or short across conductors at different voltages, particularly when live work is being carried out on electrical equipment. Some events, especially at high voltages, are caused by insulation failures after the surface of the insulation has been carbonised by current tracking across it; the tracking current is caused by effects such as moisture and dirt contamination or physical damage to the insulation. The typical consequence of the insulation failure is the expulsion from the short circuit of highly energetic arc products and hot gases, with temperatures in the plasma typically in the order of thousands of degrees Centigrade. The environment will often also contain vaporised metal.

Anybody standing in the immediate vicinity of the arc will suffer burn injuries, most commonly to the face, upper chest and hands, which are often very severe and can be life threatening or life limiting. Experience shows that burn injuries to the face heal significantly more quickly than burns to the hands, but internal burns caused by the hot plasma being ingested can cause long-term damage.

The power in the arc is determined by the fault level in the system, usually quoted in Million Volt Amperes (MVA) or thousands of amps (kA); the term 'prospective short circuit current' is also commonly used. This is a measure of the amount of current that can be fed into the fault, which is determined by the voltage and the impedance in the fault circuit. Modern systems, even at low voltage, often have very high fault levels.

Fire injuries

Electrical systems that are poorly designed or maintained, or are overloaded or have certain fault conditions, may overheat, or generate arcs and sparks, to the extent that adjacent flammable and combustible materials may be ignited, causing a fire that may cause injury or death. Incidents of this type are quite common in workplaces.

In cables, the amount of current that can be carried safely is determined by the type of conductor, typically copper or aluminium; the conductor's cross-sectional area; the type of insulation employed in the cable's construction, such as PolyVinyl Chloride (PVC), Cross-Linked PolyEthylene (XLPE) and Ethylene Propylene Rubber (EPR); and the environment in which the cable is installed, such as being buried under insulation or grouped together with other electricity cables. If the maximum current-carrying capacity is exceeded, perhaps because the cable is overloaded or because fault current is flowing, the cable may heat up to the point at which the insulation begins to melt, creating the conditions for ignition of flammable materials.

Poor or loose connections can create electrical resistance at terminals and connections in distributors' cut-outs, metering equipment, switchgear, controlgear, appliances and equipment. The amount of power generated in the form of heat at a poor connection is equal to I^2R watts, the square of the current times the magnitude of the resistance. So in locations of relatively high current, such as in cut-outs, consumer units and distribution boards, a small amount of resistance created by a poor connection can lead to the generation of thermal power measured in hundreds or thousands of watts. The heat generated can be high enough to melt insulation and lead to the ignition of flammable materials.

If material such as dust or sawdust covers cooling vents in electrical equipment, such as a motor driving machinery in a factory, the equipment can heat up to the point at which the material is ignited.

Damage to electrical systems, such as a cable being penetrated by a sharp object or insulation failing because of excessive bending and flexing, can sometimes create sparks or arcs that have the potential to ignite adjacent flammable materials and cause a fire.

Although fires caused by cooking appliances, most commonly chip pan and similar fires, dominate the statistics for electrical fires in domestic dwellings, fires caused by faults in other domestic appliances, especially tumble dryers, washing machines, dishwashers and electric blankets, are not uncommon and contribute to fatal and non-fatal incidents every year. Whereas some of these fires have a root cause of design failings that lead to product recalls, some are caused by poor installation standards and inadequate maintenance/cleaning.

A recent phenomenon has been the dramatic increase in the number of chargers used to charge the batteries in a wide range of equipment and appliances including phones, tablets, laptop computers, toys, power tools and e-cigarettes. Insurance companies report an increasing trend in fires caused by failures of these chargers, many of which have been purchased on-line and may not meet the design and construction standards stipulated in product standards. It is not unknown for the lithium ion batteries in these types of devices to ignite, as witnessed by the recent experience of Samsung who took product recall action for a new-to-market mobile phone after the batteries in a small number of devices caught fire.

Explosion injuries

There are many areas in industrial and commercial premises that have flammable or potentially explosive atmospheres. For example, petrochemical sites such as refineries, printing works with highly flammable inks, flour mills in which fine dust is created, and motor vehicle paint shops where highly flammable paints are used are premises in which there will be areas that may contain flammable and potentially explosive atmospheres. The hazardous atmospheres may not exist all the time but may be created for short periods during routine operations or in the event of a containment failure.

Electrical equipment installed in those areas may act as an ignition source, an example being a light switch in which an arc is momentarily created between the internal contacts when switched off. Electrostatic discharges can have the same effect. If the energy in the arc is above the minimum ignition energy of the atmosphere, an explosion may occur, leading to people in the vicinity suffering burns or other trauma.

Electromagnetic radiation

Electromagnetic fields are generated by electrical systems and have the potential to create harmful health effects. They can be created deliberately, as in the case of radio and radar systems, or accidentally, as in the case of an arc flashover event.

Consideration of electromagnetic radiation's health effects spans the spectrum from ultra low frequencies associated with the electric and magnetic fields emanating from overhead power lines, through the radio spectrum, up to the infrared, visible, ultraviolet, X-ray and Gamma ray end of the spectrum.

There is much controversy about the potential for ill health from exposure to power frequency radiation (50 and 60 Hz) and the radiation from cellular masts and mobile phones. A considerable amount of research has been carried out internationally to determine whether or not there is a link between these forms of non-ionising radiation and ill health. Interested readers will be able to find a considerable amount of information in research papers published by the International Commission on Non-Ionising Radiation Protection (ICNIRP).

What is well known and can be said with certainty is that overexposure to infrared, ultraviolet, X-ray and Gamma radiation causes ill health, the form of which is well documented.

In the radio spectrum, covering the band 3 MHz to 30 GHz, electromagnetic fields can cause heating of the body tissue by energy absorption, in much the same way that microwave ovens use radio waves to heat food. The eyes are particularly vulnerable to damage because of their water-like composition and because there is no blood circulation to assist in heat dispersal.

In the UK, the limits for exposure to electromagnetic radiation are set down by the Health Protection Agency's Centre for Radiation, Chemical and Environmental Hazards. The limits for exposure to non-ionising radiation are based on the ICNIRP's guidance. For ionising electromagnetic radiation, such as X-rays and gamma rays, information on the hazards, risks and protection requirements are available in the Ionising Radiation Regulations 1999 and supporting guidance material.

Whilst accepting that electromagnetic radiation is essentially an electrical phenomenon, the accidents and incidents arising from radio frequency, laser and ionising radiation applications will not be considered further in this book, which is concerned primarily with the risks from electricity.

Electrical accidents and dangerous occurrences

Introduction

This chapter explains the main types of accidents and dangerous occurrences that occur on electrical systems. Some types of electrical accidents and incidents have to be reported to the relevant regulatory bodies, so the first section explains what needs to be reported and to whom. There is also a section at the end of the chapter that provides some statistical information on incidents, mainly those that occur during work activities.

Reporting electrical accidents and dangerous occurrences

An electrical accident occurs when a person suffers an electrical injury, usually electric shock or burns, or a non-electrical injury arising from an electrical event such as an electric shock or explosion. Electrical accidents at work, in common with other health and safety incidents, must be reported to the relevant regulatory body if the legal reporting criteria are met. For most types of electrical accidents, the criteria are laid out in the Reporting of Injuries, Diseases and Danger Occurrences Regulations (RIDDOR) 2013, which places duties on employers and the self-employed to report certain types of incidents involving employees, the self-employed and members of the public.

The principal regulatory bodies who investigate RIDDOR-reportable incidents, be they accidents, ill health or dangerous occurrences, are the Health and Safety Executive (HSE), local authorities, and the Office for Rail Regulation (ORR), depending upon the type of premise in which the incident occurs. For example, an electrical accident in a shop or office would be investigated by local authority environmental health officers in whose area the premises are located, whereas an incident in a factory or hospital would be investigated by HSE inspectors of health and safety; the split in responsibility between the HSE and local authorities is described in the Health and Safety (Enforcing Authority) Regulations 1998 and associated guidance on the HSE's website (www.hse.gov.uk).

These bodies don't investigate all incidents reported to them; they have a set of incident selection criteria that they use to select which incidents to investigate, usually relating to the severity of the incident and public interest considerations, the details of which can be found on their websites.

Under the terms of the 2013 regulations, an electrical accident involving an employee or someone who is self-employed is reportable if it results in

- death;
- any of the regulations' specified injuries, including
  - serious burns which cover more than 10% of the body or cause significant damage to the eyes, respiratory system or other vital organs;
  - fractures, other than to fingers, thumbs and toes;
  - amputations;
  - any injury likely to lead to permanent loss of sight or reduction in sight;
  - any crush injury to the head or torso causing damage to the brain or internal organs;
  - any scalping requiring hospital treatment;
  - any loss of consciousness caused by head injury or asphyxia.
- the person being away from work, or unable to perform their normal work duties, for more than seven consecutive days as the result of their injury. This seven day period does not include the day of the accident, but does include weekends and rest days.

A non-fatal accident involving a member of the public being injured is reportable if the person is taken directly from the scene of the accident to a hospital for treatment to that injury. For reporting purposes, diagnostic tests and examinations do not constitute 'treatment', so someone being given an ECG examination after suffering a suspected electric shock is not receiving 'treatment' as far as the regulations are concerned. Application of a defibrillator to restart a heart that has arrested or gone into ventricular fibrillation would be regarded as being 'treatment'.

There are two types of incidents involving electrical systems that are classified as 'dangerous occurrences' and are therefore reportable under RIDDOR 2013. The incidents are

- Overhead power line incidents in which any plant such as a crane or equipment such as a ladder or scaffold pole unintentionally comes into contact with an uninsulated overhead electric line in which the voltage exceeds 200 volts; or into such close proximity with an overhead electric line that it causes an electrical discharge.
- Any explosion or fire caused by an electrical short circuit or overload (including those resulting from accidental damage to electrical plant) which either results in the stoppage of the plant involved for more than 24 hours; or causes a significant risk of death. This particular definition is quite difficult to interpret and creates considerable confusion about whether or not incidents need to be reported as dangerous occurrences. For example, most strikes by pneumatic drills and pick axes of buried live power cables cause an 'explosion' of some degree, usually in the form of arcing activity, and will have a risk of death by electrocution by virtue of the live conductors being touched by the tool, although the term 'significant risk' is wide open to interpretation. The best advice is to not automatically report such incidents and to give the risk of death some careful thought because not all explosion or fire incidents caused by short circuits or overloads have a significant risk of death. However, if in doubt about this, err on the side of caution and report the incident; the regulator should reject any reports of incidents that are not in fact reportable.

Some types of electrical incidents involving the generation, transmission and distribution of electricity are reportable under the terms of the Electricity Safety Quality and

Continuity Regulations (ESQCR) 2002, as amended. Examples are a fire in a distributor's fused cut-out and a member of the public being injured by coming into contact with an overhead power line. Such incidents are reportable on-line or by telephone to the HSE by a limited set of duty holders, as explained further in Chapter 7.

Some types of incidents involve medical devices, as defined in the Medical Devices Regulations 2002 which are made under the Consumer Protection Act 1987 and enact into UK law the requirements of European directives covering the safety of medical devices. Examples of equipment captured by these regulations include medical implants, infusion pumps, electrical profiling hospital beds, battery-powered wheelchairs, kidney dialysis machines, powered patient lifting machines and so on. The safety of these devices is regulated in the UK by the Medicines and Healthcare products Regulatory Agency (MHRA), to whom any reports on dangerous incidents involving medical devices must be submitted, such as a battery-powered electric wheelchair catching fire due to an electrical fault. If any workers are injured in these incidents, or if a member of the public is injured as a result of a work activity involving a medical device, a report would need to be submitted under the terms of RIDDOR 2013.

Electric shock accidents

Accidents involving an electric shock are commonly subdivided into two categories – direct contact and indirect contact shocks. Although the standards that will be considered later no longer use this distinction when considering preventive measures, the terminology remains meaningful and in widespread use so will be used in this book. A direct contact shock occurs when conductors that are meant to be live, such as bare wires or terminals, are touched. An indirect contact shock usually occurs when an exposed conductive part that has become live under fault conditions is touched; an example of an exposed conductive part would be the metal casing of a washing machine that should normally be earthed.

The majority of direct and indirect contact electric shock and burn accidents occur at low voltage on distribution systems or on connected equipment. There are many instances in which high voltage overhead lines are touched, which is a form of direct contact, although casualties tend to suffer severe burn injuries as well as electric shock.

Direct contact electric shock

Figure 2.1 illustrates the principles of direct contact electric shocks. The figure shows the low voltage side of a high voltage to low voltage distribution transformer feeding a 4-wire 3-phase and neutral circuit; switches and protective devices such as fuses and circuit breakers are omitted. The star point of the secondary windings is connected to earth, which is the case with all distribution transformers supplying customers of electricity supply companies in the UK. This creates an earth-referenced system, with the neutral of the system connected to this earthed star point.

In diagram A the person is experiencing a hand-to-hand direct contact shock between two 230 V phase conductors, giving a shock voltage of 400 V; this type of incident is not very common.

In diagram B the person is experiencing a hand-to-hand direct contact shock between a 230 V phase conductor and the neutral of the supply – this could be on, for example, a distribution circuit, or in a distribution board, or in equipment fed at 230 V single phase.

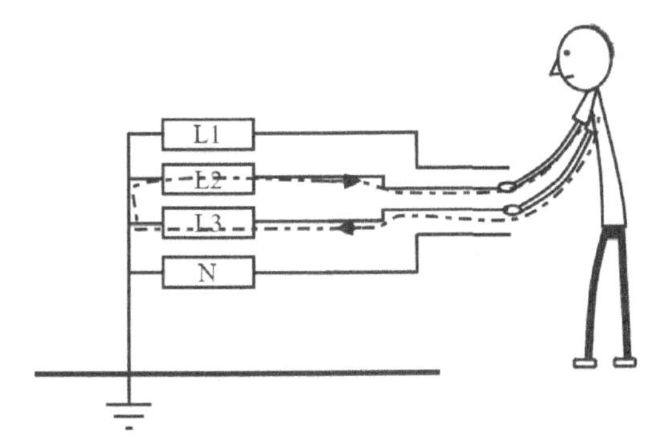

A. Phase-to-phase direct contact shock

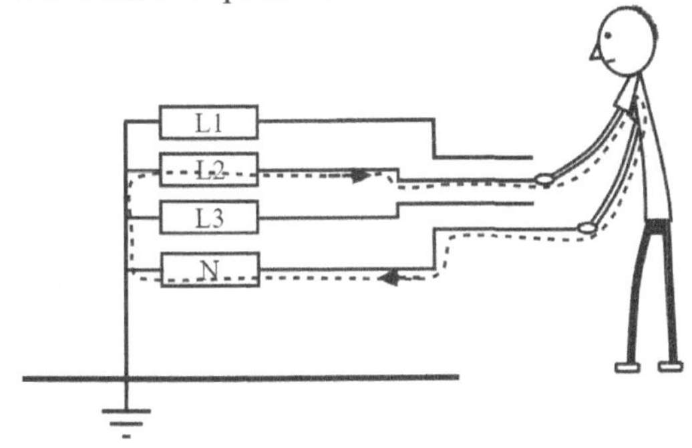

B. Phase-to-neutral direct contact shock

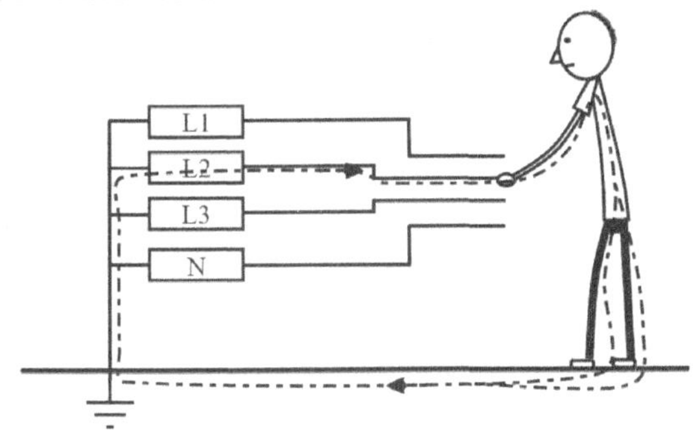

C. Phase-to-earth direct contact shock

Figure 2.1 **Direct contact electric shock**

In diagram C the person is experiencing a hand-to-foot 230 V shock between one phase of the supply and earth through a conducting floor (such as a concrete floor); recall from Chapter 1 that a hand-to-foot shock has more severe effects on the heart than a hand-to-hand shock of the same magnitude and duration.

In each case, the electric shocks will prove to be fatal if the magnitude and duration of the current flowing through the body meet the criteria explained in Chapter 1. Experience shows that the figure in diagram C is experiencing the most common form of direct contact injury: a phase-to-earth electric shock. A typical incident would be a person picking up a cable on which the sheath and insulation is damaged and touching an exposed live conductor; a shock would be experienced between the hand and feet.

Another type of incident would be an electrician working inside a live electrical panel, making inadvertent contact with one of the live terminals and experiencing a shock between the hand touching the live terminal and any other parts of his body in contact with earth. This situation occurred on a 3-phase 400 V distribution board of the type shown in Figure 2.2, into which an electrician was connecting a new circuit while the supply to the board was energised; he was electrocuted when he made direct contact with the conductor of a wire from which he was stripping the insulation after he had

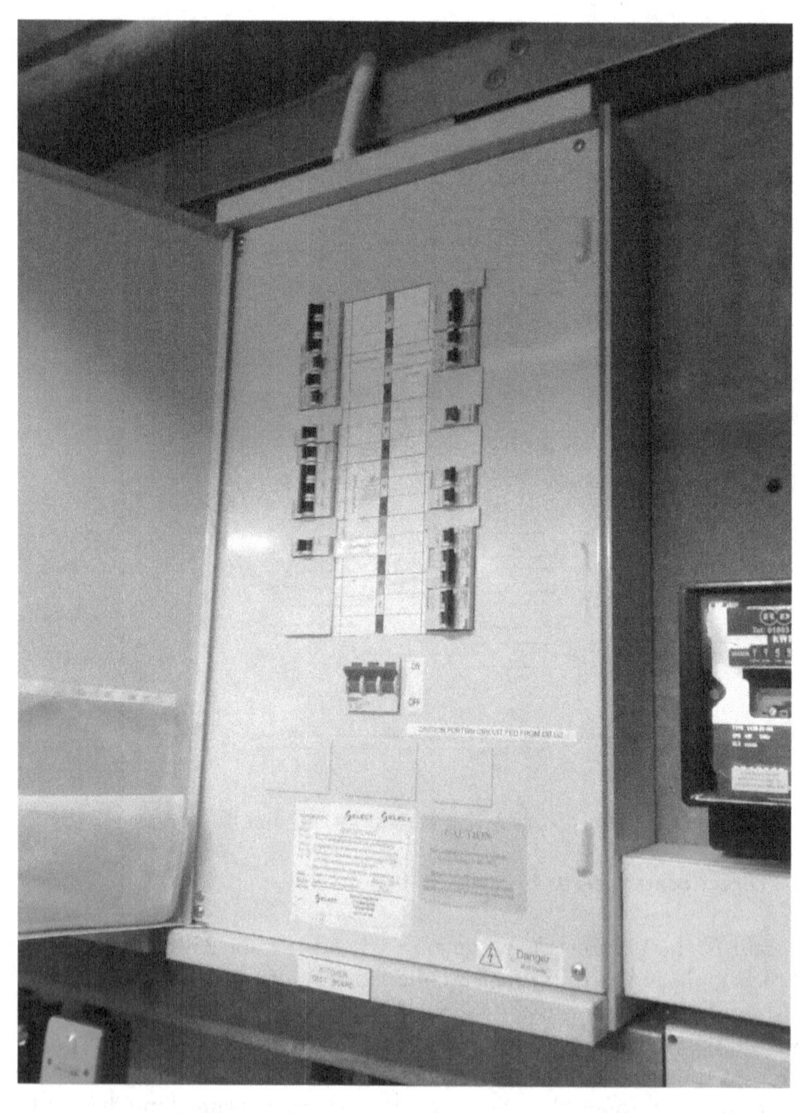

Figure 2.2 Low voltage 3-phase distribution panel of the type involved in a fatal accident

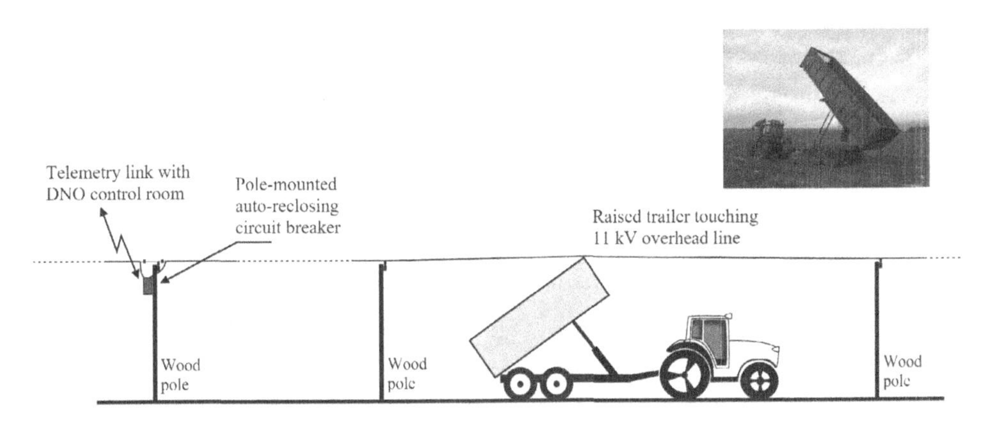

Figure 2.3 **Illustration of raised trailer in contact with 11,000 volt overhead power line**

already connected the other end into the live board and not switched the circuit's circuit breaker off.

The most common form of work-related death from direct contact is contact with an overhead power line, which over the past 30 years or so in the UK has accounted for roughly half of all work-related electrocution and electric burn fatal accidents. A typical fatal accident incident is illustrated in Figure 2.3, which shows a farm tractor pulling a raised trailer in contact with a 2-wire 11 kV overhead line with uninsulated wires; the phase voltage to earth would have been about 6.3 kV. The tractor driver had stopped his tractor in a farm field and raised the trailer to allow him to grease the trailer's brakes during a lull in harvesting activity. Unbeknownst to him, he had stopped directly underneath the power line – but the pole-mounted autoreclosing circuit breaker closest to the point of contact tripped off when the line was first touched, making the wires dead. While he was sitting in the tractor cab raising the trailer, he cannot have heard over the engine noise any arcing activity as the trailer touched the wire, so he would have had no awareness that the trailer was touching the wire. He got out of the tractor cab and started to work on the trailer's brakes, and while he was doing this, the power line was reenergised by the Distribution Network Operator's control engineer remotely switching the circuit breaker back on via a telemetry link in an attempt to restore supplies to consumers, leading to the driver's death by electrocution.

Indirect contact electric shock

Turning now to the case of indirect contact electric shock incidents, Figure 2.4 illustrates the principle by which this can happen.

The diagram illustrates that the circuit protective conductor (earth wire) in the flexible cord supplying a washing machine, which is a metal-cased appliance, has become open circuit, meaning that the metal case is no longer earthed. A second fault, an internal insulation failure inside the washing machine, has connected the metal case to the supply conductors resulting in the case becoming live.

Under the fault condition shown, the casing of the washing machine will become live at the full mains potential of about 230V with respect to earth. When the person depicted

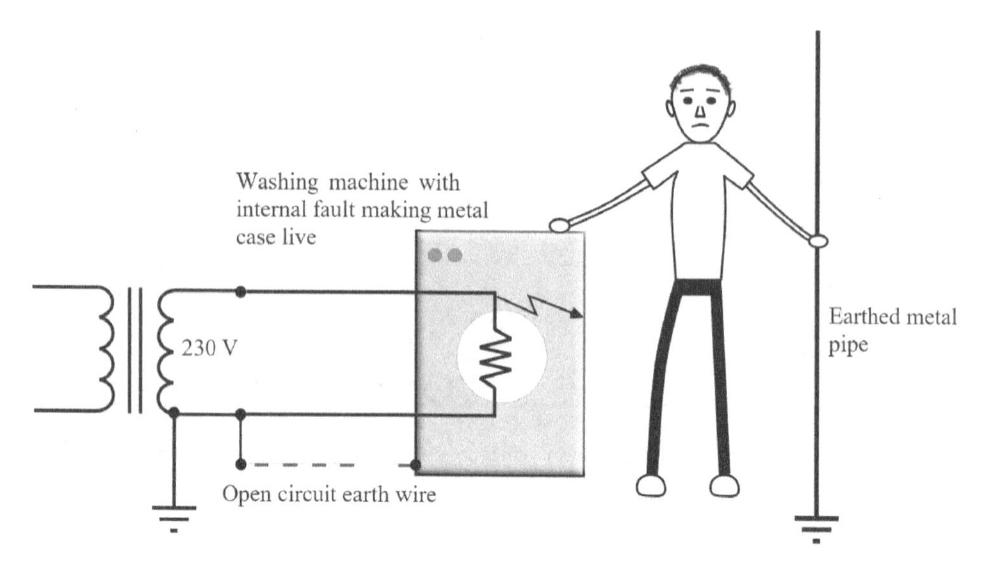

Figure 2.4 Indirect contact electric shock

simultaneously touches the live casing of the appliance and an earthed object, such as a metal water pipe, a hand-to-hand mains voltage shock is suffered; it is assumed that the casualty depicted is standing on an insulating floor or wearing shoes with insulating soles, so no current flows to earth through his feet. The flow of shock current in this case of indirect contact is depicted in Figure 2.5, and the equivalent circuit is shown in Figure 2.6.

It may seem at first sight that this type of indirect contact scenario is rather unlikely to occur. However, it is in fact a surprisingly common occurrence. What tends to happen is that a break in the earth wire, or circuit protective conductor, occurs first, maybe as a result of the connection at an earth terminal becoming loose, or the earth connection not being made correctly in the first place, or an open circuit fault occurring in the circuit protective conductor. An open circuit earth fault will not be detected unless the circuit is tested and the fault revealed; the fault can therefore exist for some considerable time before it is found. If, in the intervening interval, a phase-to-earth fault occurs as a result of an insulation failure, the conditions for an indirect contact electric shock are created. In this type of situation, the faults will not be apparent until an electric shock is experienced. This is because the appliance will operate as normal, despite the fact that the exposed metalwork, which should be at earth potential, has become live at mains voltage.

Referring to Figure 2.6, the phase voltage at the substation transformer will usually be a little higher than 230V to allow for the inevitable voltage drop in the distribution cables. In urban areas, the line/neutral and the line/earth loop impedances will be comparable and will probably be only a small fraction of an ohm, whereas the casualty's hand-to-hand impedance will be in the order of 2000 Ω. Under these circumstances, the effects of the circuit impedances can be ignored. The casualty's touch voltage will be about 230V and, for a total body impedance of 2000 Ω, the shock current would be 230/2000 = 0.11 A. This is high enough to cause ventricular fibrillation in many people, should the current flow for about 0.5 seconds.

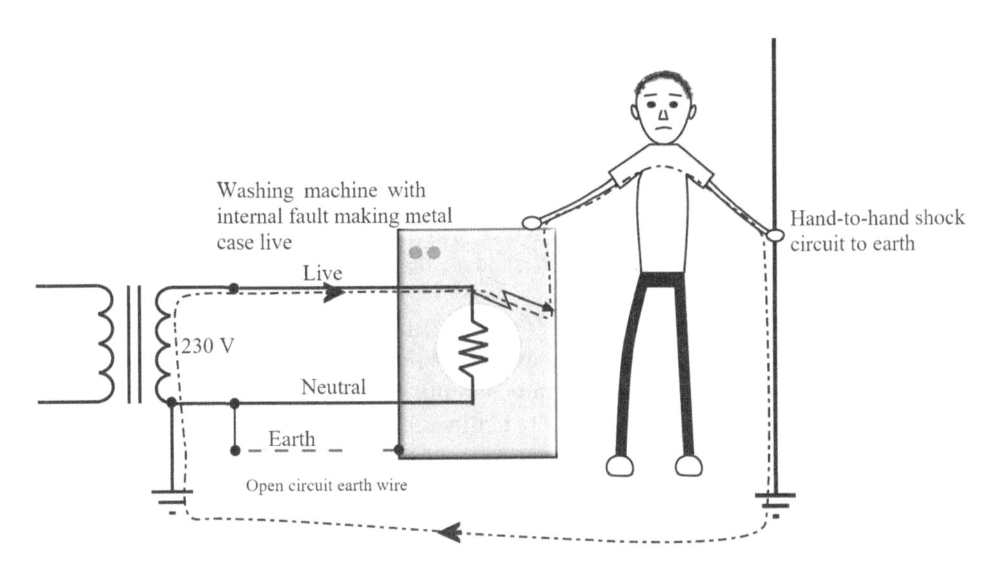

Figure 2.5 Shock current flow in case of indirect contact

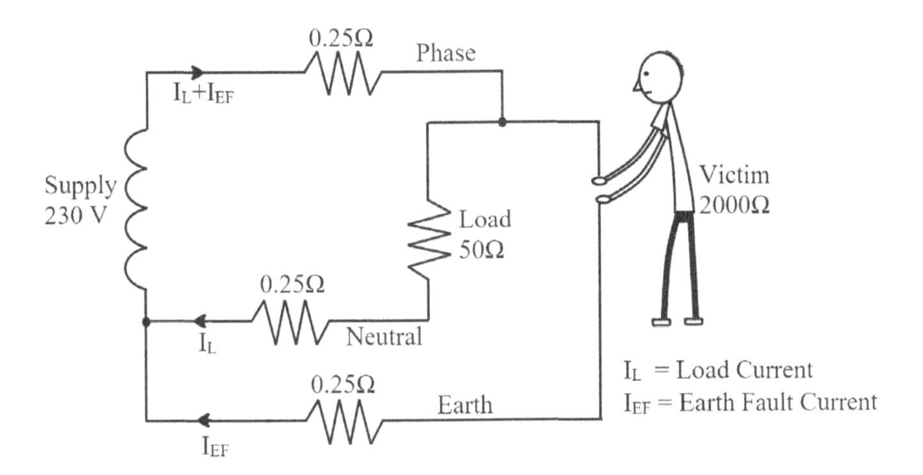

Figure 2.6 Equivalent circuit of Figure 2.5, indirect contact shock

It is interesting to observe that our electrical systems are designed in such a way that two faults such as this can lead to injury and death. It is for this reason that preventive maintenance of electrical systems is very important – if we were not to maintain our systems, the rate of occurrence of these two faults and the consequential injuries would be far higher.

The nature of the fault described means that the touch voltage is in the order of 230 V. However, many indirect contact shock accidents occur at less than mains voltage. This can be quite fortunate for the injured person because the shock current will be lower, thereby reducing the adverse effects and improving their chances of surviving the incident.

To explain how the voltage is reduced, consider the fault shown in Figure 2.7, where there is a short circuit between the centre of, say, a heating element and the appliance's metal case. If the appliance has previously lost its earth connection because, for example, the earth core in the flexible cable became detached from its plug terminal, the metal case will acquire the potential to earth of half the supply voltage, about 115 V. This will be the shock voltage for the person who is in contact with the case and the earthed water pipe. At this voltage the casualty's body resistance will be greater than it would have been at 230 V, a speculative figure being 3000 Ω, which will have the beneficial effect of reducing the shock current to an even greater extent than the proportional reduction associated with the reduction in voltage.

If the appliance has not lost its connection to earth via the circuit protective conductor, the potential difference between the appliance's metalwork and the water pipe would normally be less than 50 V, and the person would not experience an electric shock of any physiological consequence. The equivalent circuits associated with these two situations are shown in Figures 2.8 and 2.9.

Sometimes both faults occur inside a plug, when the earth wire becomes detached from the earth pin and becomes lodged underneath the live pin or the fuse holder. This is especially prevalent on portable or hand-held appliances on which excessive tension is applied to the cable where it enters the plug, resulting in the wires becoming detached from the pin terminals inside the plug. An example of this is illustrated schematically in Figure 2.10, which shows the situation associated with the electrocution of a plumber. He was using a metal-framed Class I handlamp supplied via an extension lead; the lamp's exposed protective metal frame and handle needed to be earthed for safety. As illustrated, the earth wire in the extension cable's plug had been pulled out of the earth pin, possibly because the cable's sheath had been cut back too far and the cable was not secured by the cable grip. The earth wire became lodged underneath the

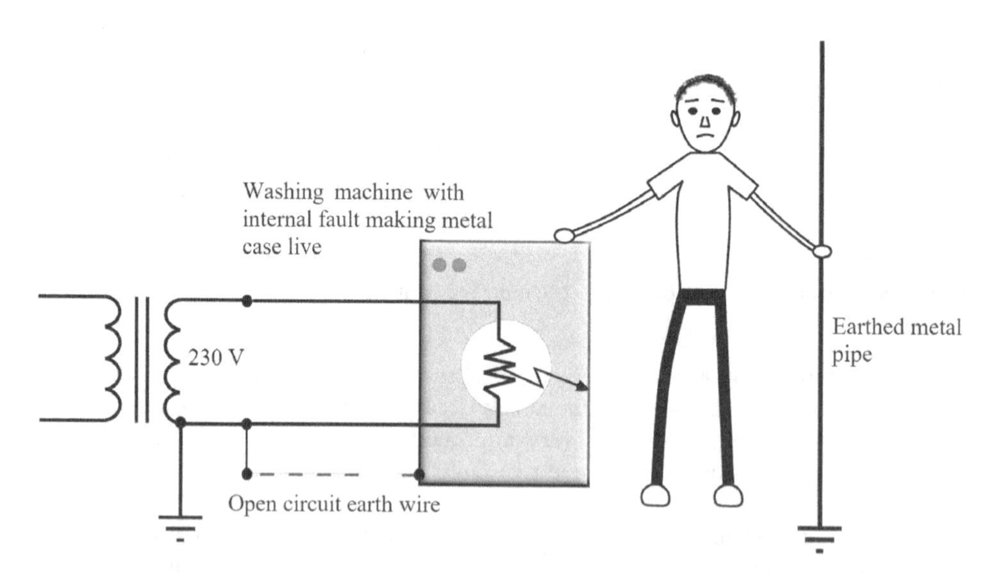

Figure 2.7 Indirect contact shock at voltage lower than mains voltage – open circuit earth wire (circuit protective conductor)

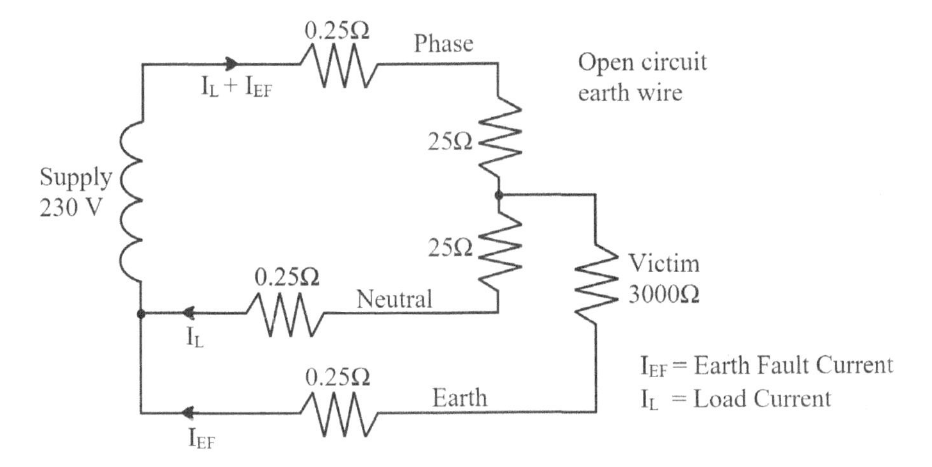

Figure 2.8 Equivalent circuit of Figure 2.7

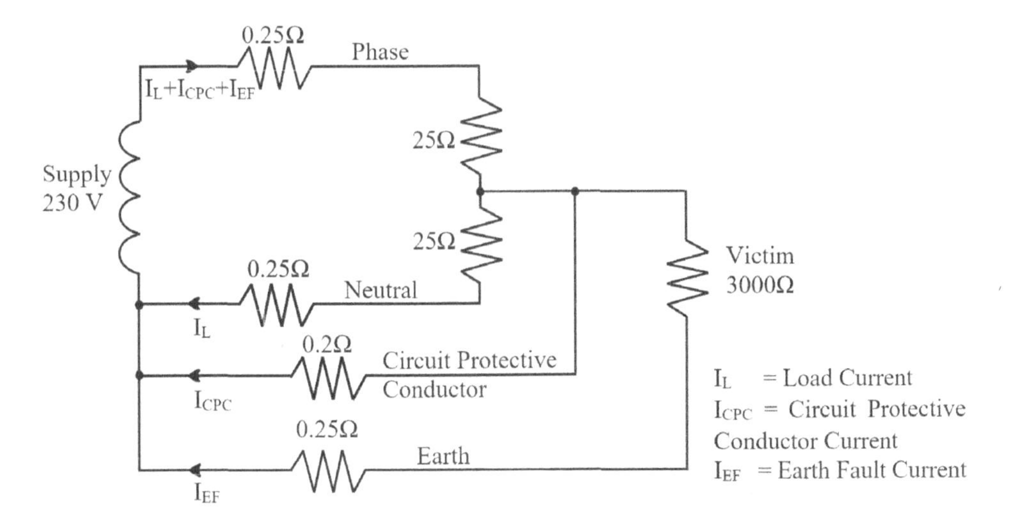

Figure 2.9 Equivalent circuit of Figure 2.7 but with the circuit protective conductor intact

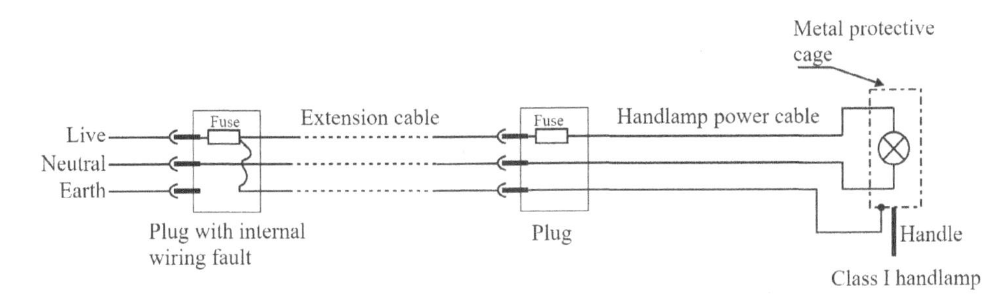

Figure 2.10 Class I handlamp with defective wiring in extension cable plug leading to indirect contact electric shock

live pin, which resulted in the lamp's metal handle and case becoming live at about 230 V when the lamp was plugged in. The lamp would have illuminated as expected, but the user of the lamp suffered a fatal electric shock when he held the live metallic handle and simultaneously touched a path to earth, believed to be an earthed central heating radiator, with his other hand.

It is important that 3-pin plugs to BS 1363, the British standard for 13 A plugs, socket-outlets, adaptors and connection units, are wired correctly. This is not just to ensure that the three wires are connected to the correct terminals (brown-live, blue-neutral, green/yellow-earth) but also to ensure that the wires are cut to the correct lengths. In particular, the earth wire should be cut longer than the live and neutral wires to ensure that should excessive tension be applied to the cable, the earth wire is not the first one to become detached from its pin. This illustrated in Figures 2.11, 2.12 and 2.13.

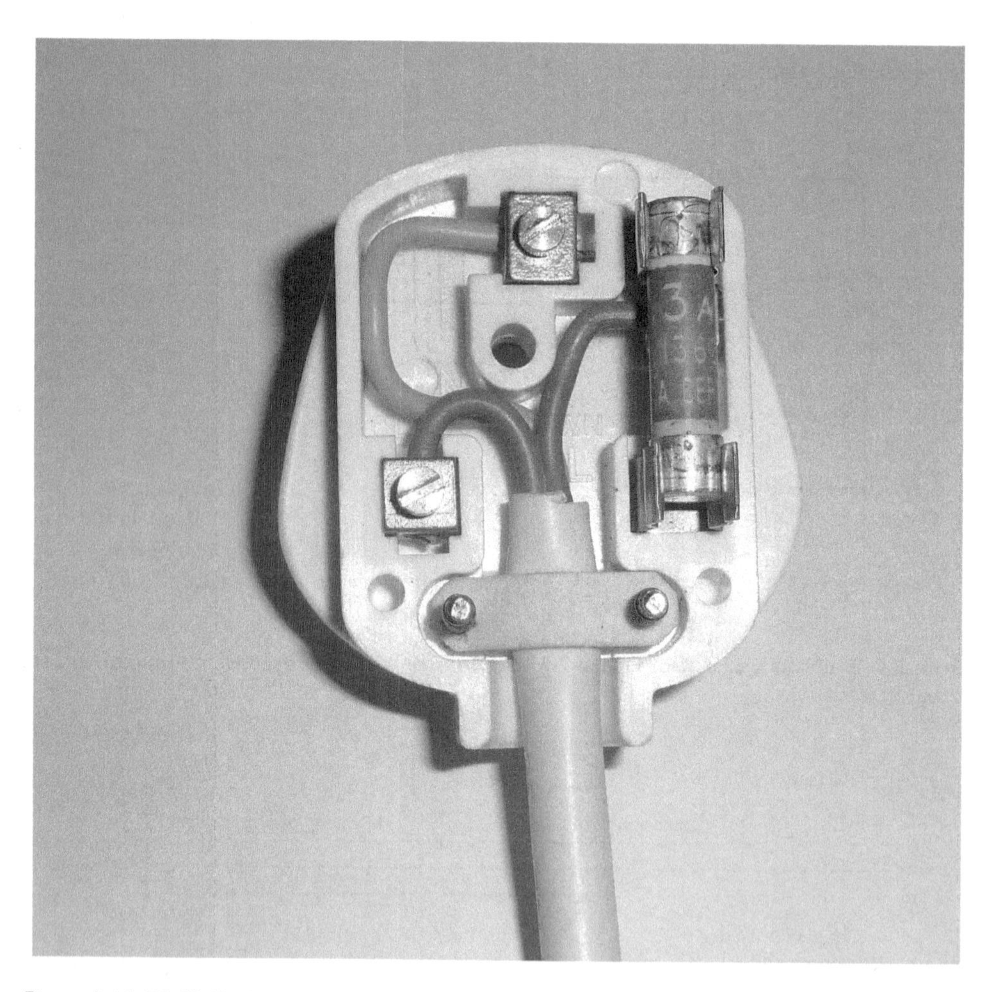

Figure 2.11 BS 1363 plug with earth wire cut to be longer than neutral and live wires, cable sheath secured by cable grip

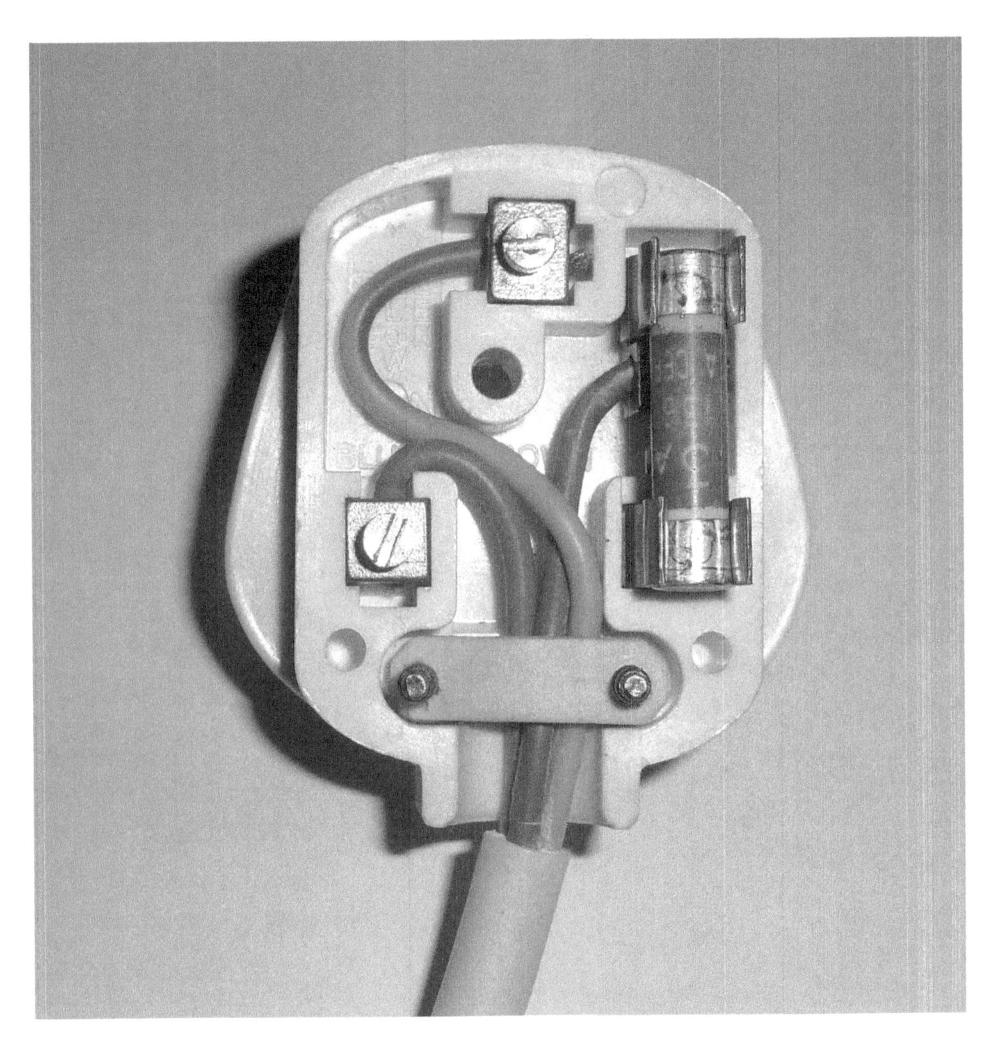

Figure 2.12 Cable pulled out of cable grip but earth wire remains connected to earth pin

Portable appliances

Although many electric shock incidents occur on power distribution systems, commonly called fixed wiring in the case of low voltage distribution systems, many occur on appliances, especially portable or hand-held appliances such as power tools, hair dryers and straighteners, and irons; the previously described fatal accident involving the handlamp is a case in point. The accidents are commonly linked to a failure of the insulation, mostly on the cables supplying the appliances. This is especially prevalent in harsh environments, such as construction sites and farms, where the risk of mechanical damage to the sheathing and basic insulation is especially high. Insulation damage also occurs where cables are constantly and excessively bent and flexed.

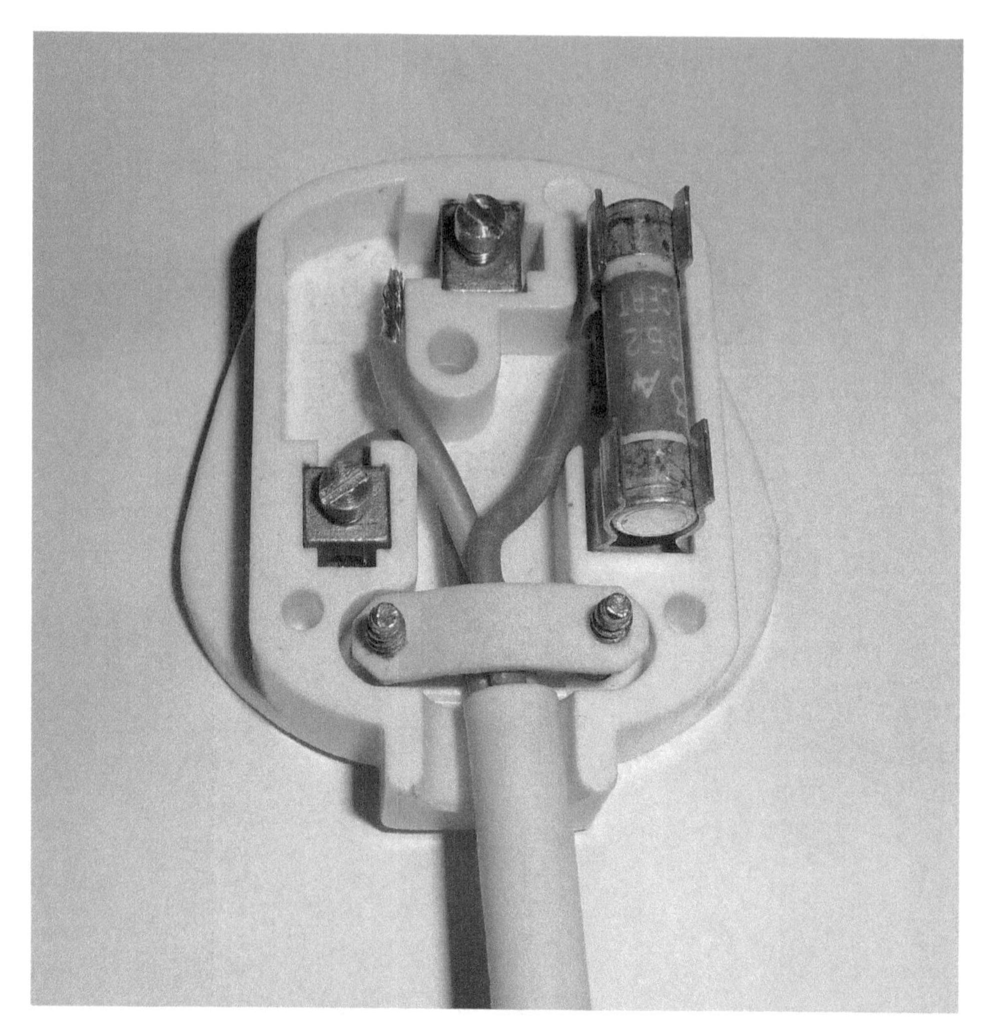

Figure 2.13 Plug wired with earth wire cut too short and becomes detached from earth pin when cable is pulled out of cable grip

The damage will often expose the live conductor in the cable. When this is touched, a hand-to-feet shock will be experienced. If the live conductor is being gripped at the instant the shock current starts to flow, the casualty may well not be able to let go.

Since portable appliances such as power tools are frequently used on ladders, scaffold structures and platforms, a relatively common result of a person experiencing an electric shock when using a power tool under these working conditions is a fall from a height, leading to impact injuries.

Extraneous conductive parts

Extraneous conductive parts are metal items that are not part of an electrical installation, such as metal gas and water pipes, reinforcing bars and window frames, but which may assume a potential, including earth potential. Sometimes they can become live at a hazardous voltage and cause electric shock accidents.

An example of this is the tragic case of a woman who suffered a fatal electric shock when she touched a scaffold that had been erected outside a shop in a busy street in Glasgow. The situation is illustrated in Figure 2.14.

The scaffold's vertical tubes, also known as standards, were standing on wooden blocks located on the pavement in front of the shop, which formed part of the ground floor of a multi-storey tenement building. Above the shop there was a narrow ledge covered in lead flashing, which one of the standards was touching. A lead-sheathed electric cable with two rubber-insulated cores (live and neutral) was lying on top of the flashing; this type of cable was used for fixed mains circuits prior to the 1950s and had probably been in situ for 30 to 40 years by the time of the accident. The cable had been used in the past to distribute power at 230 V to external lights on the tenement block but had been out of use for some time, although it was still connected into the lighting control circuit. Power to the lighting circuit, including the redundant lead-sheathed cable, was controlled by a timer switch, which switched the lights on and off, and a photoelectric switch which overrode the timer switch if the conditions were overcast and dark.

The cable's lead sheath was meant to act as the earth wire for the lighting circuit, but at some time in the past, the earth connection had broken where the cable connected into a distribution box. Also, at some time the lead sheath had split open where the cable passed across the lead flashing, possibly as a result of mechanical damage. The rubber insulation on the live wire had perished, and when it rained the rainwater provided a conducting path between the live wire and the lead sheath and hence to the lead flashing on which the cable was lying. So, when it rained and when the timer or photoelectric switch turned the power to the lighting circuit on, the lead sheath and the flashing on the ledge became live at 230V. Once the scaffold was erected and touched the flashing, it too became live at 230V.

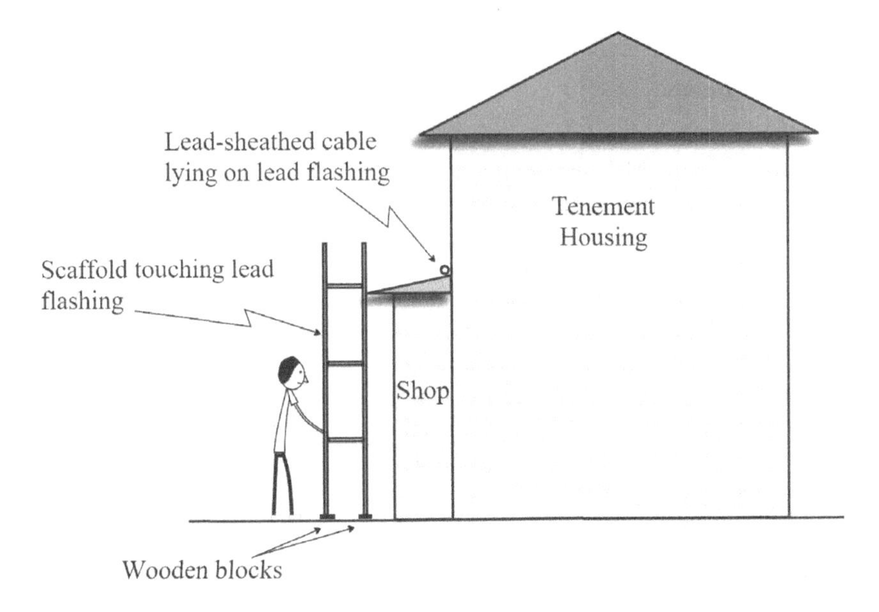

Figure 2.14 Electrocution death from live extraneous conductive part

At mid-afternoon on the overcast day of the accident, the photoelectric switch turned on the power to the lighting circuit and by doing so made the scaffold live at 230 V. The wooden blocks footing the scaffold effectively insulated the scaffold from earth; any leakage current to earth would have been below the fusing current of the 20 A fuse protecting the lighting circuit. The deceased woman had been inside the shop and on exiting had grabbed a vertical scaffold tube, effectively bridging the voltage on the scaffold to earth through her feet, and she suffered a hand-to-foot fatal electric shock. Again, the shock current to earth would have been well below the fusing current of the 20 A fuse.

This tragic accident was a case of indirect contact shock involving extraneous conductive parts. It is perhaps surprising that no one else on the busy street suffered a similar fate, but the most likely explanation for this is that the victim had grabbed the scaffold tube and had become gripped-on, unable to let go. Other people who might have brushed against the scaffold might have suffered a potentially painful momentary shock, but would not have been gripped onto the tube.

This incident serves to illustrate a number of points, as follows:

- Many electrical accidents result from a number of different failures or faults or events occurring at the same time, each of which on its own is unlikely, making the overall circumstances improbable and such accidents infrequent, indeed rare.
- Redundant cabling should always be removed.
- Overcurrent protection such as a fuse does not always provide protection against electrical injury.
- An electrical system should be inspected and tested periodically to ensure that it is in a safe condition.

Some of these observations are addressed in more detail in later chapters.

Burn accidents

Electrical contact burns

Many electric shock incidents such as those described are accompanied by minor burn injuries at the points of contact. These are frequently characterised by small blisters at the entry and exit points. More severe burn injuries arise when large currents flow through the body, such as when a high voltage conductor is touched.

An example of this type of accident occurred on a croft in the Outer Hebrides, where a 3-wire 11,000 V overhead line supported on wooden poles crossed one of the croft's fields. The line was struck by lightning, which caused damage to the insulators on the cross-arm at the top of one of the poles. Leakage current flowed down the pole to earth over a period of days or even weeks, burning the pole to the extent that it eventually broke at a point about 4 metres above ground level. The upper section of the broken pole fell vertically down and became embedded in the marshy ground, with the conductors now at a height of about 2 metres above ground level; the three conductors did not clash as the pole fell and they remained live. The crofter, a man in his seventies, went into the field to inspect the damage and to collect the porcelain shard remains of the insulators which had shattered as the pole fell. He had not realised that the conductors were live. His head touched one of the high voltage conductors, and he suffered severe burns when current flowed to earth through his feet. Remarkably, he survived the consequential shock and burns, but his burn injuries, particularly to his legs, resulted in one of his legs having to be amputated.

Arc burns

Arc burn injuries, also known as flashover or arc flash injuries, are mostly sustained when work is being carried out on live electrical equipment. Some incidents occur when switchgear such as motor control switches or circuit breakers or transformer tap changers are operated and there is a disruptive high energy failure, although these types of failures are now quite rare. Most incidents occur during live fault finding and testing activity or when equipment is being added to or removed from existing live equipment.

For example, a recent incident occurred when some electricians were removing a faulty power factor correcting capacitor assembly from a live low voltage distribution board. The electricians knew about the risks of live working and had attempted to insulate the assembly from the live busbars inside the board before they started to slide it out of the board. However, after they had unbolted the assembly from its fixings and as they were pulling it out of the board, they lost control of it and it slipped down into the board and touched the internal live 3-phase 400 V rising busbars, providing a short circuit path between the busbars and earthed metalwork. There was a high energy arc drawn between the busbars and the earthed metalwork, and the two electricians suffered serious burns to their upper bodies from the consequential explosion.

As explained in Chapter 1, the effect of a short circuit is to cause a very large current to flow, the magnitude of which depends upon the circuit voltage and impedances but which may be in the order of thousands of amps, especially if the short circuit occurs close to the distribution transformer. An arc drawn in air is hot enough to vaporise the metal with which it is in contact. The net effect is an explosion of extremely hot gaseous products and flames accompanied by a blast wave that evolves from the point of the short circuit, enveloping anybody in the near vicinity. The explosion will last until either the circuit protective device operates or the current is extinguished by the destruction of the circuit or the arc self-extinguishes.

The typical injuries are burns to the face, neck and chest. The backs of the hands are also frequently badly burned, as they are quickly raised to protect the face. If the hot gases are breathed in, they can cause serious life-limiting internal burns. In some rare cases, the burns are sufficiently deep and extensive to be fatal, although in most cases a recovery is made.

The short circuit can be caused by any conducting material. For example, in another case, a number of electricians were installing a new cable on a cable tray adjacent to a 400 V distribution board. They were using single strands of wire to secure the cable to the tray, with the wire lengths typically in the order of 0.75 m. As one of the electricians was manipulating one of the pieces of wire, unbeknownst to him one end of it entered a small gap in the cover of the live distribution panel. It created a short circuit between the earthed cover of the panel and a 230V terminal on an internal air circuit breaker; the fault level at that point was 15 megavoltamperes (MVA), where the term 'fault level' refers to the amount of power or current that would be fed into a short circuit fault and which is quoted in either MVA (power) or kA (current). The ensuing explosion blew the cover off the panel and the flames severely burned the electrician on his back.

Another accident happened while an electrician was working on the internal part of a low voltage motor control panel, which had been isolated apart from a set of internal busbars and 3 fuse carriers that formed the 3-phase supply to the panel. These parts were meant to be covered and protected by a Perspex cover, but unfortunately the cover wasn't fitted correctly and there was a small gap. A nut that the electrician was handling dropped through the gap onto the live busbars, causing a short circuit and a high energy flashover that vaporised metal and resulted in him suffering burn injuries, mainly to his unprotected face and hands.

Short circuits or flashover events such as this have also been caused by uninsulated tools which, when carelessly used, have bridged between live parts or between a live part and earthed metal.

Another common cause is the use of inadequately insulated instrument probes. The bare section of the metal probe, if carelessly used, may bridge between a live part and earthed metalwork or between live parts of different polarity, again causing a short circuit with the consequential arcing and burn injuries.

Fires

It is well known that electrical systems have the potential to cause fires, for the reasons explained in Chapter 1. However, it is often difficult to attribute the cause of a fire to an electrical fault or defect because, in many cases, the physical evidence is destroyed by the fire. Fires tend to consume all the combustible materials in electrical equipment such as plastic insulation and paint, as illustrated in Figure 2.15, which shows the burned-out remains of a metal single phase distribution board that had been involved in a fire.

Figure 2.15 Burned-out remains of single phase metal distribution board involved in a fire

The forensic investigator has to sift through the metallic remains of cables, terminals, busbars and switchgear, identify them, work out the design of the distribution system, search for any telltale signs of short circuits and arcing activity such as globules of melted copper and make judgements about the likelihood of the fire starting in or around the equipment. It might be appreciated that this is not a straightforward task.

Although in many cases it is obvious that a fire has started in electrical equipment or apparatus, fire investigators will often need to exclude other potential sources, such as arson or a discarded cigarette, before concluding that a particular fire must have been caused by an electrical fault on the basis that there is no other credible alternative. Conclusions are often reached 'on the balance of probabilities' rather than because the cause of the fire has been determined 'beyond reasonable doubt'.

According to the UK government's fire statistics for the year April 2013 to March 2014, published by the Department of Communities and Local Government on 29 January 2015 (ISBN 978-1-4098-4669-7), there were 22,200 fires in premises that were not domestic dwellings. Of these, about 75% were classified as being accidental, with the main cause being faulty appliances and leads, representing 25% of all accidental fires. Another main source of ignition in accidental fires was reported to be electrical distribution, accounting for nearly one-fifth of accidental fires. Other key sources of ignition were cooking appliances and other electrical appliances. With the ever increasing number of electrical appliances used in homes, there is a correlated increasing concern, particularly in the insurance industry, about the fire risks. This is especially with regard to lithium ion batteries and their chargers associated with mobile phones, tablet computers, personal computers, e-cigarettes and the like; some of these chargers are counterfeited items and are of dubious manufacturing quality.

Wiring and connection faults

Faulty wiring and connections are the cause of a significant number of fires in buildings, commonly due to insulation failures or bad high resistance connections. It is also often blamed for fires where the cause is unknown, possibly because fire investigators have to state the probable cause of the fire in their investigation report. If the fire has destroyed the evidence and there is no evidence of any other cause such as an accelerant associated with arson, there is a temptation to attribute the fire to a cigarette end or faulty wiring. However, there is no doubt that wiring defects do cause fires.

One of the most serious fires in the UK in recent years, in terms of the number of fatalities, occurred in 2004 at the Rosepark Care Home in Scotland. There was a very lengthy and complex fire investigation, followed by a public inquiry in the form of a Fatal Accident Inquiry held between November 2009 and August 2010; the full FAI report was published in May 2011 and is available on the Scottish Courts and Tribunals website.

The cause of the fire, which led to the deaths of 14 elderly residents, was attributed to a wiring defect in a metal single phase distribution board mounted on the wall inside a storage cupboard which had contained a number of pressurised cans of air freshener.

The fire investigation concluded that an earth fault had occurred where a 2-core and earth PVC-insulated and sheathed cable, commonly known as flat-twin-and-earth cable, with solid copper conductors with cross-sectional area of 6 mm^2, passed through a hole in the back plate of the distribution board. The hole, known as a knockout, had sharp edges which should have been covered by an edge protector known as a grommet to prevent damage to the cables that passed through it; however, no grommet had been fitted at the time of installation some 12 years before the fire occurred. The cable distributed power from the distribution board to two washing machines which were in regular use.

The knockout had a number of cables passing through it, and the 6 mm^2 cable had been pressed against the edge of the knockout. Over a period of years, the sharp metallic edge of the distribution board's back plate had migrated through the PVC insulation on the cable's live core to the point at which arcing activity occurred between the live conductor and the earthed back plate, generating hot sparks. The investigation concluded that these sparks escaped from the distribution board and either ignited solid combustible materials in the storage cupboard or, potentially, ignited a flammable atmosphere that may have been generated by highly flammable propellant leaking from a pressurised can of air spray stored in the cupboard. The resulting fire quickly spread, with tragic consequences.

Lead-sheathed vulcanised-India-rubber insulated cable

Before World War II, a favoured method of house wiring used lead-sheathed, twin vulcanised-India-rubber (VIR) insulated conductors which were concealed beneath the wall plaster, run below the floorboards and in the roof space; this is the same type of cable as the one involved in the fatal electric shock accident depicted in Figure 2.14. For earth continuity at junctions, bridging conductors and clamps on the lead sheath were used. In the course of time, these clamps tended to slacken and the joint to corrode, which increased the impedance of the protective circuit. The natural rubber insulation perished, hardened and cracked and separated from the conductor if disturbed. Under these conditions, dampness or conducting foreign matter would cause leakage between the phase conductor, earthed sheath or neutral, or the phase conductor, if disturbed, might contact the lead sheath. The leakage currents might be insufficient to cause immediate fuse failure but might well cause sparking or very high temperatures at the bad contact points which could, in turn, ignite anything flammable in the vicinity and cause a fire.

Most of this wiring has been replaced, but there is still a fair amount installed in older buildings.

Tough-rubber-sheathed and PVC cable

Lead-covered wiring was superseded by the tough-rubber-sheathed (TRS) system where the VIR insulated conductors were contained in a tough rubber sheath with an uninsulated protective conductor. This solved the earth continuity problem, but the wiring became a fire hazard when the rubber perished. After World War II, polyvinyl chloride (PVC) was introduced and replaced rubber for both insulation and sheath on low voltage cables; apart from some teething troubles in its early years, it now has a good safety record and a very long service life. As it burns less readily than rubber it is a lesser fire hazard, however, it contains halogens that can create toxic smoke and fumes, principally hydrogen chloride, when the material is heated or burns.

The toxicity of PVC smoke/fumes has led cable manufacturers to develop insulation materials that are classed as Low Smoke and Fume (LSF) and Low Smoke Zero Halogen (LSZH) (or Low Smoke Halogen Free – LSHF). These polymeric materials should be specified for applications where the toxicity of PVC in fire conditions cannot be tolerated, such as public buildings and enclosed spaces.

Installing PVC-insulated cables in a metal conduit or trunking offers improved protection against mechanical damage and a lower fire risk, because in the event of a fault the metal enclosure acts as a fire barrier.

Other insulation materials

In addition to PVC, which is a thermoplastic material, and the LSF and LSZH materials mentioned previously, there are other types of plastic materials used for cable insulation. The main ones are the thermoplastic material polyethylene (PE) and the thermosetting compounds cross-linked polyethylene (XLPE) and ethylene propylene rubber (EPR). Broadly speaking, they do not contain the halogens that are characteristic of PVC. XLPE is widely used on high voltage cables, largely because it has better temperature, dielectric strength and mechanical characteristics than PVC.

Another type of insulation in widespread use, especially on high voltage distribution cables, is oil-impregnated paper. It does not have the toxicity problems of PVC, but its high weight and costs, together with the complexity of jointing, means that it is being phased out in favour of the thermosetting compounds. However, paper is still used as an insulator in many applications, such as in the windings of transformers, motors and generators.

Mineral-insulated, metal-sheathed cable

Mineral-insulated, metal-sheathed (MIMS) cable has conductors placed inside a metal tube which is filled with magnesium oxide powder that acts as an insulator. If the tube is made of copper, the cable is known as mineral-insulated copper-sheathed cable (MICC). In many applications, the outer metal tube is covered in a plastic outer sheath, commonly coloured red, orange, black or white, which can be a low smoke and fume material. MI cable is commonly used where fire and heat resistance is required, such as for fire and security alarm circuits.

Whereas this type of cable is suitable for use in most environments, it is not so suitable for use where it may be subjected to vibration which can cause embrittlement of the copper sheath, which can crack and break.

Poor connections – high resistance joints

Poor connections in terminals, typically due to loose screws and crimps and poor contacts in socket receptacles or fuse carriers, can lead to localised resistive heating and micro-arcs or sparking which, in turn, can cause corrosion, very high temperatures of conductors and possible ignition of anything flammable nearby.

An example of this effect which has been of some concern in the UK over recent years is fires at electrical fused cut-outs in dwellings, at the electrical intake. A fused cut-out is the component into which an electricity distributor connects its cable where it enters a property, the outgoing cables usually connect into an electricity meter and thence to the distribution board or consumer unit. It typically comprises a fuse in the phase conductor, or fuses in 3-phase applications, and a brass terminal block commonly known as the neutral block. The whole assembly is covered in plastic insulation such as glass-filled polyester or, in the case of older assemblies, phenolic compounds. The cut-out assembly belongs to the distribution network operator whose cable connects into the property, and who is responsible for its safety (see Chapter 7). An example is shown in Figure 2.16, which shows a single phase fused cut-out supplying a consumer unit via an electromechanical induction watt-hour meter and a 2-pole switch disconnector.

The fused cut-out carries the full load current for the premises, so any resistance at its terminals has the potential to create heating effects that may lead to fire; the same is true of terminal connections at meters and main switchgear devices in distribution boards and consumer units where full load current is carried.

One concern is that, over the years and particularly in recent decades, the electrical load taken by domestic dwellings has increased considerably in line with the increasing number of electrical devices in homes – dishwashers, tumble driers, washing machines, microwave ovens, ceramic/induction hobs, IT and audio/TV devices and so on. Higher levels of current imply higher temperatures, which in turn have the potential to increase chemical decomposition and corrosion effects that can eventually lead to higher electrical resistance and thermal failure or fire. The principles behind this failure mode are reasonably well understood, but with the current state of the art it is not possible to predict with any certainty which connections will fail, why they will fail, and how long it will take them to fail. Figure 2.17 shows a cut-out that has been subjected to overheating and Figure 2.18 shows the heating effect concentrated on one of the prongs of the fuse holder.

East Sussex Fire and Rescue Service have reported (*Investigation Report into Fires Originating in Electrical Intakes;* East Sussex Fire and Rescue Service; July 2010) that in the 12-month period from May 2009 to May 2010, it attended 35 fires originating at electrical intakes. Anecdotally, this figure matches the experience of other fire and rescue services and probably reflects the national picture for this type of fire incident.

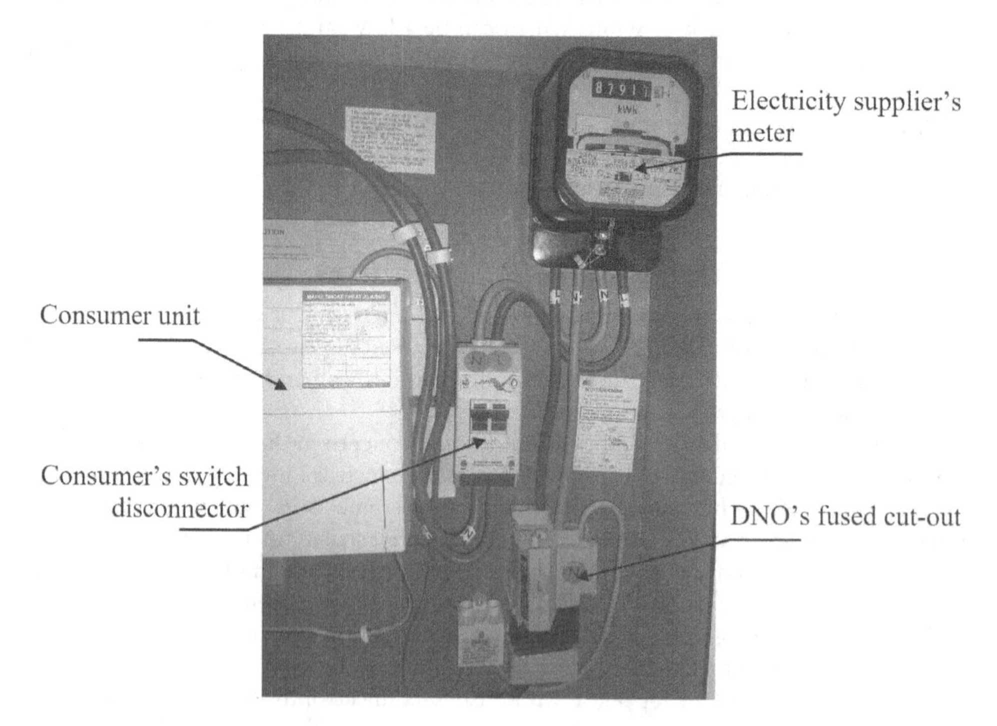

Figure 2.16 Typical domestic electrical intake with single phase fused cut-out

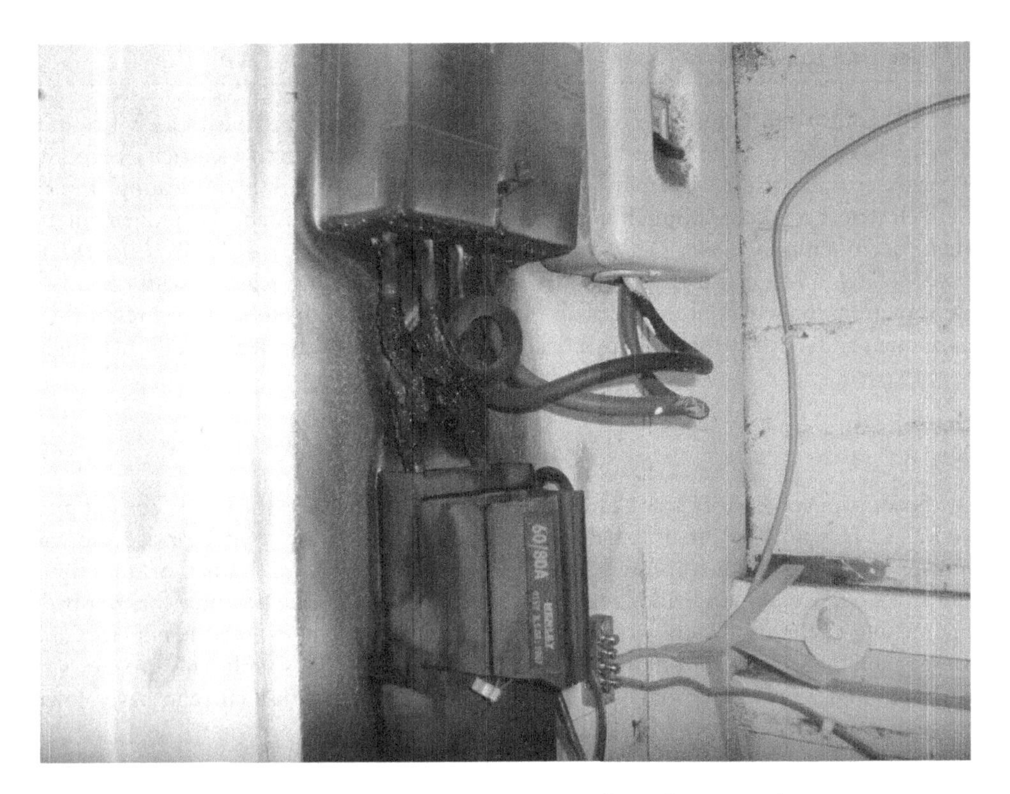

Figure 2.17 Single phase fused cut-out showing signs of significant overheating

Figure 2.18 Remains of overheated fuse assembly from cut-out

In 2012 the High Court (Technology and Construction Court) heard a civil case in which 5 test cases were brought by domestic and small commercial electricity consumers against an electricity distributor for negligence relating to resistive heating fires in their electrical intakes. Although each of the claims was dismissed, the judgement by Mr Justice Akenhead (*Smith & Others v South Eastern Power Networks Plc [2012] EWHC 2541 [TCC]* [17 September 2012]) makes for very interesting reading with regard to fire causation, the legal duties of distribution network operators to prevent fires at their equipment so far as is reasonably practicable and the preventative techniques that might be employed.

Overloading

Overloading of conductors is not a frequent fire causation because the excess current protection, such as a fuse or circuit breaker, is usually matched to the conductor size and will operate before the conductors are seriously overheated. Of course, if a nail or something similar is used in a plug in lieu of a properly rated fuse, there is the potential for overloading, melting of the insulation, arcing and sparking of the newly exposed conductors and consequential fire. It is not unknown for people to replace fuses with nails or screws in BS 1363 plugs supplying welding sets, for example, in which in-rush currents have blown the fuse.

Overheating

Electrical equipment and appliances generate heat and, if not properly designed, installed and used, they have the potential to cause excess heating that can lead to fire. Localised overheating can occur from the radiated and conducted heat from luminaires and in hot environments where cables have not been derated or suitably insulated for the high ambient temperature.

An example of this was a fire at a new spherical storage tank at the BP Chemicals plant at Grangemouth in Scotland, which caused the death of one man and severe injuries to three others. Some scaffolding had been erected to allow polyurethane insulating foam to be sprayed onto the large tank. It is believed that the fire started when a halogen lamp rated at 110 V 250 W located on a wooden scaffold board tipped over such that its face was resting on the board. The radiant heat from the lamp eventually ignited the wooden board, and the men working on the scaffold became trapped inside plastic sheeting that had been erected around the sphere.

The increase in the use of thermal insulation to reduce heat losses in buildings can be a problem, as cables covered by thermal insulation have to be derated to avoid overheating. Any overheating of conductors has the potential to damage most insulating materials and enhances the risk of fire.

The winding insulation in electric motors of the enclosed ventilated type will overheat and may break down if the ventilation is impaired, e.g. by foreign matter such as sawdust or flour dust clogging the air vents and ducts. The resultant short circuit between turns or to earthed metal may cause an ignition, but may not cause a sufficient increase in the supply current to operate the excess current protection.

In one fire incident, a convector heater had a badly designed resistance wire element. Some of the coils touched, resulting in part of the element being short circuited. The increased current in the rest caused overheating of the wire and a progressive break-up of the element. Some of the hot pieces of wire fell through the slots in the base plate on to the carpet below and set the house on fire. Although the increased current was sufficient to burn out the element, it was not enough to rupture the high breaking capacity (HBC) fuse in the plug top.

Explosions

There is sometimes difficulty in differentiating reported explosions from flashover-type incidents, because the latter can easily be interpreted as explosions. However, reported explosions include events such as underground cables being struck and generating explosive failures; the ignition of explosive atmospheres by electric arcs and sparks; and explosions in oil-filled control and switchgear, such as high voltage ring main units and oil-filled transformers.

Oil-filled switchgear failures

Explosive failures of oil-filled switchgear tend to have quite destructive power, largely because the fault level is high; in the UK, the fault level is usually a maximum of 250 MVA on the 11 kV distribution system. For example, an explosion occurred in the current transformer chamber of an 11 kV switchboard while engineers were inside the substation, a small brick building; the explosion was caused by an insulation failure on the bushings inside the oil-filled chamber in which the current transformers were located. Fortunately, the engineers heard the switchgear 'fizzing' moments before the explosion and had the presence of mind to depart the substation in something of a hurry. The blast wave hit them just as they were running through the door, resulting in minor burns to the backs of their necks and considerable sighs of relief.

Historically, much of the earlier oil-filled switchgear had what is known as dependant manual operating mechanism. In this type of mechanism, the moving contacts are directly linked to the operating handle, such that the closing speed and pressure of the contacts is directly related to the effort being exerted by the engineer operating the switchgear. This should be compared with modern switchgear in which the closing mechanisms are spring-assisted.

If a switch, for example, has dependant manual operation and is being closed onto a fault, and the closure is hesitant, the fault current will cause arcing at the contacts. In some designs, the electromagnetic forces will force the contacts apart, making the arcing worse. Alternatively, the operator may sense the increasing pressure and instinctively reverse the closing operation, again making the arcing worse. The rapid oil vaporisation and consequential internal pressure rise may burst the tank and result in the operator being enveloped in burning oil. This reversal of movement can be prevented by using an anti-reflex handle.

Although the number of oil-filled dependant manual switches and circuit breakers is decreasing, there are some still in service, on both low voltage and high voltage systems. The HSE has been concerned for some time that today's engineers may not understand the particular hazards associated with using this type of switchgear on modern high voltage distribution systems, the need to operate it quickly and positively, and the risks associated with reversing the operation when closing onto a fault.

Another concern is that some of the very old circuit breakers that are still found in industrial premises have plain break contacts, without the aids of turbulators and other techniques used to assist in the extinguishing of the arc that is drawn when the contacts open.

There is also concern that some of the older switchgear may be underrated, although most of this switchgear has now been removed. In underrated switchgear, the fault breaking capacity, which may have been adequate when it was first designed, may be lower than the very high fault levels that exist on today's distribution networks. This creates the potential for the circuit breaker to fail explosively when clearing a short circuit fault beyond its design capacity.

Finally, there is concern that the switchgear may not be properly maintained because, for example, skills and knowledge have been lost, or manufacturer's manuals have been discarded.

Although dependant manual and underrated circuit breakers are a significant hazard, they are not the only cause of accidents involving oil-filled switchgear. In another tragic accident, in 1997, two electrical engineers were killed when they were working on an oil filled 11 kV ring main unit (RMU) that was part of a power company's high voltage distribution network. An RMU is a device containing two switches and a tee-off fed through a fused switch or a circuit breaker; the tee-off circuit normally connects to a transformer which steps the 11 kV supply down to 415 V for local power distribution. In oil-filled RMUs, of which there are many thousands in use in the UK, the switching and power distribution components are immersed in oil. A photograph of a typical 11 kV oil-filled RMU (a Long & Crawford T3GF3 unit) in an external substation is in Figure 2.19.

The engineers had been using a set of test probes to connect onto the cable contacts of one of the oil switches in the unit, which was installed inside a brick-built secondary substation building adjacent to a primary substation. This is a commonly used means of testing the condition of one of the cables connected into the unit, usually as part of a fault finding routine or to check the insulation characteristics of the cable. It can be done with the other oil switch closed and the tee-off section still live to ensure continuity of supply to the local consumers.

As the engineers were withdrawing the test probes from the RMU, a steel guide pin became detached from the test probe assembly and fell into the oil tank. It dropped through the oil to the base of the tank, where it shorted one of the 11 kV busbars to earth. The arc that developed caused a gas bubble containing acetylene (a thermal breakdown product of mineral oil) to be formed in the oil and the increasing pressure forced the oil out of the tank. The mist of oil was ignited, possibly by the arc, and the engineers suffered fatal burn injuries. Despite the fact that the fault was cleared in less than 0.5 sec, nearly 90 MJ of energy had been fed into the oil tank.

Distribution substations are now commonly being installed as packaged units inside weatherproof enclosures with pressure-relief panels, an example of which is shown in Figure 2.20, and oil-filled RMUs are progressively being replaced by units that incorporate vacuum circuit breakers or circuit breakers with sulphur hexafluoride (SF_6) insulation, rather than fuses and contacts immersed in oil, an example of which is seen in Figure 2.21.

Explosive atmosphere explosions

Much emphasis is placed on ensuring that electrical equipment in explosive atmospheres, be they gaseous or dusty in nature, is so constructed that it cannot act as a source of

Figure 2.19 High voltage oil-filled ring main unit in an external secondary substation

Figure 2.20 Typical modern packaged distribution secondary substation – the switchgear is installed inside prefabricated weatherproof enclosures

ignition, and the details of this are explained later in this book. Notwithstanding the use of explosion-protected equipment in many places, incidents do occur.

For example, two men were killed and twenty-six injured in August 1996 when a flammable atmosphere composed of a highly flammable solvent was ignited at the Scottish Adhesives Ltd factory in Glasgow, most probably by an electrical arc or spark or an electrostatic discharge. This was despite much of the electrical equipment being housed in flameproof enclosures that are designed to prevent electrical arcs or sparks that routinely occur in electrical switchgear and controlgear from igniting flammable atmospheres. The ignition caused a flash fire from which the two deceased were unable to escape.

Electrostatic discharges are a common mechanism for explosive atmospheres to be ignited, potentially leading to fire or energetic and destructive explosions. Many industrial processes associated with the handling and transporting of chemicals generate electrostatic charges, including blowing powders through pipework and agitating, spraying, and splashing liquids. If a discharge occurs in the presence of a gas/air mixture or dust/air mixture

Figure 2.21 Distribution secondary substation with Schneider Electric Ringmaster 11 kV SF6-insulated switchgear

that is between the lower and upper explosive limits and the energy in the discharge is greater than the minimum ignition energy for the atmosphere, an explosion will occur.

Liquid hydrocarbons, such as petroleum and many common solvents, produce flammable vapours at typical ambient temperatures. They also have high electrical resistivity, which allows them to accumulate high levels of electrostatic charges. The combination of these two features makes them susceptible to ignition by electrostatic discharge.

It is worth keeping in mind that a person wearing insulating shoes is a common example of an insulated conductor that, with a capacitance to earth in the order of 150 pF (picoFarads), can become charged to a level up to 30 kV. A person charged to this level can experience quite painful electrostatic discharges, for example when a hand approaches an earthed metal object. The discharge may cause arcing if the electrical field strength exceeds the dielectric strength of air, which can have serious consequences if the person is working with flammable or explosive substances or in flammable conditions.

Accident statistics

It is helpful to review accident statistics in order to gain an appreciation of the frequency at which electrical accidents occur. In some ways, the use of the word 'accident' to describe incidents is unfortunate because it tends to hide the fact that many of the incidents in which people are injured are entirely preventable, although it is accepted that most are unexpected and unintentional.

Workplace accidents in the UK have been on a welcome downward trend for many years. In the year 2014/15, HSE's statistics (*Statistics on fatal injuries in the workplace in Great Britain 2015 Full-year details and technical notes*; HSE; July 2015) show that 142 people were killed at work, 43 of whom were self-employed and 99 of whom were employees. Broadly speaking, the number of employees killed at work has halved in a period of 20 years, although the accident rate has tended to level off in recent years. The figure is still too large, as each death represents a tragedy not just for the deceased but also for his or her family and friends.

In the same period of 2014/15, there were just over 76,000 non-fatal injury accidents reported to the HSE, although the HSE makes the point that there is significant underreporting of accidents. This is not too surprising, as many less scrupulous employers, and the self-employed, are understandably reluctant to report accidents that may lead to a regulatory investigation and some type of enforcement action. It is also the case that many employers are ignorant of their duty to report accidents or ill health to the appropriate agency.

Of the 142 fatal accidents, 4 were attributed to "contact with electricity", which means death by electrocution or electric burns. Two of the fatal accidents involved employees, and two involved self-employed workers. It is possible that some of the 142 fatalities were attributable to falls consequential to an electric shock or electrically initiated fires and explosions; indeed, during the year, there were three explosion incidents that killed eight people in total, with electricity being a possible ignition mechanism in two of the incidents. However, the publicly available statistics do not allow firm conclusions to be drawn on this.

Of the reported non-fatal accidents, 266 were attributed to employees and 12 to self-employed workers suffering "contact with electricity", which essentially means that the injured persons received treatment after suffering an electric shock or electric burns, and/ or were unable to work for 7 days or more as a consequence of their injuries.

To put electrical accidents into the broader context, over half of the total number of fatal injuries to workers were of three kinds: falls from height, contact with moving machinery, and being struck by a vehicle. Falls and slips and trips, combined, account for over 30% of employee injuries. Electrical accidents account for a small percentage of accidents, somewhat less than 5%.

In the same period of 2014/15, five members of the public were killed as a result of incidents on the electrical supply system; these are incidents that are reportable under the terms of the Electricity Safety Quality and Continuity Regulations 2002 (ESQCR). The most common form of fatality in this context is a member of the public coming into contact with the live wires of an overhead power line owned by a distribution or transmission company. In historic terms, the five deaths were at the low end of the spectrum – in 2012/13, for example, twenty members of the public were killed in ESQCR-reportable incidents.

Contact with overhead power lines is a common cause of electrical deaths. As explained earlier in the chapter, for many years, both in the UK and internationally, such incidents have represented in the order of 50% of work-related electrical fatalities.

Turning to accidents in the home as opposed to at work, in 2014 in England and Wales 27 people died from the effects of electric current in their homes (*UK Office of National Statistics; Mortality Statistics: Deaths Registered in England and Wales [Series DR], 2014;* 9 November 2015) and one person died in Scotland from the same cause (*National Records of Scotland: Vital Events Reference Table 6.4;* 2014). This means that more people died from electrical accidents at home than at work. This should not be too unexpected – more people are exposed to electricity at home than at work, electrical installations and apparatus in the home tend not to be maintained as well as those at work, more do-it-yourself work is carried out in the home than at work, and there are more vulnerable people such as the very young and the elderly at home than there are in workplaces.

It is worth keeping in mind that many people both at home and at work experience electric shocks that do not cause fatal, permanent or reportable injuries. The campaigning charitable organisation Electrical Safety First has reported survey results claiming that, in the UK, some 2.5 million people over the age of 15 experience a mains voltage electric shock each year, of whom 350,000 experience some form of serious injury (http://www.electricalsafetyfirst.org.uk/news-and-campaigns/policies-and-research/statistics/). These figures need to be treated with caution and some scepticism, but, if true, when linked to the national fatality statistics they do illustrate that the probability of being killed as a result of an electric shock is very low. If the probability were to be high, the death rate would be so high as to make the use of electricity unacceptably dangerous.

When considering the two types of electrical dangerous occurrences that have to be reported under RIDDOR 2013, the HSE's statistics show that in 2014/15 there were 122 incidents of contact with overhead lines and 275 incidents of fire and explosions caused by electrical short circuit or overload. These incidents did not lead to RIDDOR-reportable injury, but would have had the potential to cause fatalities. Interestingly, the frequency of these types of incidents does not match the long-term downward trend of the accident statistics.

3

Principal safety precautions

Introduction

The hazards and risks associated with electricity were described in Chapter 1, and the types of electrical accidents and incidents that occur were covered in Chapter 2. The aim of this chapter is to introduce the principles of the most common precautions that are available to prevent accidents. It is useful to have an appreciation of the techniques before the legal requirements for electrical safety are addressed, although the techniques for particular applications will be described in more detail in later chapters, with reference to relevant standards.

Precautions against electric shock and contact burn injuries

The precautions against electric shock described in the following paragraphs also provide protection against the burn injuries that often occur when contact is made with live conductors. The measures can conveniently be grouped into a small number of categories, as follows:

- Techniques known as 'basic protection' that aim to prevent live conductors being touched, including
    - the use of enclosures and insulation; and
    - placing conductors out of reach.
- Techniques that aim to limit the amount and/or duration of current that can flow when a conductor is touched, including
    - the use of reduced and extra-low voltages;
    - the use of electrically separated and unreferenced systems;
    - the use of techniques that limit energy and current to nominally safe levels; and
    - the use of residual current devices as supplementary protection.
- Techniques known as 'fault protection' to ensure that electrical systems become disconnected from their sources of energy in the event of faults that may lead to danger.

These techniques are summarised in the following text.

Enclosures and insulation

Direct contact with live parts is most commonly prevented by covering live conductors with suitably rated insulating material or by placing them inside an enclosure. In the latter case, unless the live conductors inside the enclosure are insulated or placed behind barriers to prevent them from being touched when the door is open, additional precautions must be taken. Typically, these will require that the enclosure door must only be capable of being opened using a key or a tool. Alternatively, the door must be interlocked in such a way that either the conductors must be made dead before the door can be opened or the act of opening the door reliably makes the conductors dead. Additionally, an electrical danger sign of the type shown in Figure 3.1 should be fixed to the door to highlight the danger.

The majority of fixed apparatus is of Class I construction, as defined in BS EN 61140:2002+A1:2006, *Protection against electric shock. Common aspects for installation and equipment*. This standard defines four protection classes, as described in Table 3.1; the

Figure 3.1 Electrical danger sign (black lettering on yellow background)

Table 3.1 Protection classes for enclosures – as defined in IEC 61140:2002

Protection class	*Description*	*Symbol*
Class 0	The equipment is not earthed and has only one layer of insulation covering live parts.	
Class I	The electrical apparatus is mounted inside a metal enclosure which protects the internal wiring and components from damage and prevents direct contact. The metal enclosure must be connected to earth via a separate circuit protective conductor to ensure that the metalwork cannot become energised at a dangerous potential under fault conditions.	⏚
Class II	The equipment is commonly referred to as 'double insulated', and is usually constructed with two layers of insulation or with reinforced insulation covering live parts. A connection to earth is not required for safety purposes.	⧈
Class III	A Class III appliance is designed to be supplied from a separated/safety extra-low voltage (SELV) power source.	◇ III

Table 3.2 The Ingress Protection (IP) code system

First number: protection against solid objects	
0	No protection
1	Protection against solid objects over 50 mm
2	Protection against solid objects over 12 mm
3	Protection against solid objects over 2.5 mm
4	Protection against solid objects over 1 mm
5	Protected against dust – limited ingress (no harmful deposit)
6	Totally protected against dust
Second number: protection against liquids	
0	No protection
1	Protection against vertically falling drops of liquid
2	Protection against direct spray up to 15 degrees from the vertical
3	Protection against direct spray up to 65 degrees from vertical
4	Protection against direct spray from all directions
5	Protection against low pressure jets from all directions
6	Protection against strong jets
7	Protection against the effects of immersion up to 1 metre for short periods
8	Protection against long periods of immersion under pressure

symbols shown can usually be found on the manufacturer's plate attached to the equipment or apparatus.

The extent to which an enclosure prevents the ingress of water and dust is defined using an Ingress Protection (IP) code described in BS EN 60529:1992+A2:2013 *Specification for the classification of degrees of protection provided by enclosures.* The IP code uses two numbers; the first number shows the degree of protection against solid objects and the second number shows the degree of protection against liquids. The salient features of this code are set out in Table 3.2

So, for example, an enclosure with an IP rating of IP55 would be suitable for external use, protected against the ingress of rainwater and dirt. A common code is IP2X, which means that the enclosure is fingerproof but has no rating for protection against the ingress of moisture. The plug pin holes in domestic power sockets to BS 1363 are IP2X. It is common practice to make the internal construction of electrical panels meet the IP2X standard so that the enclosure door can be opened without allowing anybody directly to touch live parts inside the enclosure.

Insulating plastics are commonly used for wiring accessories such as switches, socket-outlets and ceiling roses which usually have no accessible conductive parts. Such enclosures provide complete protection against electric shock provided they are kept dry and clean and are undamaged.

Portable apparatus may be of Class I construction, in which case the exposed metalwork is earthed by a circuit protective conductor, also commonly called the earth wire, which is a separate core in the flexible supply cable. However, modern portable and hand-held apparatus such as power drills and hair dryers are commonly of Class II construction, having no protective conductor terminal, as any exposed metalwork does not need to be earthed. It is not advisable to use ventilated Class II apparatus in wet environments unless it has an appropriate IP rating, because moisture may penetrate and provide a conductive film between the touchable surfaces and internal live conductors.

Safe by position

Uninsulated conductors energised at dangerous voltages can, in principle, be made safe by placing them out of reach. Perhaps the most obvious examples of this technique are the ubiquitous power distribution and transmission overhead lines, energised at 3-phase voltages of 400 V up to 400 kV, that exist throughout the UK. For example, 11 kV overhead lines must be suspended at heights no lower than 5.2 m above ground level, or 5.8 m where the line crosses a road (see Chapters 7 and 14 for more detail), on the basis that this is presumed to be high enough to prevent inadvertent contact. Since 10 to 15 people are killed each year by making contact with these power lines, it's reasonable to observe that the technique is not entirely successful. However, the cost of implementing engineering measures to prevent these deaths, such as insulating all the conductors, increasing their height or replacing them with buried cables, would be prohibitive and not reasonably practicable.

An example of this technique in factory environments can be found in bare conductors running at height along a wall which are used to provide power to an overhead travelling crane; the crane has power pick-offs that run along the conductors as the crane moves to provide power to the drive motors. This technique is satisfactory so long as there is no easy access to the bare conductors from the crane structure or, for example, from ladders placed up against the wall. Modern crane power pick-offs are partially insulated, or shrouded, through most of their circumference so as to reduce the probability of inadvertent contact with the live wires.

Reduced low voltage system

Increased safety is attainable by reducing the potential shock voltage to well below mains voltage. In the UK the most common reduced voltage system operates at 110 V three-phase or single phase. In the former, the star point of the supply generator or double-wound isolating transformer is earthed, resulting in a voltage of 63.5 V between a phase conductor and earth or an earthed neutral. In single phase systems, the centre point of the output winding is earthed, thereby reducing the phase-to-earth shock voltage to 55 V. Fuses or circuit breakers are placed in each phase conductor for overcurrent protection, and a residual current device (RCD) can be used for fault protection; see later in this chapter for a description of RCDs.

Reduced low voltage apparatus is commonly used to supply Class I and Class II portable and hand-held tools and temporary lighting on construction sites, but it is also used by peripatetic workers such as joiners, plumbers, kitchen fitters and electricians, as well as in some factories and similar environments. Although there is no specific legal requirement for reduced voltage systems and apparatus to be used in these environments, they are now so commonly used and available that their use has become standard practice and it is generally considered that they reduce electrical risks so far as is reasonably practicable; for that reason, the regulators are prepared to enforce their use.

The distribution system and apparatus for construction sites are described in Chapter 13, but for factory use indoors the wiring is commonly fixed and the apparatus does not have to be weatherproof or as robust as that employed on construction sites, where the risks of mechanical damage tend to be higher.

110 V systems can be recognised by their yellow step-down transformers, plugs, sockets and connectors, the use of the colour yellow being restricted by BS EN 60309–1:1999+A2:2012 *Plugs, socket-outlets and couplers for industrial purposes – Part 1 General*

Table 3.3 Prescribed colours of accessories for different operating voltages

Rated operating voltage V	*Colour*
20 to 25	Violet
40 to 50	White
100 to 130	Yellow
200 to 250	Blue
380 to 480	Red
500 to 1000	Black

requirements to accessories with an operating voltage in the range of 100 to 130 V. The full colour range prescribed by the standard for power frequency applications (50 and 60 Hz) is shown in Table 3.3.

Safety extra-low voltage (SELV)

For an even greater degree of safety, lower voltages may be used together with techniques to prevent the extra-low voltage conductors becoming energised at higher voltages under fault conditions. Opinions differ as to what constitutes a safe low voltage, but in normal dry environments it is generally considered that no harmful shock will result from handling live parts where the voltage does not exceed 25 V a.c. or 60 V ripple-free d.c. between conductors or between any conductor and earth. In wet and/or in confined conductive locations where the normal body impedance is likely to be reduced, lower voltage limits are needed to avoid danger.

For both normal and abnormal environments, it is essential to preserve the integrity of the extra-low voltage circuit and prevent it from becoming live at a higher voltage from another system. Therefore, in an SELV system, any exposed conductive parts must not be connected to, or be in contact with, the protective conductor of another system, nor with extraneous metal which could be energised by another system.

Where a step-down transformer is used to provide the safe extra-low voltage, it should be a safety isolating transformer to BS EN 61558–2–6:2009 *Safety of transformers, reactors, power supply units and similar products for supply voltages up to 1100 V. Particular requirements and tests for safety isolating transformers and power supply units incorporating safety isolating transformers* or BS EN 61558–2–8:2010 *Safety of transformers, reactors, power supply units and combinations thereof. Particular requirements and tests for transformers and power supply units for bells and chimes.* The arrangement is illustrated in Figure 3.2. Alternative sources of supply are batteries, motor generators and electronic power supplies.

As the safeguard is the extra-low voltage, which results in there being virtually no shock risk, no enclosures are required. Also, if the voltage exceeds 12 V a.c. or 30 V d.c., only basic insulation is employed between live parts and between live parts and earth, although double or reinforced insulation or other separation methods are required between an SELV circuit and any circuits that are not SELV. The lack of enclosures, however, increases the risk of burn injuries from handling live parts in the event of arcing or overheating consequent on a short circuit fault.

Typical examples of SELV equipment at 25 V a.c. and 60 V d.c. are bells, low voltage strip lighting and equipment used in wet environments such as swimming pools.

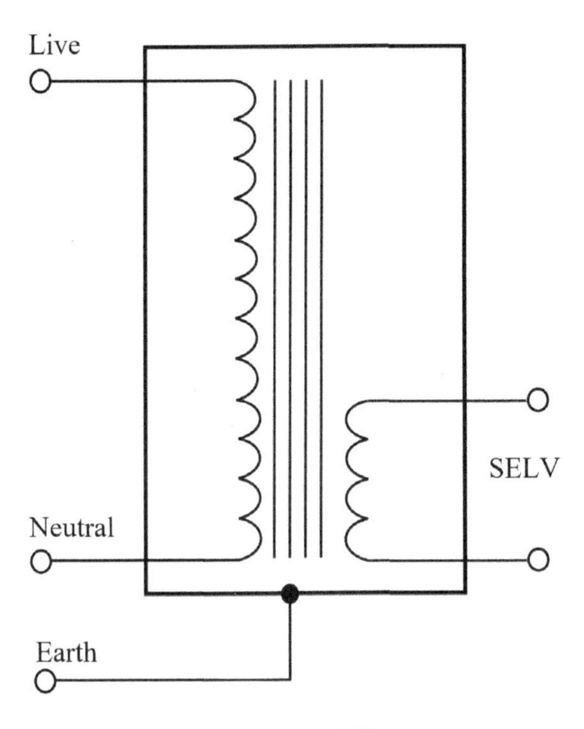

Figure 3.2 Safety isolating transformer supplying SELV circuit

Above 25V a.c. and 60V d.c. up to 50V a.c. and ripple-free 120V d.c., there is a shock risk, albeit a relatively small one in normal dry environments, so some additional precautions are required. Direct contact with live parts should be prevented by insulation, barriers or enclosures. No part of the circuit should be earthed, and exposed conductive parts should neither be connected to a protective conductor nor otherwise earthed. This ensures that two faults would have to occur to create a shock hazard. Again, if exposed conductive parts have to be in contact with extraneous conductive parts, the latter must not be capable of attaining a voltage exceeding that of the SELV circuit. In abnormal environments, reductions in the upper voltage limits, relative to the degree of risk, are necessary for the same level of protection.

Extra-low voltage other than SELV

BS 7671:2008 *Requirements for electrical installations. The IET Wiring Regulations*, 17th edition, covered in Chapter 12, recognises two other forms of extra-low voltage: protective extra-low voltage (PELV) and functional extra-low voltage (FELV). These protection techniques are also recognised in the standard BS EN 61140:2002 (IEC 61140:2001) *Protection against electric shock. Common aspects for installation and equipment.*

PELV differs from SELV only by having its circuits earthed at one point, as illustrated in Figure 3.3 which, when compared with Figure 3.2, illustrates the main difference between the two techniques. The difference means that PELV systems can be rendered unsafe by earth faults in other systems that may lead to the circuit protective conductor in the PELV system becoming energised at a hazardous voltage. This hazardous fault condition does not exist in SELV systems.

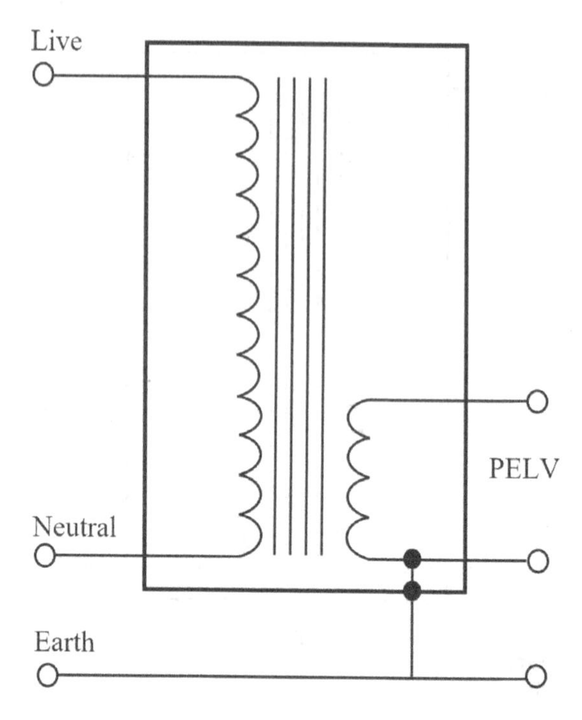

Figure 3.3 Safety isolating transformer supplying PELV circuit

PELV's main application is in low voltage power supplies for IT and similar apparatus.

FELV systems have voltages not exceeding 50 V a.c. or 120 V d.c. but do not have the same level of isolation from other systems as SELV and PELV. This means that a transformer-based supply would not need to be a safety isolating transformer. The insulation should be improved so that it will withstand the same test voltage as applies to the primary circuit, and exposed conductive parts should be connected to the protective conductor of the primary circuit. A typical application of an FELV system would be the low voltage control system of a machine.

Limitation of energy

There are a number of applications where limitation of energy is used as the method of protection. As an example, a common application in the agricultural sector for stock control, and increasingly in the security sector for guarding applications, is electric fences. As explained in Chapter 1, these consist of uninsulated wires, often running to lengths measured in kilometres in the case of stock control fences, which are connected to one or more energiser units. The energiser transmits pulses onto the fence wires, with the peak voltage of each pulse being in the order of 5–10 kV, with a pulse duration in the order of 1 millisecond and pulse repetition frequencies of about 1 Hz. The design aim is to provide a sufficiently severe electric shock to an animal (or intruder in the case of security fences) to deter the animal from touching the fence, whilst limiting the amount of energy delivered to humans touching the fence to a level below that which will cause fibrillation effects. The relevant standard, BS EN 60335–2–76:1999 *Specification for safety of household and similar electrical appliances. Particular requirements. Particular requirements for electric fence energizers*, prescribes an energy limit of 5 Joules into a non-inductive 500 ohm load as being safe – the 500 ohm figure being chosen to represent the typical lowest value of human body resistance at the fence operating voltages. The 5 Joule figure is estimated to be below the energy needed to cause cardiac fibrillation effects in the large majority of the population.

Another example of the use of limitation of energy is found in the test probes used to carry out high voltage insulation resistance tests on cables and appliances; these probes may be energised at voltages in excess of 1 kV. The amount of current that can be delivered by the test set supplying power to the probes is commonly limited to 5 mA to ensure that anybody touching the probes will not suffer electrical injury. In situations where current, and therefore energy, limitation is used in this way, the means of limiting the current must be reliable and such that foreseeable single faults will not cause an increase in current flow.

Although not strictly associated with preventing electric shock or burn injury, a very important application of the energy limitation technique is the intrinsic safety concept used in explosion protection. The technique, designated Ex i and explained in BS EN 60079–11:2007 *Explosive atmospheres. Equipment protection by intrinsic safety "i"* and in Chapter 15 of this book, ensures that any electrical equipment installed inside flammable atmospheres cannot generate sufficient energy during normal operation and under fault conditions to ignite the atmosphere. This allows instrumentation to be installed in areas such as process vessels where a flammable atmosphere is likely to be present during routine operation.

Non-conducting location

Non-conducting location as a form of protection is not much used in the UK because of the practical difficulties of providing it and then ensuring its integrity is preserved. It has an application in the testing of electronic apparatus, where the test area is specially designated for use by authorised personnel with the knowledge and skill for successful operation (see Chapter 16).

The essence of the concept is the prevention of direct and indirect shocks by contact between a live part, or a conductive part made live by a fault, and earth. To this end, insulating walls and floors have the minimum amount of touchable conductive and extraneous conductive parts. Any such parts that could become live under fault conditions have to be spaced as to prevent anyone touching two of them at the same time. The area has to be earth-free, so no protective conductors are employed and exposed conductive parts are not earthed. Extraneous conductive parts such as metal pipes which are in the location and outside it, need protection by insulation or barriers or to be placed out of reach so that, in the event of a fault, they cannot transmit a potential, including earth potential, in either direction.

Earth-free local equipotential bonding

Earth-free local equipotential bonding as a form of protection is for special applications only and might be used in a test area, as described in the preceding section, to connect metal instrument cases and thus prevent a potential difference arising between them, which would be an indirect shock hazard.

Electrical separation

In the electrical separation technique, the supply is not earthed or otherwise referenced – such systems are commonly known as 'isolated' or 'unreferenced' supplies. The source for this type of system is usually an isolating transformer or a small motor generator set, and the circuit voltage is limited to a maximum of 500 V.

The principle, as illustrated in Figure 3.4, is that anybody simultaneously touching one pole of the unreferenced supply and earth will not experience an electric shock because there is no complete circuit back to the point of supply. However, if there were to be an insulation failure that inadvertently connected one pole to earth, the system would become an earth-referenced supply, and anybody touching the live pole and earth would be at risk, as illustrated in Figure 3.4. For that reason, it is very important that measures are taken to prevent earth faults, such as protecting cables against damage and routinely testing for such insulation failures. Figure 3.4 does not show the overcurrent protective devices (fuses or circuit breakers) that must be installed in both poles of such unreferenced systems.

These types of unreferenced systems are common in applications where small single phase generators, typically up to about 4 kVA, are used to supply power tools, mobile food outlets, external lighting and the like. The general advice is that they should be limited to single loads and that cables should be kept as short as possible and protected against mechanical damage so as to prevent earth faults. Additional precautions are required if more than one load is to be supplied; these include ensuring that exposed conductive parts of the separated circuits, such as metal enclosures, are connected together by insulated bonding conductors. However, the bonding conductors must not be connected to earth, any circuit protective conductors or exposed conductive parts of any non-separated circuits, or to any extraneous conductive parts such as water or gas pipes. More information can be found in Section 418.3 of BS 7671:2008.

Earthed equipotential bonding and automatic disconnection of supply

Earthed equipotential bonding and automatic disconnection of supply, abbreviated EEBADS, is the most common technique used for protecting against indirect contact electric shock. It is used in situations where the electricity supply to the premises is referenced to earth, which is the case with all supplies provided by distribution network operators.

There are five main types of earthing arrangement on electricity supplies, as illustrated in Figure 3.5. For simplicity only single phase installations are shown, although the principle applies to 3-phase systems as well. Also, distribution boards, consumer units, metering, and protective devices such as fuses and circuit breakers are not included in the schematics.

In the diagrams, the letter 'T' means 'terra' or earth or ground, the letter 'N' means 'neutral', the letter 'I' means 'isolated', the letter 'C' means connected and the letter 'S'

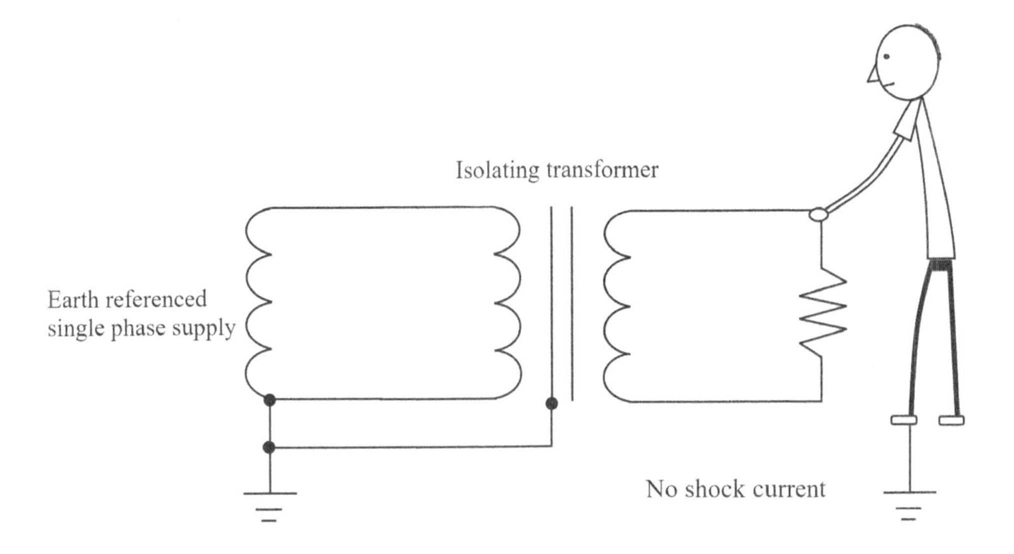

Electrically separated secondary circuit providing shock protection

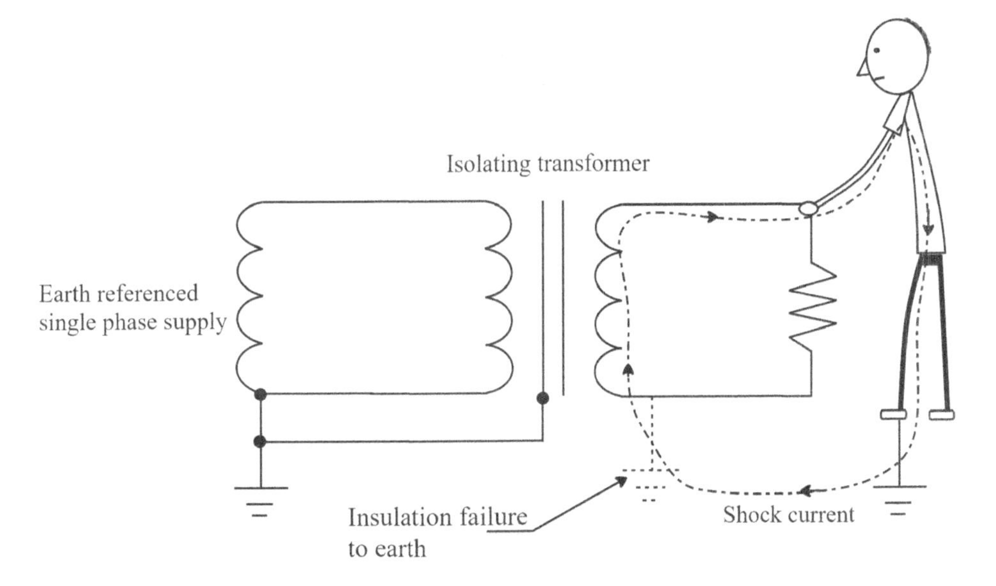

Insulation failure references secondary circuit to earth, removing the protection against electric shock

Figure 3.4 Principle of electrical separation

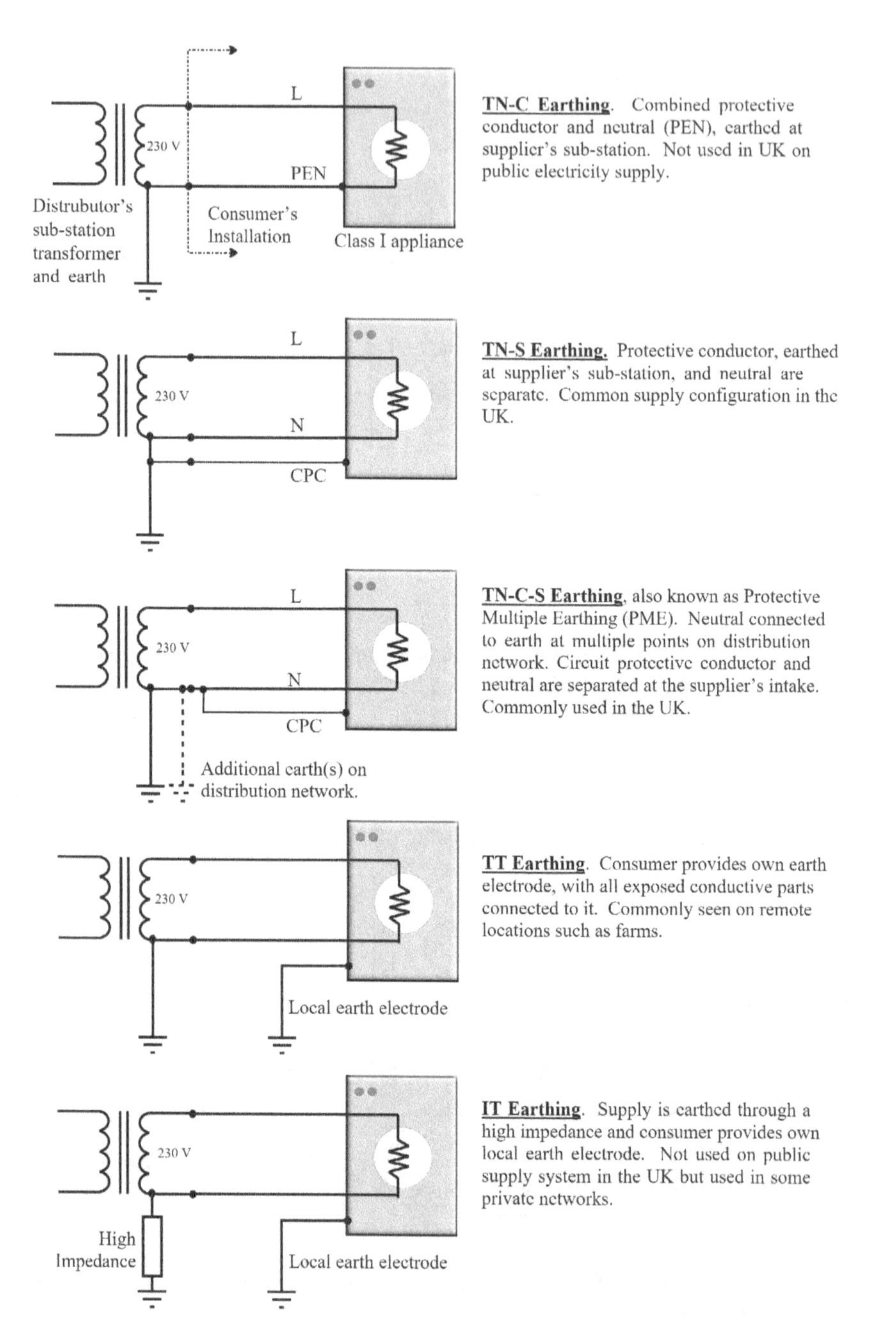

Figure 3.5 Illustration of the principles of the five main earthing techniques on power supplies

means 'separated'. In TN and TT earthing systems, the first letter indicates the connection between earth and the source equipment, be it a generator or transformer, as follows:

T: Direct connection with earth
I: No connection with earth, or connection through a high impedance

The second letter indicates the connection between earth and the load electrical equipment, as follows:

T: Direct connection to earth using an earth electrode
N: Direct connection to neutral at the origin of installation, which is connected to the earth

The principle of the EEBADS protection technique is illustrated in Figure 3.6: although the illustrated overcurrent protective device is a fuse, it could also be a circuit breaker.

Extraneous conductive parts, being metalwork such as gas and water pipes and structural steelwork that does not form part of the installation but which can introduce a potential (usually earth) into the installation, must be bonded together and connected to earth. This is usually achieved using insulated green/yellow bonding conductors to connect water, gas and oil pipes (and any other metallic services) to the main earth terminal. The circuit protective conductors of distribution cables, and hence the exposed conductive parts of any connected equipment, are also connected to the main earthing terminal, usually through an earth bar in the consumer unit or distribution board.

Connecting together the exposed and extraneous conductive parts in this way creates an equipotential zone, which is normally held at earth potential. In areas of higher risk, such as bathrooms and swimming pools, supplementary bonding conductors are used to connect together all the exposed metalwork unless the supplies into these zones are protected by 30 mA RCDs; buildings in which livestock is housed must have supplementary bonding conductors installed.

When an earth fault occurs, connecting a live conductor to exposed metalwork, the fault current finds its way back to the earthed point(s) of the supply system via the circuit protective conductor. The impedance of the fault circuit, called the earth fault loop impedance (Z_s), must be low enough to ensure that the fault current is high enough to blow fuses or trip circuit breakers quickly enough to disconnect the supply so as to prevent danger, so the speed at which these protective devices operate under fault conditions is an important consideration.

A fuse consists of a length of wire (fusing element) that starts to melt when an overcurrent or short circuit current flows through it. As it melts, arcing occurs across its length; eventually the arcs extinguish and the current falls to zero. There are essentially two types of fuses: rewireable fuses in which the fuse wire has to be replaced after operation, and cartridge fuses in which the cartridge is replaced after operation; some cartridges are filled with sand to allow for a high breaking capacity. The speed at which a fuse operates is determined by the fuse element's material and cross-sectional area and the magnitude of the current flowing through it – the smaller the cross-sectional area or the larger the current, the faster will the fuse blow. There is a wide variety of types of fuses and thermal characteristics, but a typical time-current relationship is illustrated in Figure 3.7, for a semi-enclosed rewirable fuse to BS 3036 with a rated current of 30 A.

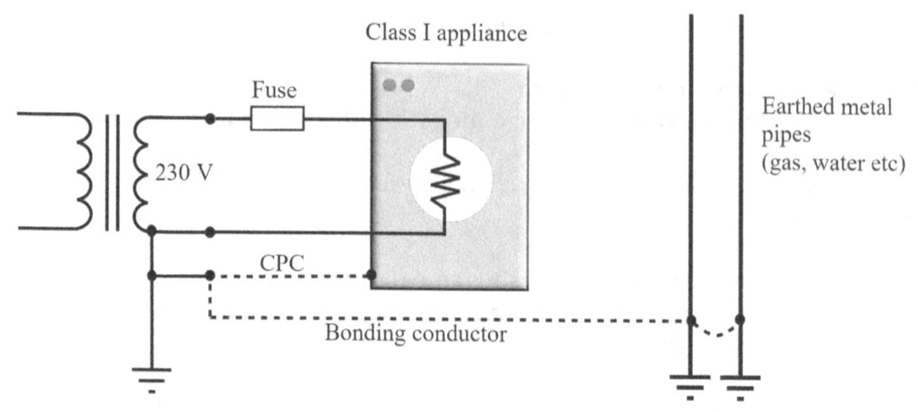

Class I appliance earthed, exposed and extraneous conductive parts bonded, overcurrent device (fuse/circuit breaker) in phase conductor. TN-C-S earthing system illustrated.

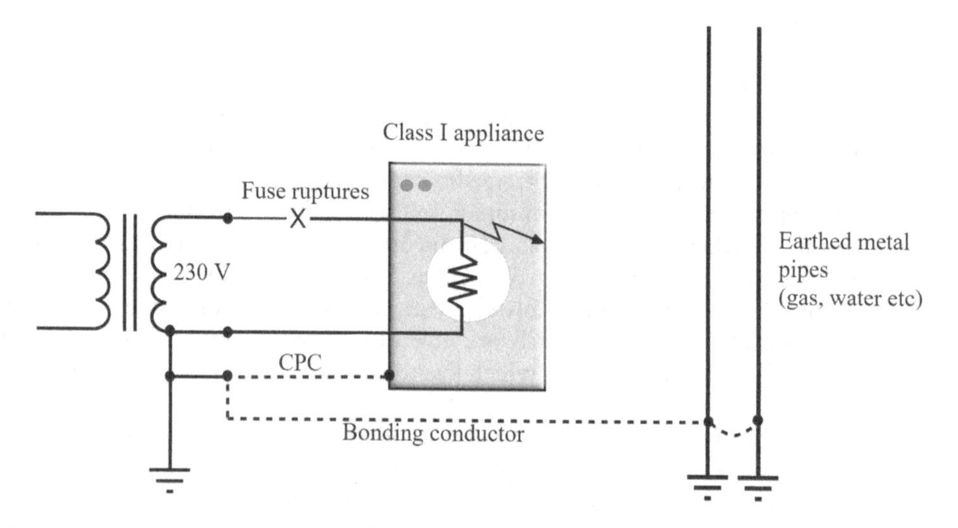

In event of fault, fault current via CPC causes fuse to blow, disconnecting supply. All exposed metalwork is at same voltage during fault, negating risk of electric shock.

Figure 3.6 **Principle of Earthed Equipotential Bonding and Automatic Disconnection of Supply (EEBADS)**

In contrast to fuses with their fusing elements, circuit breakers are mechanical resettable switching devices with a set of fixed contacts and a set of movable contacts able to carry normal load current and to make and interrupt fault current. Figure 3.8 shows the internal parts of the type of circuit breaker commonly known as a miniature circuit breaker.

Circuit breakers have a means for extinguishing arcs that occur when the contacts open, especially when breaking fault current. The stack of copper plates seen in the upper left quadrant of Figure 3.8 form part of the arc chute that lengthens and extinguishes the arc drawn from the opening contacts seen to the right of the stack.

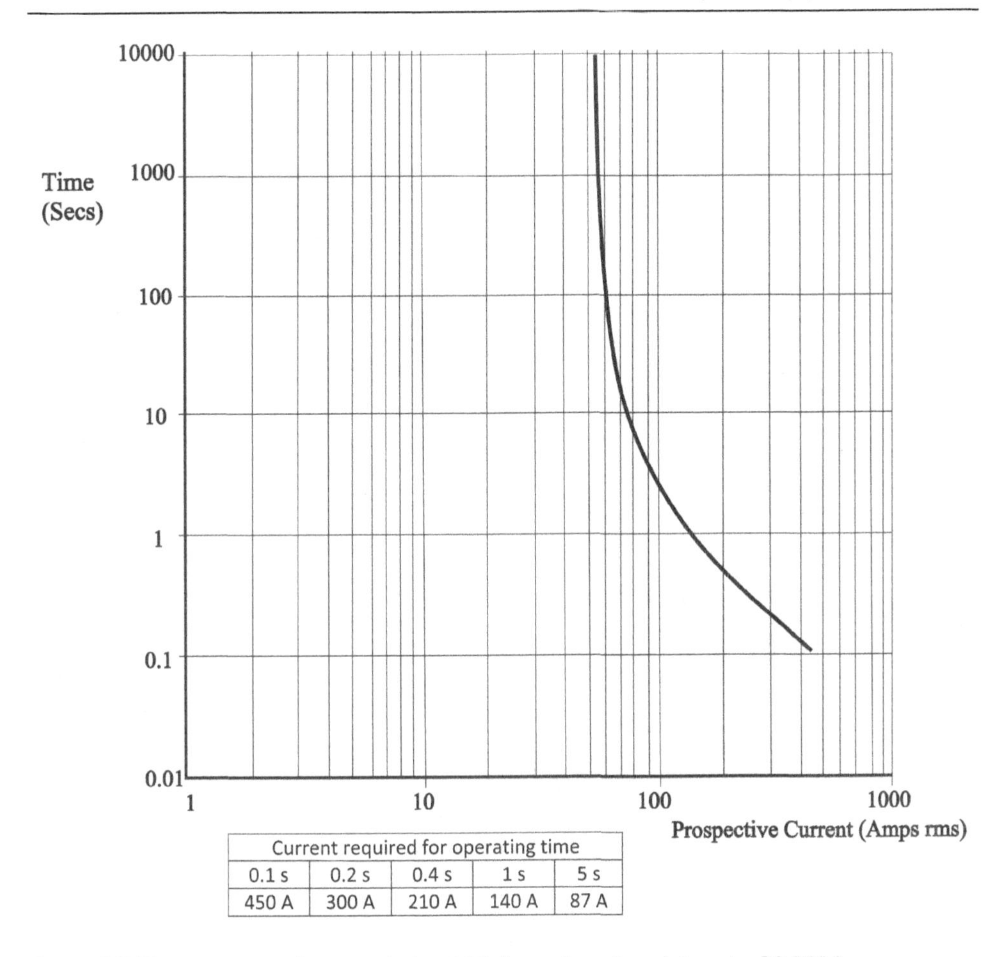

Current required for operating time				
0.1 s	0.2 s	0.4 s	1 s	5 s
450 A	300 A	210 A	140 A	87 A

Figure 3.7 Time-current characteristic of 30 A semi-enclosed fuse to BS 3036

In many applications, the tripping mechanism that initiates the opening of the contacts comprises both a thermal time-current characteristic similar to that of a fuse, commonly based on the operation of a bimetallic strip (seen below the coil of the circuit breaker in Figure 3.8), and an instantaneous characteristic that operates under high current short circuit faults and which is commonly based on the operation of an electromagnetic device such as a solenoid (the coil seen below the arc chute in Figure 3.8). The characteristics of a 32 A Type B circuit breaker to BS EN 60898 are shown in Figure 3.9 as an example of a typical circuit breaker time-current curve; it can be seen that for fault currents higher than 160 A the circuit breaker will trip instantaneously, but for fault currents lower than this the tripping time will be determined by the inverse time curve. There are many different types of circuit breakers for use at low, high and extra-high voltages with different time-current characteristics which, in many devices, can be altered to suit the protection requirements of the system in which they are installed.

The ability to set the time-current characteristic is especially useful when designing discrimination into distribution systems that have protective devices connected in series,

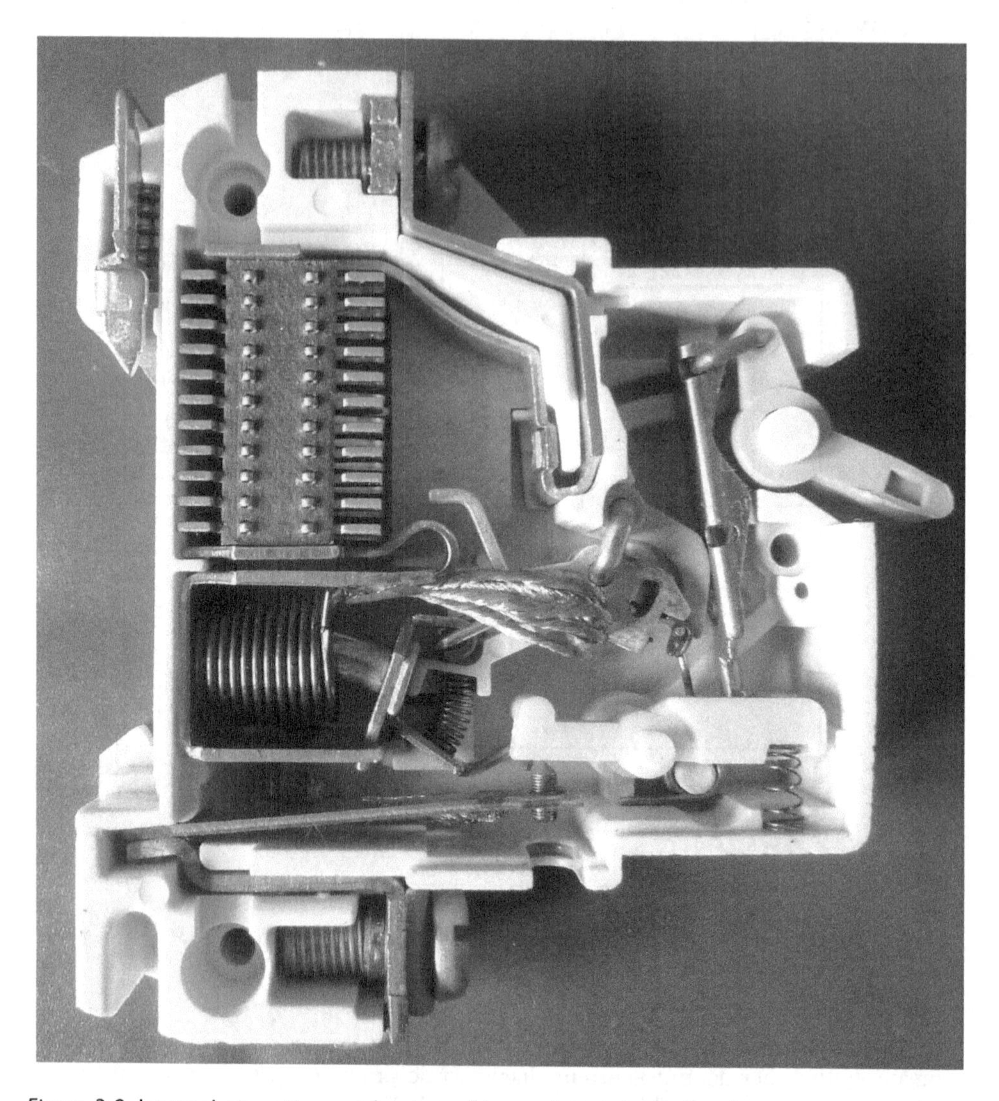

Figure 3.8 Internal operating mechanism of low voltage circuit breaker

as most systems have. Discrimination aims to ensure that the device nearest the fault operates in preference to other devices, thereby minimising the extent of the system that is adversely affected by the fault.

In 230V a.c. systems with TN earthing, BS 7671 stipulates that final circuits not exceeding 32 A, such as those supplying socket-outlets, spur units and lighting circuits, must be disconnected by fuses or circuit breakers in 0.4 seconds; this figure reduces to 0.2 seconds in a.c. systems with TT earthing unless the system contains equipotential bonding and the disconnection device is a fuse or circuit breaker. The required disconnection times for distribution circuits, which are circuits supplying distributions boards and switchgear, is 5 seconds for systems with TN earthing and 1 second for systems with TT earthing.

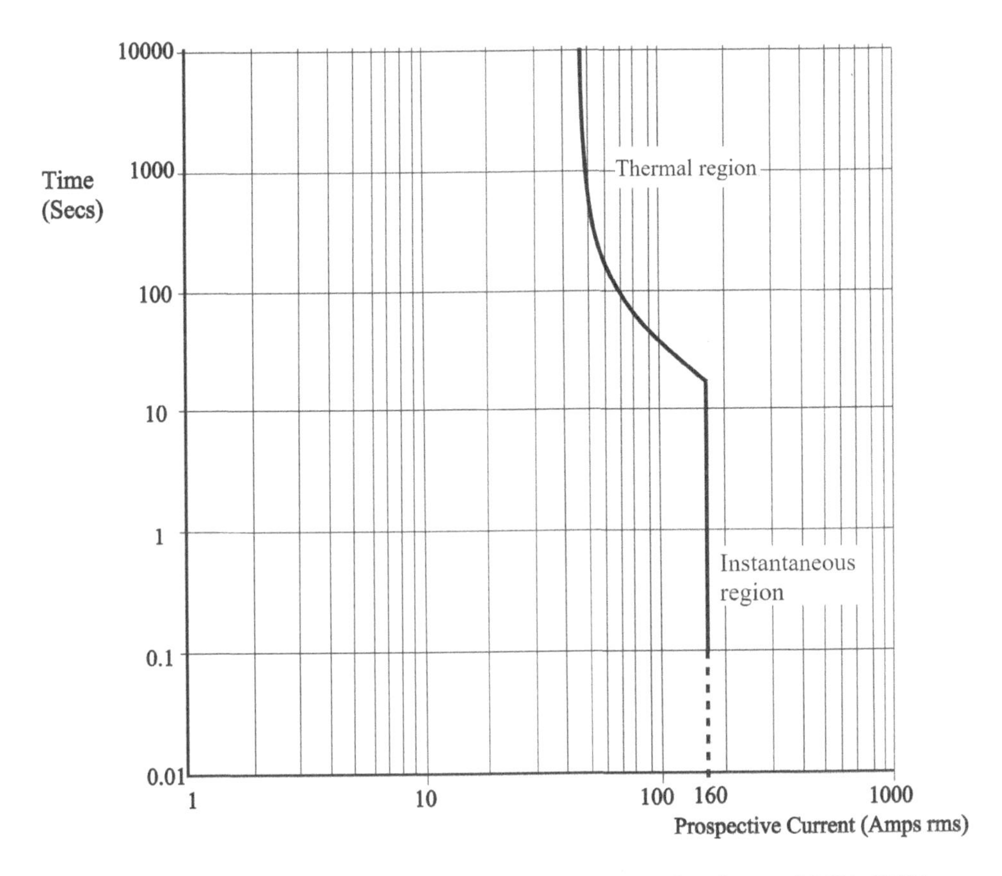

Figure 3.9 Time-current characteristic of Type B 32 A circuit breaker to BS EN 60898

The limiting values of earth fault loop impedance to achieve the required disconnection times for the different types and ratings of fuses and circuit breakers are listed in BS 7671, which also provides the detailed requirements for EEBADS systems (see Chapter 12). In circumstances where sufficiently low values of earth fault loop impedance cannot be achieved, such as on many TT systems, the disconnection device may need to be an RCD, as described in the next section.

For the duration of the fault, until it is cleared by operation of the protective device, the exposed conductive parts will rise to a voltage above earth potential, determined by the magnitude of the fault current and the resistances in the fault circuit; the design requirement is that this voltage should be no greater than 50 V in a.c. systems. Bonding the exposed and extraneous metalwork together to create the equipotential zone has the effect that, for the duration of the fault, all the metalwork in the installation that may be at a potential will rise to the same potential, minimising the risk of electric shock injury to anybody who may be simultaneously touching different metalwork. This is the purpose of creating the equipotential zone.

The integrity of the earthing system is obviously of considerable importance in this type of system, so it must be periodically tested to detect any faults before they lead to danger.

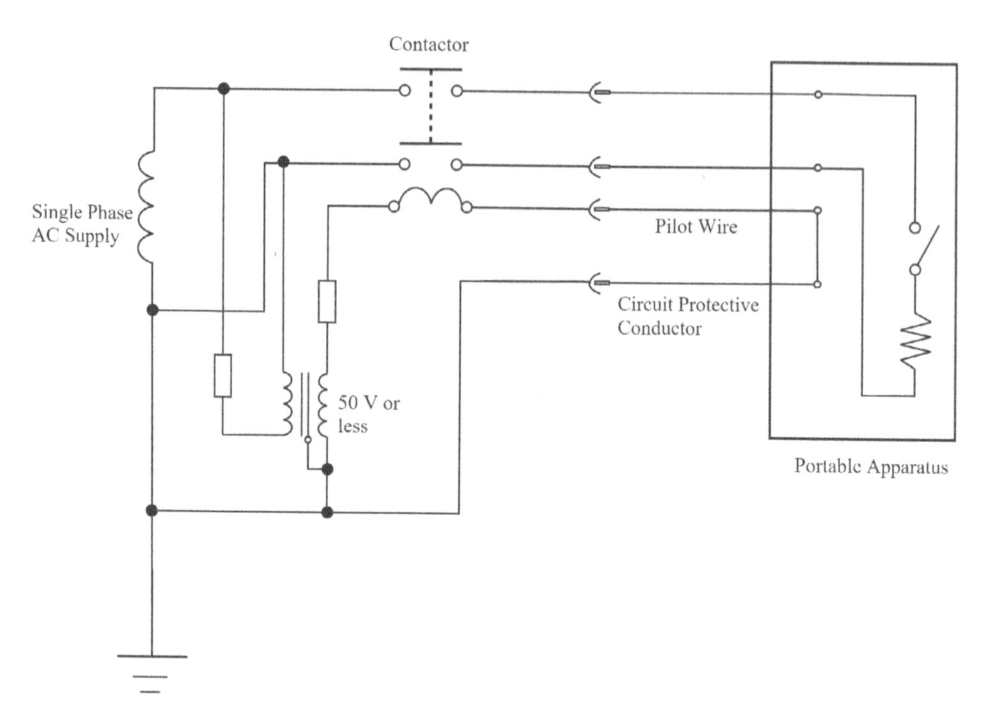

Figure 3.10 **Basic circulating current earth monitoring**

In some circumstances, such as when there is constant flexing of the cable carrying the protective conductor, it may be necessary to take measures to increase the earthing integrity. There are techniques available for this. For example, a second protective conductor core will reduce the likelihood of the loss of earthing. Alternatively, the integrity of the earth system can be continuously monitored by, for example, the circulating current technique.

The basic circuit of a circulating current earth monitoring system is shown in Figure 3.10. It entails an extra core in the flexible cable. The step-down transformer's secondary winding circulates a current of a few milliamps at PELV through a contactor coil, the pilot core in the flexible cable, the protected apparatus and back to the transformer via the protective conductor. Any break in this circuit causes the contactor to open and to switch off the supply. The system can also be applied to multiphase systems. If a braided armoured flexible cable is used, the armour can be employed as the pilot conductor instead of an extra core in the cable. In this case, the armour should have an overall insulating sheath to avoid fortuitous contacts with conducting materials providing a parallel path.

This type of system is becoming less common as a result of the increasing use of reduced voltage (110V CTE) and even battery-operated Class II hand tools in construction sites and other harsh environments. It does, however, have a role in ensuring the integrity of earthing systems where there are high levels of protective conductor current (see BS 7671 Section 543.7).

Earth leakage protection including residual current devices (RCD)

Circuit breakers and fuses are installed in circuits to operate in the event of excess current arising from overload conditions and faults. The most common type of fault is an earth fault, but it is frequently the case that the fault current that flows in the event of an earth fault is too low to operate the overcurrent protection devices, commonly because there is some electrical resistance in the fault. In addition, the overcurrent protective devices will not operate in the event of somebody making direct contact with a live conductor – the current which flows through the body to earth will be too low to operate the devices but will often be high enough to cause serious or fatal electric shock and burns. These two problems can be obviated by the use of earth leakage protection devices.

There are two generic types of devices used for earth leakage detection: those that are voltage operated and those that are current operated. The voltage operated devices are no longer used in the UK but, for completeness, they consisted of a coil connected in series in the earthing conductor or between the metalwork of the installation and an auxiliary earth electrode. The device sensed a voltage rise in the metalwork with respect to earth and, when this occurred, tripped the circuit breaker.

The current operated devices are called Residual Current Devices (RCD) and work on a different principle, as illustrated in Figure 3.11 for a single phase system. When the circuit is fault-free the current flowing in the phase conductor (I_{ph}) will be the same as the current flowing in the neutral (I_n). If there is an earth fault, some current (I_{ef}) will flow back to the source via the earth path, creating an imbalance between the current flowing through the phase and neutral conductors. It is this imbalance that is detected, usually by passing the phase and neutral conductors through a core balance transformer or by electronic circuits. Any current imbalance produces a resultant magnetic flux which is picked up by the sensing coil and which, if it reaches a predetermined level, will cause the trip coil to operate and switch the circuit off.

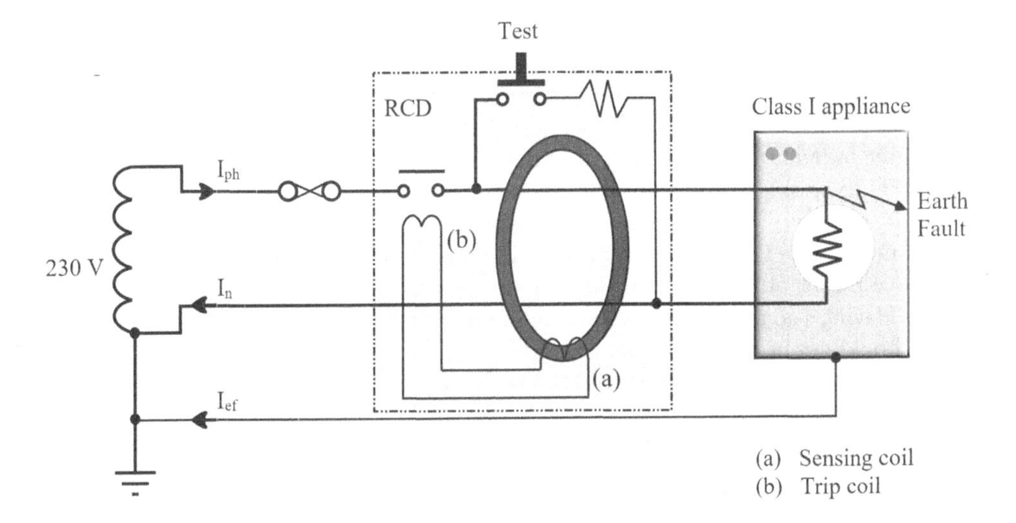

Figure 3.11 Basic RCD in single phase circuit

The current imbalance needed to operate the device varies according to the application. However, when the RCD is provided for protection against electric shock, it should have a rated residual operating current (i.e. the current imbalance that causes the device to operate) not exceeding 30 mA and an operating time not exceeding 300 milliseconds at the rated residual operating current and 40 milliseconds when the residual current is 5 times the rated residual operating current, i.e. 150 mA. The figure of 30 mA is chosen as a compromise between, on the one hand, ensuring that the value is below the threshold of ventricular fibrillation and, on the other hand, preventing nuisance tripping caused by earth leakage current arising from the normal operation of the loads.

RCDs with higher rated residual operating currents (commonly 100 mA but up to 300 mA) are available and are used for fire protection rather than prevention of electric shock.

Earth leakage protection is not restricted to single phase systems or to low voltage systems. The technique is commonly employed on low and high voltage 3-phase systems. On high voltage systems, the core balance method is not the only one used. For example, another way to detect earth fault current is to monitor the amount of current that flows in the earthing conductor at the point of supply, using a current transformer. If the amount of current exceeds a particular value, a circuit breaker will operate to cut off the supply.

Every residual current circuit breaker on low voltage supplies has a test button which, when pressed, creates an imbalance in the phase and neutral conductors passing through the transformer. This allows the tripping mechanism to be tested, although it does not provide a test of the magnitude of the residual operating current or the tripping time; proprietary test equipment is available for this purpose. It is very important that the test button is operated periodically to confirm the RCD's serviceability because RCDs are sensitive devices and it is not uncommon for them to fail to danger (i.e. they fail in a way that means the contacts are closed but the device will not operate on demand to open the contacts).

This failure characteristic means that an RCD should not be relied upon as the sole means of protecting against injury from direct contact. Another reason for this is that, for the RCD to operate in the event of direct contact, current of at least 30 mA must flow through the 'victim'. RCDs rated at 30 mA do not prevent electric shock or limit the amount of current flowing through an individual who may have touched a live conductor. They minimise the amount of time that current greater than 30 mA flows through the body to a nominally safe value as far as effects on the heart are concerned, but they do not prevent the individual experiencing the painful sensation of an electric shock and may not prevent injury arising from the muscular contraction – such as falling off a ladder, or being thrown against a wall. Since the Electricity at Work Regulations require the prevention of injury, and since an RCD may not prevent an injury in the event of direct contact, its use as the sole means of protection against direct contact injury would be unlikely to satisfy the law. Having said that, the device's value in providing supplementary protection against injury should not be underestimated.

There are several different types of RCDs, the main ones being

- Residual Current Operated Circuit Breaker without Integral Overcurrent Protection (RCCB);
- Residual Current Operated Circuit Breaker with Integral Overcurrent Protection (RCBO);
- Socket-Outlet incorporating a Residual Current Device (SRCD);

- Fused Connection Unit incorporating a Residual Current Device (FCURCD);
- Portable Residual Current Device (PRCD); and
- Circuit Breaker incorporating Residual Current Protection (CBR).

These RCDs are subdivided into those that operate 'instantaneously' and those that incorporate a time delay that allows discrimination between RCDs connected in series, the principle being that the RCD nearest the fault should trip first so as to maintain supplies to loads on circuits other than the one immediately affected by the fault.

A further subdivision relates to the RCD's sensitivity to direct current and is of importance if an RCD is protecting a circuit with switched and non-linear loads, such as found in circuits associated with photovoltaic generators and switched mode power supplies; the subdivisions are as follows:

- Type a.c. RCDs that operate for residual a.c. currents;
- Type A RCDs that trip for residual a.c. currents and pulsating d.c. currents; and
- Type B RCDs that trip for residual a.c. currents, pulsating d.c. currents and smooth d.c. currents.

The use of RCDs has become much more common in recent years, and the latest editions of BS 7671 have increased the requirements for their use. Taking BS 7671:2008's recommendations and the general legal duty to minimise risks, RCDs and other earth fault protection devices should be considered for use in the following circumstances:

- On all socket-socket-outlet circuits with a rated current not exceeding 20 A (Exceptions to this are labelled circuits or socket-outlets supplying 'essential' equipment, such as a freezer, and in workplaces, where a risk assessment indicates that the lack of RCD protection would be acceptable.)
- Where mobile equipment with a current rating not exceeding 32 A is to be used outdoors
- On all circuits within bathrooms and shower rooms
- Where there are concealed cables in thin walls fewer than 50 mm deep and the cables are not mechanically protected and not run in safe zones
- Where there are concealed cables within metal partitions and the cables are not mechanically protected
- Where required disconnection times for fault protection cannot be achieved using overcurrent devices such as fuses and circuit breakers
- In situations where there is an increased risk due, for example, to the presence of water (this would include the power supplies to power washers.)
- Where 230 V hand tools and power tools are being used, especially in harsh work environments such as construction sites and workshops
- In special locations such as caravan parks, swimming pools, saunas and marinas
- In high voltage distribution systems, especially those that incorporate overhead lines, where the protection will contribute to risk reduction (this is known as sensitive earth fault protection.)

Many circuits and appliances generate leakage currents to earth through, for example, filters. This means that in larger systems there can be quite a substantial amount of earth leakage current flowing through the protective conductors under normal operating conditions. In these types of installations, a 30 mA RCD installed at the origin can be subject to nuisance tripping, so RCDs should be installed closer to the loads to minimise the frequency of nuisance trips.

Combined techniques

There are circumstances where a combination of the above techniques is used, a good example being welding sets which, by virtue of their design and function, present a risk of direct contact with the live uninsulated welding electrode.

The most common welding process is manual metal arc welding, also known as shielded metal arc welding, where flux-coated stick electrodes are used to strike an arc between the consumable electrode and the piece being welded. The heat created by the arc melts the weld metal and thereby allows the parts to be joined together. The other main welding techniques are

- MIG (metal inert gas) and MAG (metal active gas) systems in which the welding electrode is a gas-shielded uncoated wire, fed from a gun held by the operator
- TIG (tungsten insert gas) in which a tungsten electrode which does not come into contact with the work piece is used (The arc is struck and maintained by ionising the gas between the electrode tip and work piece with a high voltage which is sometimes at a high frequency.)
- Plasma arc welding and cutting

Some welding work is carried out using automated welding machines, particularly in heavy fabrication work, but the majority is carried out using manual techniques in which the operator holds the welding electrode. In manual metal arc welding, the power source is usually constant current and is commonly either a transformer with inductive control of the welding current or a rectifier/inverter system; the two types of power source are illustrated in Figure 3.12.

The open circuit voltage on the welding stick is commonly in the order of 80 V, either a.c. or d.c. depending upon the type of welding set being used, falling on load to some 20 to 40 V depending on the load current. From an electrical hazard perspective, the use of d.c. welding voltage is safer than using a.c. voltage. When used in environments with an increased risk of electric shock, such as confined spaces, BS EN 60974–1:2012 *Arc welding equipment. Welding power* stipulates that the no-load voltage must not exceed 113 V d.c. or 68 V a.c. (peak) and 48 V a.c. (RMS). Welding current ranges between 50 A and 600 A.

This is a form of reduced low voltage system, although the shock voltage may be greater than the nominally safe voltage of 50 V in dry conditions. Because the operator is exposed to the hazard of direct contact with the uninsulated electrode energised at a potentially hazardous voltage, measures additional to the reduced voltage must be taken to prevent injury. The risk of electrical injury from direct contact with the live uninsulated electrode is controlled in the first place by using insulated electrode holders and by the operator wearing heavy duty insulating gloves. However, the risks may not be controlled sufficiently should the gloves become wet. For this reason, the operator instructions issued with welding sets emphasise the importance of keeping the gloves and other clothing dry.

In order to prevent the electrode becoming energised at higher voltages under fault conditions, the welding circuit must be isolated from the mains supply to the same extent

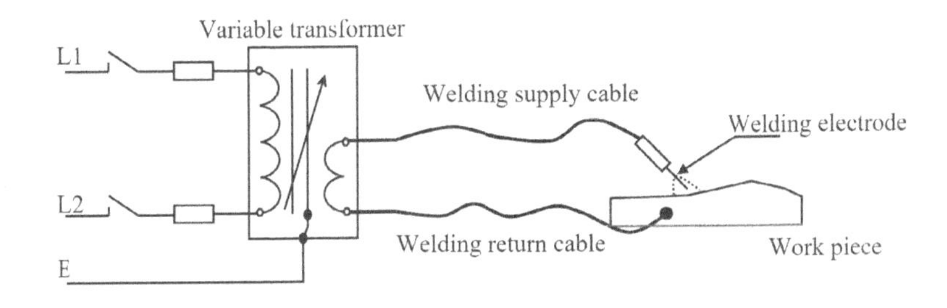

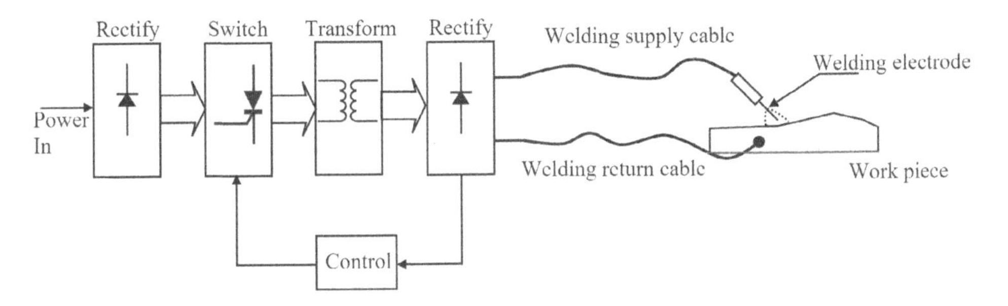

Figure 3.12 **Constant current transformer welding power source (a.c.) and modern semi-conductor inverter power source (d.c.)**

as achieved by a safety isolating transformer. As long as this standard is achieved, there is no need to earth the work piece or the welding return circuit because it will not become energised at voltages above the electrode voltage under foreseeable fault conditions. Indeed earthing the work piece may introduce additional hazards because the high levels of welding return current may return to the power source by various routes other than the return cable, creating the risk of fire.

The welding set itself may be of Class I construction, in which case the enclosure should be earthed, or of Class II construction.

Use of socket covers

Socket covers are plastic assemblies with a face plate and three pins with rectangular cross sections that are meant to mimic the construction of a 3-pin plug to BS 1363. For many years they have been marketed as safety devices that, when plugged into a socket-outlet, prevent people such as children and the elderly interfering with the socket-outlet in a fashion that may be dangerous. The main markets have been dwellings, nurseries, primary schools, care homes and the like.

In recent years there has been a campaign against their sale and use resulting in some organisations such as NHS Trusts in England and various local authorities throughout the UK banning their use. The campaign's main safety-related objections have been as follows:

- Some designs of socket cover have oversized pins aimed at ensuring that, when plugged in, the cover's pins are securely gripped by the socket-outlet's sockets, thereby making it difficult for the socket cover to be withdrawn from the socket-outlet by children and the elderly. The oversized pins have the potential to loosen the socket-outlet's internal brass sockets, leading to the possibility of loose connections between a plug's live and neutral pins and the sockets with a consequential risk of overheating and fire.

- Some designs of socket cover have had face plates that are too small, leading to the live and neutral sockets in the socket-outlet being exposed when the socket cover is plugged in. This would allow thin metal objects to be pushed into the sockets while the socket-outlet's internal shutters are open, leading to the risk of electric shock and burn injuries.
- It is possible for the socket cover to be inverted and the plastic earth pin to be pushed into a socket-outlet's earth socket. This would force open the shutters in the socket-outlet's live and neutral sockets, allowing metal objects to be inserted, leading to the possibility of electrical injury.

This author's opinion is that the only significant safety advantage of socket covers is that, when plugged into a socket-outlet, they deter children and others from plugging potentially hazardous appliances into the socket-outlet, switching the appliances on, and playing with them in a fashion that may cause injury; examples would be hair straighteners, hair dryers, kitchen appliances, and power tools that a child, for example, may be able to get a hold of in a typical domestic environment. This advantage should be realised only by socket covers that have been designed with correctly dimensioned pins and face plates.

The campaign is right to argue that socket covers with oversized pins and undersized face plates should not be placed on the market and should not be used. The hazards which the campaign has identified could be removed by ensuring, firstly, that socket cover pins have the same dimensions as the plug pins specified in BS 1363 and, secondly, that the face plate has at least the same length and width dimensions as a BS 1363 plug. The face plate can be designed in such a way that it is more difficult to grip and remove the cover when compared to a standard plug. In this regard, it would be very helpful if a new British standard were to be produced or an addendum were to be added to the existing BS 1363 to describe, among other things, the dimensional limitations and preferred materials of construction for these devices.

The campaign is also correct to argue that some designs of socket covers can be inverted and the socket-outlet's shutters opened by pushing the cover's earth pin into the earth socket. However, it is equally correct to argue that this can be done using inverted BS 1363 plugs and any other implement with dimensions that allow it to be inserted into the earth socket, such as pencils, pens, screwdrivers and so on. Socket covers are not uniquely hazardous in this respect and only become hazardous if a child, for example, is determined to open the socket-outlet's shutters and poke something into the live socket.

Precautions against burn injuries

Contact burns

Contact burn injuries, either through direct or indirect contact, can be prevented by the techniques explained in the preceding section. However, flashover or arc burn injuries occur without direct or indirect contact with live parts, and the next section describes the preventive techniques available.

Arc burns

As explained in Chapter 2, many burn injuries are caused by the effects of arcing resulting from insulation failure and short circuit incidents, especially on systems with high fault levels. The main hazards are

- extreme temperature;
- explosive forces (shock wave) and fast moving debris;
- high noise level;
- very bright light including ultraviolet light;
- emission of plasma;
- toxic smoke and fumes; and
- hot/burning liquid (in oil-cooled or -insulated material).

Explosive arcing failures of switchgear during operation are now thankfully rare, but the risk of injury can be minimised by

- ensuring that the equipment is suitably rated for its duty (e.g. ensuring that it is designed to interrupt the peak prospective short circuit current on the system in which it is installed);
- where possible, reducing the system fault level and/or altering protection settings so as to minimise the amount of energy released under fault conditions;
- ensuring that the equipment is properly maintained in accordance with the manufacturer's instructions;
- ensuring that switchgear does not have dependent manual operation where the speed of closing the contacts is dependent only on the speed at which the operator operates the closing mechanism;
- operating the switchgear remotely, possibly using a lanyard or other form of remote control device;
- ensuring that off-circuit devices such as transformer tap changers are not operated on-circuit;
- replacing oil-filled switchgear with vacuum or SF_6 switchgear; and
- ensuring that any operating restrictions promulgated by manufacturers, trade bodies or other users are adhered to.

Whenever possible, the risk of flashover during work on live equipment should be avoided by working on systems that have been made dead and securely isolated. If this is not possible, then detailed preplanning of the task should be done to minimise the risk, culminating in a suitably detailed risk assessment. Insulating screens between live parts of different polarity and between live parts and earthed metalwork can be used and an insulating mat or stand provided for the operator. Tools should be insulated and can be magnetised to assist in the safer positioning or withdrawal of ferrous parts. Since bare instrument probes have caused short circuit incidents and consequential injury, only the contact points should be bare and have a maximum length of exposed metal of 4 mm, with the rest of the probe insulated. For some applications, probes with retractable contacts can be used with advantage. Advice on the safety standards for electrical instruments for use by electricians is published in the HSE's guidance note GS38 *Electrical test equipment for use by electricians*.

During the past decade there has been an increasing take-up in the UK and the rest of Europe of a preventive measure which originated in the United States known as arc flash protection, where the term arc flash has the same meaning as flashover. The measure requires that an arc flash analytical study is carried out to determine the incident energy exposure that personnel working on or near energised equipment could receive if an arc flash occurs. The study builds on the short circuit or fault level studies that should be carried out on distribution systems, especially high voltage systems, to determine fault levels, protection settings for fuses and circuit breakers and the requirements for adequate discrimination between protective devices (also known as grading studies).

For each item of equipment the analysis identifies the flash protection boundary, which is the distance at which the onset of second degree burns could occur in the event of an arc flash incident. It also identifies the incident energy at assigned working distances throughout the electrical system, where incident energy is measured in Joules/cm^2 (J/cm^2) or calories/cm^2 (cal/cm^2). Second degree burns occur when the incident energy on bare skin is 5 J/cm^2 or 1.2 cal/cm^2. Incident energies are calculated using methods described in either the standard IEEE 1584–2002, *Guide for Performing AC Arc Flash Calculations*, or the guidance document published by the National Fire Protection Association, NFPA 70E–2000, *Standard for Electrical Safety Requirements for Employee Workplaces*. Both of these are American codes and matched to American legislative requirements for electrical safety.

The standards require that arc flash hazard labels are attached to the equipment included in the study. The label must include information such as nominal system voltage, arc flash boundary and guidance on the required level of Personal Protective Equipment (PPE).

The results of the arc flash calculations are used to allocate risk categories and thereby identify a level of arc flash PPE that personnel should wear when working on or near the equipment. A balance has to be drawn between, on the one hand, the type of PPE needed to provide sufficient protection to prevent second degree burns and, on the other hand, the need to avoid providing more protection than is necessary, as hazards may be introduced by the garments themselves, such as poor visibility and restricted movement. The American codes describe five levels of risk category, which are shown in Table 3.4. This is the minimum level of PPE, quoted in cal/cm^2, required to protect the worker from the thermal effects of an arc flash at a distance of 18 inches from the source of the arc.

Table 3.4 Arc flash risk categories and minimum levels of thermal protective PPE

Min incident energy, J/cm^2	*Max incident energy, J/cm^2*	*Risk category*	*Min level of PPE, cal/cm^2*
0	5	0	2
5.001	16.74	1	4
16.741	33.47	2	8
33.471	104.6	3	25
104.601	167.36	4	40
167.361	And above	5	100

Manufacturers of arc flash PPE supply a variety of items manufactured from thermally rated materials and which are designed to meet the various risk categories; examples are shown in Figure 3.13.

This protection philosophy has obvious benefits, such as requiring companies to assess the arc flash hazards and risks from their systems. However, companies in the UK, or the

Figure 3.13 Two standards of arc flash PPE (by kind permission of J & K Ross Ltd) – 8 cal/cm^2 PPE and 40 cal/cm^2 PPE

Figure 3.13 Continued

rest of Europe, that intend to adopt this arc flash protection technique need to do so with some caution. In particular, companies should not use the availability of thermal protective PPE as an excuse to carry out hazardous live work where there are alternatives to carrying out the live work or where the risks are not reasonable (see Chapter 6 and the discussion on Regulation 14).

There is no legal duty in the UK for companies to carry out arc flash surveys of the type described in IEEE 1584 and NFPA 70e, nor is there a duty to display arc flash hazard notices and signs; that is not to say that these measures should not be taken, but just to observe that the UK's law does not specifically require them to be done. Furthermore, it is important to appreciate that the wearing of PPE to protect against thermal effects of an arc flash is at the lower end of the preferred hierarchy of protective measures and that European legislation requires that, where practicable, effective risk control measures are considered and taken before allowing people to carry out hazardous work where they rely on PPE for their protection (see Chapter 8 and the discussion on the Management of Health and Safety at Work Regulations 1999). The hierarchy of risk control measures to be taken before relying on PPE are as follows:

(a) Where possible and reasonably practicable, avoid the risk altogether by taking the engineering measures listed earlier in this chapter.
(b) Evaluate arc risks that cannot be avoided by carrying out a risk assessment and recording its significant findings.
(c) If, in the final analysis after all other protective and preventive measures have been considered and adopted, work has to be done that presents a risk of flashover, then arc flash control measures involving competent people wearing appropriate PPE must be implemented.

Precautions against fire

A low voltage installation designed and constructed to the requirements of BS 7671, described in Chapter 12, is not a fire hazard if it is properly maintained. The risk of ignition from the several possible causes listed in the following subheadings can be minimised by taking the described precautions.

Overloading

Overloading occurs when the current demanded by a load exceeds the safe and rated current-carrying capacity of the cables and switchgear in the circuits supplying the load. The potential consequence is overheating of the cables and switchgear, leading to the risk of fire. The risks arising from overloads can be prevented by taking the following precautions:

- Cables should be of adequate cross-sectional area for the load, having regard to environmental conditions, e.g. under or in thermal insulation or in hot environments, and the extent to which they are grouped with other cables.
- Cables should be protected by correctly rated excess current protective devices – a fuse or circuit breaker with rated operating current less than the current-carrying capacity of the cable being protected but greater than the expected load current.
- Apparatus such as motors which may be subject to overloading should be provided with their own excess current protection and, where necessary, single phasing protection.

Earth faults

Protection against the effects of earth faults should be taken where prescribed by the standards and described in the preceding section on earth leakage protection. It is increasingly common for 100 mA and 300 mA RCDs to be installed to enhance protection against fire.

Arc fault current

Not all arcs are caused by earth faults. Phase-to-phase or phase-to-neutral arc faults that have the potential to initiate a fire will not be detected by RCDs, and the fault current may be too low to operate overcurrent protection such as a fuse or circuit breaker. One option that may be considered is that of the arc fault detection device (AFDD), also known as an arc fault circuit interrupter. This is a unit about the size of a standard 2-pole disconnector that analyses voltage and current waveforms and, when the waveforms have the characteristics associated with the presence of an arc fault, will interrupt the affected circuit. Such devices are widely used in the United States and other countries, and although their use is not currently a requirement of British standards, it is likely that it will become so in the coming years.

Short circuits

In order to prevent danger from short circuits, ensure that circuit breakers and fuses are adequately rated for the potential fault level, or prospective short circuit current, and can safely and rapidly interrupt the short circuit current.

Wiring, accessories and equipment

The type of wiring used in circuits should be suitable for the environmental conditions and its use. Examples of methods for ensuring this are the provision of

- adequate protection against mechanical damage and the ingress of dust and moisture (the IP rating must be appropriate for the environmental conditions); use of armoured cables where appropriate; installation of cables in trunking, conduit, ducts etc; cable tiles or tape above buried cables;
- consumer units made of non-combustible material or installed in non-combustible enclosure;
- fire sealing, stopping and protection for any cables or accessories that penetrate fire-rated walls;
- heating cables such as under-floor heating installed and protected in accordance with the manufacturer's instructions;
- MIMS cable in very hot locations such as on furnaces;
- corrosion-resistant materials in polluted atmospheres;
- flexible conductors where there is cable movement; and
- explosion-protected apparatus and wiring in flammable and explosive atmospheres.

Luminaires

Luminaires, especially of the recessed variety, have the potential to ignite combustible materials if not suitably selected and installed. Luminaire manufacturers provide fitting instructions to minimise the fire risk; special attention should be given to

- ensuring that the luminaire has an IP rating appropriate for the environment in which it is installed;
- the amount of heat produced by the luminaire, including the maximum rating of the lamps that should be fitted in it;

- the risk of adjacent combustible materials being ignited by radiant or conducted heat;
- the recommended clearances to structures and materials such as thermal insulation;
- the requirements for heat-resistant supply cables and external wiring; and
- the advisability of fitting a non-combustible box around the recessed parts of the luminaire.

Poor workmanship

A properly designed installation can be a fire risk if it is badly constructed. Only properly trained and competent operators should be employed on electrical installation work, and there must be adequate skilled supervision on site to ensure

- compliance with the specification;
- a high standard of workmanship; and
- that all the relevant initial verification inspections and tests are done and that the results are satisfactory.

Poor maintenance

Inadequate and poor maintenance enhances the fire risk so a planned fault reporting, periodic inspection, test and servicing programme is necessary to prevent the occurrence of most faults and the early discovery and rectification of those that do occur. The fault reporting system should identify faulty apparatus between inspections. The inspections should look out for signs of overheating and should reveal such potential fire hazards as

- loose connections;
- loose cable grips;
- dirty, misaligned or damaged contacts;
- contaminated oil in transformers, switches and control gear;
- physical damage to plugs, sockets, apparatus, enclosures and wiring;
- dust ingress into enclosures;
- corrosion; and
- cooling vents on motors and apparatus covered in dust, cardboard, cloth and other detritus.

One option increasingly being used in industrial and commercial premises is thermal imaging, where a camera that operates in the infrared part of the spectrum is used to detect hot spots. Whereas the technique has advantages and positive benefits, it is not without its problems and limitations. This topic is covered in greater detail in Chapter 18.

The testing programme should prove the integrity of the protective and bonding conductors, that the insulation values are adequate, and that the protective devices operate correctly.

A problem with periodic inspection programmes is that they will rarely detect incipient failure of connections before they overheat and create the conditions for a fire. This is because the final breakdown mechanism stemming from long-term resistive heating tends to occur in the space of hours, possibly days. This thermal runaway effect can occur years, even decades, after the equipment was installed, as witnessed in the cut-out fires described in Chapter 2. The chances of a periodic inspection detecting the thermal breakdown before it leads to dangerous overheating are quite slim. This emphasises the importance of ensuring the equipment is installed in the first place to a high standard using good quality components.

Additional precautions in high fire risk locations

In timbered buildings, factories and warehouses containing flammable materials, oil depots and refineries and the like where the economic consequences of a fire merit extra precautions, the installation should be designed to minimise the fire risk consequent on the occurrence of faults. The following recommendations should be considered:

- Use air-insulated, vacuum or SF_6-insulated switchgear and controlgear instead of oil-insulated equipment. If oil-filled equipment such as transformers is installed internally, there should be a bund to collect oil spillage, and they should be in a separate fire-protected area with fire detection and suppression systems.
- Use air-insulated rather than compound-filled busbar chambers.
- Use metal-clad switch and controlgear.
- Use a metal-clad wiring system. MIMS cables should not be plastics-covered. Insulation on cables should be selected for its fire propagation and smoke/fume properties.
- Install RCD protection not exceeding 300 mA on distribution circuits, other than where mineral-insulated cables or busbar trunking or powertrack systems are used. The RCD's rated residual operating current should be reduced to 30 mA if a resistive fault that may cause a fire is foreseeable.
- Consider installing arc fault detection devices in final circuits.
- Wiring should be routed as far as possible clear from flammable materials and secured to non-flammable surfaces and situated where it is not vulnerable to damage.
- A fireman's switch should be provided on the outside of a building so that in the event of a fire, the installation therein can be made dead and the fire dealt with safely and expeditiously.

Precautions against explosions

The precautions described in the preceding sections on burns and fires also serve to prevent the explosions caused by short circuits in high fault level circuits. Reference should also be made to Chapter 15 for the relevant measures to avoid the explosions that can occur when an underground cable is damaged during excavation work.

Oil-immersed apparatus needs to be properly maintained if explosions are to be avoided, as referred to in the preceding section on fires, 'Poor maintenance'. Refer also to BS 6423:2014 *Code of practice for maintenance of low-voltage switchgear and controlgear* and BS 6626:2010 *Maintenance of electrical switchgear and controlgear for voltages above 1 kV and up to and including 36 kV. Code of practice for low voltage and high voltage switchgear and control gear maintenance*, and BS EN 60422:*2013 Mineral insulating oils in electrical equipment. Supervision and maintenance guidance.* The latter code describes the deterioration of switchgear mineral oil that occurs in service, oil sampling techniques, and the frequency and types of testing required. It is advisable to examine the contacts for damage and test the oil of an oil circuit breaker after it has operated to interrupt a high energy short circuit fault.

Another precaution against the risks associated with the use of oil-filled switchgear is to replace it with switchgear that uses an alternative insulating medium; some manufacturers supply circuit breaker assemblies that can be used as a retrofit replacement for legacy oil circuit breakers. Options include sulphur hexaflouride (SF_6), air-insulated and vacuum circuit breakers. The SF_6 variety is beginning to lose favour because of the environmental and toxic hazards that the gas poses, so vacuum and air-insulated gear is now preferred by many high voltage users and distributors.

Another advantage that modern high voltage switchgear has over its predecessor designs is that it is mostly of the fixed pattern design without having withdrawable components. This means that features such as busbar and cable spout shutters are not needed, with a consequential increase in safety arising from the fact that engineers operating the switchgear do not need to work close to uninsulated conductors. Figure 3.14 illustrates a modern distribution board comprising gas-insulated fixed pattern high voltage switchgear.

Electrical equipment installed in flammable or dust-laden potentially explosive atmospheres needs to be selected so that it cannot act as an ignition source. This requires hazardous areas to be classified according to the likelihood of a flammable or explosive atmosphere being present, and then suitable equipment being selected, installed and maintained by competent persons. Detailed information on the requirements is contained in Chapter 15.

Electrostatic discharges capable of causing explosions can be prevented by designing the process to avoid the accumulation of electrostatic charges and by earthing objects to allow any charges to dissipate to earth. If the accumulation of charge cannot be prevented, then measures must be taken to prevent discharges causing hazardous events.

PD CLC/TR 50404:2003 *Electrostatics. Code of practice for the avoidance of hazards due to static electricity* provides comprehensive guidance on preventing static electricity hazards. Techniques include avoiding contact between materials that might generate static charge through friction or separation; reducing flow rates of materials; preventing splashing when

Figure 3.14 Distribution board with SF6-insulated switchgear

pouring liquids; earthing conducting materials; inerting the atmosphere inside vessels in which flammable vapours might otherwise accumulate using, for example, nitrogen; and using static eliminators designed to neutralise or safely discharge any static electricity that may be generated.

Safe systems of work

The safety precautions described so far have concentrated on hardware and technical design solutions. However, nowadays the majority of electrical accidents result from people working unsafely on electrical systems, so the adoption of safe systems of work will have benefits in accident reduction. Safe systems of work are frequently referred to as 'software' measures, in comparison with the hardware measures already covered.

Employers who expect their employees to work on electrical systems have to carry out risk assessments covering those work activities, as explained in Chapter 8. A risk assessment should identify the hazards arising from the activity, their severity (usually expressed as high, medium or low), the likelihood of the hazard occurring and the likelihood of being able to avoid the hazard. These factors then need to be combined to derive an overall assessment of the risk. Having done that, those activities that are judged to have unacceptably high levels of risk must have action taken to bring the risk down to a tolerable level. Risk assessments are usually undertaken by teams of people, often under the control of a health and safety adviser; it is important that the people who will actually be undertaking the work should participate in the risk assessment so that they have a sense of ownership of them and are therefore more likely to implement the control measures.

The output of the risk assessment is a set of control measures that if implemented diligently by competent people will reduce the risks to an acceptable level. In industrial concerns, especially in the electrical supply industry and in premises that have high voltage or extensive low voltage distribution systems, it is common for these safety procedures to be set out in electrical safety rules or similar documents. These rules, which should be drawn up by somebody familiar with the electrical systems at the premises, the nature of the work being done on them, and the principles of electrical safety, should explain the employer's policies for achieving safety from the electrical system. The amount of detail should be commensurate with the complexity of the systems and the level of risk.

In reality, there are two main areas to be considered; firstly, isolation procedures and precautions on equipment that has been made dead; and secondly, precautions during live working. These are described below. There are other work activities that will need to be covered in the safety procedures, examples being switching operations, testing, non-electrical work such as painting and cleaning being carried out near electrical equipment such as in substations, and excavation works near buried cables.

Safe isolation procedures

It should be the norm that work is carried out on systems that have been made dead and on which precautions have been taken to prevent them being reenergised while work is going on. The generic safe isolation procedure is as follows:

(i) Identify the circuit or apparatus on which work is to be done.
(ii) Identify all possible points of supply.

(iii) Disconnect the supply by, for example, switching off an isolator, withdrawing a plug or tripping to 'off' a circuit breaker that has a contact gap that can positively be seen to be open and which has a gap large enough for isolation purposes.

(iv) Secure the point of isolation by applying a padlock with a unique key, by locking an enclosure door or by any other equally effective manner. Ensure that access to the key used to lock the isolation device is under the control of the person doing the work; this reduces the possibility of inadvertent reenergisation while the work is being carried out. Locking off the means of isolation must always be preferred, but where it is not possible to lock off the means of isolation, apply proprietary electrical danger warning tape to the device used to achieve the disconnection, whilst recognising that this may not be secure enough if it is at all foreseeable that the tape may be removed or ignored. Do not use insulating tape to secure points of isolation – it can easily fall off or be removed.

(v) Post caution notices at the point(s) of isolation to warn that work is being done on the disconnected circuit(s). Figure 3.15 illustrates a two-way distribution board on which one outgoing circuit has been locked off using a universal locking kit and a padlock has been attached to a circuit breaker, on which a caution notice has been posted.

(vi) At the point of work, prove the conductors are dead using a suitable voltage indicator such as test lamps. The voltage indicator must be proven to be serviceable

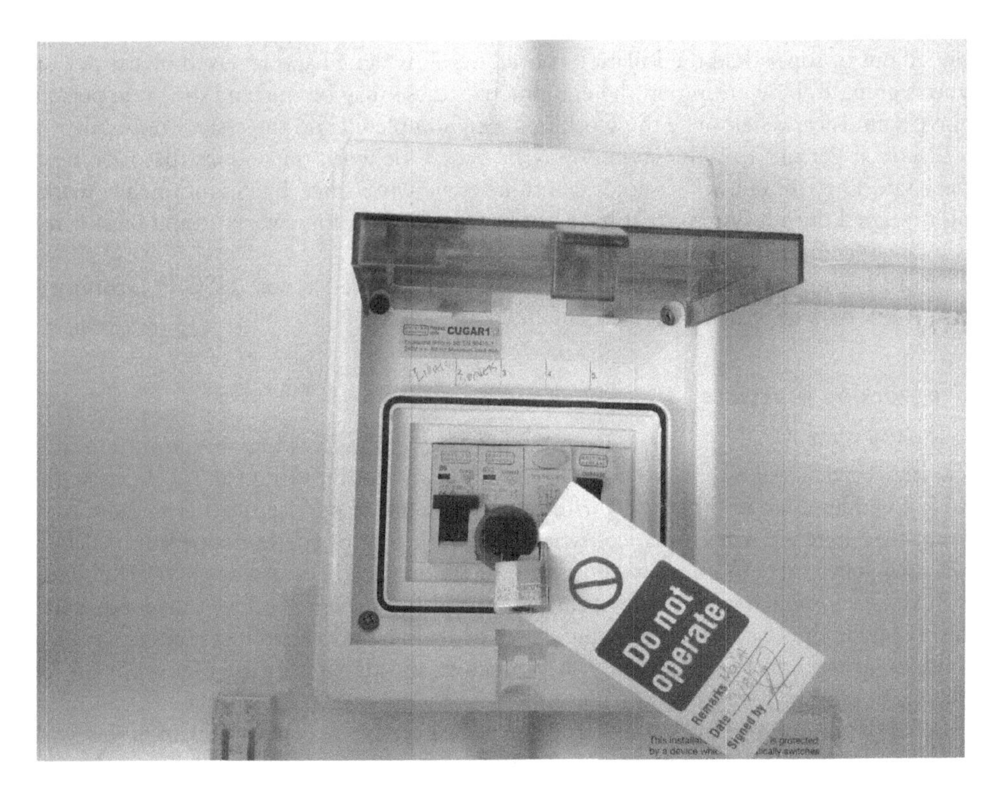

Figure 3.15 Distribution board with outgoing circuit securely isolated

immediately before and after the conductors have been tested for voltage. A proving unit should be used for this; alternatively another live circuit may be used.

(vii) In the cases of high voltage equipment, and some higher risk low voltage systems, applying earths to the conductors on which work is to be done enhances safety. This is to ensure that all dangerous electrical energy is dissipated and to maintain the conductors at earth potential. The earth can be applied using a circuit breaker or switch designed for the purpose, or using earthing leads.

(viii) Post danger notices on adjacent live equipment and circuits.

(ix) Ensure that the persons carrying out the work on the isolated systems are aware of the scope and limitations of the work to be done.

Safe isolation procedures such as these should be carried out only by people, such as electricians and technicians, who have been trained to implement them and who are familiar with the equipment involved. See Chapter 19 for more information on competence and authorisation procedures.

In the cases of high voltage systems and high fault level low voltage systems, it is accepted practice that the safe isolation procedures are formalised through safety documents known as Permits to Work. Essentially, an electrical Permit to Work is a document, normally over two sides of A4 paper, which describes the work to be done and identifies the equipment to be worked on. It lists the points of isolation and identifies where any earths and caution/danger notices have been applied. It is good practice to attach to the permit a switching schedule that identifies the sequence of actions needed to make the system safe. The permit is signed and issued by an authorised person who has actually carried out or supervised the isolation procedures. It is signed and received by the person who is going to be working on the equipment – these may be one and the same person. The permit is cancelled once the work has been completed and the system restored.

Electrical Permits to Work should only be issued for work on systems that have been made safe. They should not be used to authorise live work since, by definition, live working is carried out on systems that have not been made safe. Live work should be covered by a task-specific risk assessment.

Model Permits to Work are published in the HSE's guidance note HSG85 *Electricity at work – Safe working practices* and in BS 6626:2010.

Live working practices

Live working is frequently carried out during fault finding, testing and commissioning activities, particularly on low voltage systems. In many cases it is unnecessary because the work could be done in other ways with the equipment dead, but nonetheless there is no doubt that there are many instances where live work is justified. Some specialised activities, such as live cable jointing, live line working on overhead power lines, and phasing out high voltage conductors, are carried out by specialists who have received in-depth training, usually in the electricity supply industry. The main hazards are direct contact with live conductors and flashovers caused by short circuits, and the aim of the precautions to be taken is to prevent injury from such events.

Before hazardous live work is undertaken, the responsible person in control of the work must be satisfied that it is unreasonable to make the conductors dead and, if it is, that the risks are acceptable when precautions against injury are taken; this is described in more detail in Chapter 6 on the Electricity at Work Regulations 1989.

The most important precaution is to ensure that the people carrying out live work are competent for the task or are being closely supervised by someone who is competent. This means that they must have been trained in the task and been assessed as being competent, must understand the system on which they are working, and must be provided with the appropriate tools, test equipment and personal protective equipment. Live work should only be undertaken after a risk assessment has been conducted, the risks identified and the appropriate safety precautions have been determined.

The types of precautions that can be taken are as follows:

(i) Ensure that the area of the work activity is barriered off or otherwise delineated and protected to prevent the workers being distracted or disturbed. It will also ensure that non-competent or unauthorised personnel are kept away from the live conductors. Danger notices can be used to highlight the danger from the exposed conductors. In permanent test areas it is common practice to have flashing warning lights to highlight when testing work is being done.
(ii) Use tools and test equipment suitable for the job. Suitably rated insulated tools such as screwdrivers and spanners should be used. Test equipment should conform to the guidance in the HSE's guidance note GS38 on electrical test equipment. This means, for example, that probes should be fused, should have finger guards to stop the fingers slipping down the probe onto the live conductors, and there should only be 2–4 mm of metal exposed at the tips of the probes. Care should be taken to ensure that the test equipment itself, such as Class I oscilloscopes and signal generators, does not introduce additional hazards.
(iii) Where necessary, shroud off metalwork in the vicinity of the work that may be at other potentials, including earth. For example, when live work is being carried out inside an equipment enclosure or panel, the metal structure of the panel will be at earth potential; it should be shrouded to reduce the possibility of a live-to-earth shock being experienced. This can be done using flexible insulating sheeting made from materials such as neoprene and polythene. If the floor is conducting, use insulating rubber mats to remove the shock path to earth.
(iv) Where it will contribute to risk reduction, arrange for the live worker to be accompanied. The benefits are twofold; firstly, the second person may be able to offer a second opinion or to detect unsafe practices before they lead to injury; and, secondly, he will be able to take emergency action and render first aid if things do go wrong. Accompaniment is not always essential, but it should be considered.
(v) Use appropriate personal protective equipment. Antiflash clothing such arc flash PPE and eye protection will be appropriate where there is a risk of flashover or arcing. Insulating rubber gloves will be appropriate when live conductors are being handled. The PPE must be maintained in good condition.
(vi) Follow any approved procedure that has been created for the task. For example, DNOs have cable jointing manuals which detail the live jointing procedures for different types of low voltage cables; these procedures should be followed by the cable jointers, and in situations in which a procedure cannot be followed, the presumption should be that the work will not be done live.

Some companies, particularly in the electricity supply industry, use a safety document called a Sanction for Test to manage the safety of testing work on high voltage apparatus, including cables. This is because test work often requires the removal of earths that have been applied under a Permit to Work.

4

The legislative framework

Introduction

The UK's criminal law covering electrical safety is spread across a number of acts and regulations put before Parliament by different government departments, or secretaries of state, with different 'audiences' in mind. The target audiences are, generally speaking, those who supply electrical equipment for work and/or domestic use; those who use electrical equipment and systems; and those who generate, supply and distribute electrical energy. The relevant legislation is explained in detail in the following chapters, but this chapter looks at the general framework.

This book is primarily concerned with the criminal law, which is the law based on statutory instruments that define what the government expects duty holders to do, usually to prohibit harm to property, public health and safety, and welfare. The law covers the offences that would occur in the event of non-compliance and sets out the punishments that the government expects the courts to impose on those found guilty of non-compliance. This contrasts with civil litigation, in which people can use the courts to sue for compensation in circumstances where, for example, they have suffered injury as a consequence of someone else's negligence. Whereas civil litigation contributes to the body of knowledge and case law in the field of health and safety, the legislation described in this book is part of the criminal law.

Strictly speaking, the acts and regulations (often referred to as primary and secondary legislation) referenced in this and other chapters apply in Great Britain; i.e. in England, Scotland and Wales. Northern Ireland has its own primary legislation in the form of the Health and Safety at Work (Northern Ireland) Order 1978 and its own enforcing authority, the Health and Safety Executive for Northern Ireland, which is an executive non-departmental public body sponsored by the Department of Enterprise, Trade and Investment (DETI) of the Northern Ireland Executive. However, its legislation and enforcement policies tend to mirror those used in Great Britain. Where this book refers to the UK it should be kept in mind that Northern Ireland's arrangements are separate from those of Great Britain.

The main statutory instrument covering health and safety in GB is the Health and Safety at Work etc Act 1974, described in the next chapter, which is concerned with ensuring the health and safety of employees, the self-employed (other than those excluded by statute, as described later) and those who may be affected by work activity. There is a range of subsidiary regulations made under the Health and Safety at Work Act, such as the Electricity at Work Regulations 1989. Some of these regulations derive from European directives targeted at worker safety, important examples being the Management of Health and Safety at Work Regulations 1999 and the Provision and Use of Work Equipment Regulations 1998. The Health and Safety Executive, which is a non-departmental public body sponsored by the Department for Work and Pensions, usually takes the lead on the development of this type of legislation.

After the UK population's momentous decision on 23 June 2016 to leave the EU, this legislation will remain in force while the exit is negotiated and until such time as it occurs, on the assumption that it does in fact occur. There is a good chance that it would be retained as domestic legislation after exit, but it would no doubt have to be reviewed to determine what, if any, changes would be needed to reflect the new order. Some changes would definitely be needed to remove references to the authority of EU bodies and compliance with EU legislation, while other changes could be made to take the opportunity to reduce the perceived burdens on business created by EU-inspired legislation. In this latter regard, one extreme option would simply be to repeal the European legislation and return to the situation that existed before it was transposed into UK law.

Other legislation, such as the Electrical Equipment (Safety) Regulations 1994, is derived from European directives whose main aim is to ensure the free movement of goods throughout the European Union, but which have a subsidiary but very important safety content. The Department for Business, Innovation and Skills (previously the Department of Trade and Industry and, with effect from July 2016, the Department for Business, Energy and Industrial Strategy) has taken the lead on the development of this type of legislation which is made under consumer protection legislation, the Consumer Protection Act 1987, rather than the Health and Safety at Work etc Act 1974. It is in this area of European-derived safety-related legislation that most changes may be needed in response to the UK's departure from the EU, although it is likely that the fundamental safety principles expressed in the essential health and safety requirements of the various regulations would be retained in some form or another. Moreover, elements of the compliance procedures that form an important part of the product safety legislation may need to be retained if the UK were to negotiate access to the common market as part of a new trade agreement, perhaps as a new member of the European Free Trade Association alongside Iceland, Liechtenstein, Norway and Switzerland. UK businesses selling into Europe will continue to need to meet the EU's legal requirements covering the safety of products placed on the EU's internal market.

The Electricity Safety Quality and Continuity Regulations 2002 (as amended) are the responsibility of the Department for Business, Energy and Industrial Strategy (BEIS), which in July 2016 emerged from the merger of the Department of Energy and Climate Change (DECC) and the Department for Business, Innovation and Skills (BIS). They are aimed at ensuring that generators, distributors, meter operators and suppliers of electricity meet certain performance standards, including those covering electrical safety. This legislation is made under the Electricity Act 1989, but the safety-related requirements are enforced by the HSE.

In addition to this health and safety legislation, employers also need to be aware of the Corporate Manslaughter and Corporate Homicide Act 2007. This legislation allows companies and organisations to be prosecuted for corporate manslaughter in England and Wales, or corporate homicide in Scotland, if it can be proved that a gross breach of a duty of care occurred as a result of serious management failings. This important legislation is explained in greater detail in Chapter 10.

Goal-setting and prescription

A feature of modern UK-originated legislation in the field of health and safety is that it tends not to be prescriptive in describing what duty holders need to do to comply with the law. It is more goal-setting in nature, describing in broad terms what needs to be achieved but not how to achieve it. Law derived from European directives tends to be slightly more prescriptive, as exemplified by the detailed Essential Health and Safety Requirements that form part of the Machinery Directive and its most recent enactment in the UK, the Supply of Machinery (Safety) Regulations 2008.

The lack of prescription in the legislation is meant to be offset by the provision of comprehensive guidance and the setting of benchmark standards by bodies such as the HSE and the BSI, a good example being British Standard 7671 *Requirements for electrical installations* which is the subject of Chapter 12. Compliance with these standards and guidance normally offers a presumption of conformity with the relevant legislation. Trade associations and other similar organisations have an important role in this respect, and guidance material published by bodies such as the Institution of Engineering and Technology (IET), the Electrical Contractors' Association (ECA) and the BEAMA is extremely useful.

One means at the disposal of the HSE to be more prescriptive is the development and publication of an Approved Code of Practice (ACOP) to support a set of regulations. There is, for example, an ACOP supporting the Provision and Use of Work Equipment Regulations 1998. An ACOP is something of a peculiar object because it has the status of neither a regulation nor guidance. The Health and Safety at Work Act, Section 17, stipulates that failure to follow an ACOP's provision is not an offence in itself. However, where criminal proceedings allege the breach of particular regulation for which there is a relevant provision in an ACOP, and it can be proved that the ACOP provision was not followed, this can be taken as proof of the contravention of the regulation. The defendant may escape a guilty determination, however, if he can prove to the satisfaction of the court that he took alternative steps that had the same level of safety as the ACOP's provisions. Neither guidance material nor British Standards enjoy this level of legal status.

It has to be said that the goal-setting nature of the legislation causes problems for hard-pressed managers in small- and medium-size enterprises who may not have ready access to health and safety professionals; many of them would much rather be told precisely what they need to do to comply with the law than have to spend time and energy searching for and interpreting guidance and standards. There is an overriding need for simple, straightforward and unambiguous guidance for these organisations.

Reasonable practicability

A fundamentally important concept embedded in much of the health and safety legislation is that of reasonable practicability. Many of the regulations require action to be taken so as to reduce risks 'so far as is reasonably practicable', often spoken or written about in terms of reducing risks to a level that is as low as reasonably practicable (the so-called ALARP principle). The most often quoted definition for reasonably practicable was set out by the Court of Appeal (Edwards v. National Coal Board, [1949] 1 All ER 743) as

> 'Reasonably practicable' is a narrower term than 'physically possible' . . . a computation must be made by the owner in which the quantum of risk is placed on one scale

> and the sacrifice involved in the measures necessary for averting the risk (whether in money, time or trouble) is placed in the other, and that, if it be shown that there is a gross disproportion between them – the risk being insignificant in relation to the sacrifice – the defendants discharge the onus on them.

This means that there should be a balance between, on the one hand, the level of risk reduction being sought against, on the other hand, the cost in terms of money, time or trouble needed to reduce the risk. So, for example, if there is a disproportionate cost needed to minimise still further an insignificant risk, the further risk reduction measures could be judged not to be reasonably practicable and therefore do not need to be taken.

The concept of 'risk' relates the probability of a hazard causing harm with the severity of the harm caused. So, if a particular hazard such as electricity can cause serious injury or death, and the probability of serious injury or death occurring in a particular set of circumstances is high, then the risk will be assessed as high. On the other hand, if the probability of death occurring is very low, the risk may be medium or low, despite the consequence of the accident being very serious.

The use of the term 'so far as is reasonably practicable' implies that an assessment of the risk must be carried out before a judgement can be formed on whether or not a measure has reduced risks so far as reasonably practicable. Whereas the need for a risk assessment is not explicitly laid out in electrical safety legislation, it is explicitly a requirement of the Management of Health and Safety at Work Regulations (see Chapter 9). Thus risk assessment practices and procedures are a cornerstone of electrical safety management practice.

Whereas it is obviously better if risks to health and safety can be eliminated altogether, there are many circumstances in which this is simply not possible, or is impracticable, or is not reasonably practicable. We therefore have to tolerate the presence of risk and be aware of it and manage it. The concept of the 'tolerability of risk' is addressed in one of the HSE's most important publications, *Reducing risks, protecting people*, published in 2001. The publication explains the HSE's decision-making process when making regulatory decisions on risk control and confirms and emphasises the importance of risk assessment when duty holders are seeking to put in place reasonably practicable measures to manage risks in their workplaces.

In recent years there have been a number of important court judgements clarifying the concepts of risk and reasonably practicable as they relate to the Health and Safety at Work Act. These are briefly explained in Chapter 5.

In general terms, any company or other entity defending a criminal health and safety charge needs to be able to prove that it was not reasonably practicable to do more than was in fact done or that there was no better practicable means than was in fact used to have prevented an accident or removed a risk to health and safety. Having said that, the defendant only needs to be able to prove the point on the balance of probabilities, rather than using the more onerous 'beyond reasonable doubt' test that applies to the prosecution, although that is not to belittle the difficulty of presenting such a defence.

Some regulations do not have the so far as is reasonable practicable caveat and have an absolute duty for compliance; for example, the Electricity at Work Regulations, Regulation 10, requires that every joint and connection in a system shall be mechanically and electrically suitable for use. In cases such as this, a person charged with an offence under the regulation has the defence of due diligence available.

Enforcement

The different items of legislation are enforced by a variety of agencies, including HSE inspectors of health and safety, inspectors from the Office of Rail Regulation, and environmental health officers (EHO) and trading standards officers employed by local authorities.

The police will be involved if there has been a fatality at work because they will investigate the circumstances to look for any evidence of a serious criminal offence other than a health and safety offence, such as murder or manslaughter or corporate manslaughter/homicide. In reality, the police are normally the first agency to attend the scene of a fatal accident and will usually deal with the immediate aftermath in conjunction with the ambulance and fire and rescue services.

A police force's role during a fatal accident investigation, and its relationship with the other enforcing authorities such as the HSE, is governed by a protocol known as the Work-Related Deaths Protocol; there is a protocol for England and Wales and a separate one for Scotland, although the principles in the two documents are the same. The protocol requires that the various agencies work in cooperation with each other, but its essential significance is that the police will lead the investigation and have primacy until such time as a criminal offence such as murder or manslaughter can be discounted, at which point the relevant health and safety enforcing authority would assume primacy. It has been the case over the past few years that some fatal accidents at work have led to prosecutions under both the corporate manslaughter legislation and health and safety legislation.

Which health and safety authority enforces which legislation depends upon the particular legislation and the location or type of premise in which the electrical system is installed or is operating. Here are some examples:

- The BEIS's engineering inspectors enforce the quality and continuity requirements of the Electricity Safety Quality and Continuity Regulations. HSE's specialist electrical inspectors enforce the safety content of the regulations.
- The HSE's inspectors enforce health and safety legislation in factories, hospitals, construction sites, nuclear and offshore sites, local authority premises and so on. The legislation is enforced in shops and offices and similar locations by EHOs. The split between the agencies is governed by the requirements of the Health and Safety (Enforcing Authority) Regulations 1998.
- Trading standards officers enforce the Electrical Equipment (Safety) Regulations for electrical equipment used in the domestic market, whereas the regulations are enforced by HSE inspectors for equipment used in premises for which the HSE has enforcement responsibilities.

HSE inspectors and EHOs exercising their powers under the Health and Safety at Work Act 1974 have considerable discretionary powers. For example, they are empowered to do the following:

- Enter work premises at any reasonable time.
- Make examinations and investigations, including taking measurements and photographs, as necessary in the exercise of their powers.
- Serve improvement notices requiring action to be taken within a prescribed timescale. This type of notice can be appealed at an Industrial Tribunal.

- Serve prohibition notices requiring immediate action to obviate an imminent risk. This type of notice can be appealed at an Industrial Tribunal.
- Take dangerous articles or substances into possession and have them examined and tested.
- Arrange for samples to be taken and tested.
- Take statements from witnesses. Anyone suspected of breaching a legal requirement must be cautioned before being interviewed.
- Take prosecutions under the relevant legislation. In Scotland, the decision on whether or not to prosecute is taken by the Crown Office and Procurator Fiscal Service, not by an HSE inspector or EHO.

Penalties

Individuals or companies found guilty of criminal charges under the health and safety legislation are liable to various penalties. To some extent these depend on the date that the offence was committed because the Health and Safety (Offences) Act 2008 increased penalties for some offences committed after 16 January 2009. In England and Wales, Section 85 of the Legal Aid, Sentencing and Punishment of Offenders Act 2012 (which came into force on 12 March 2015) had the effect of increasing the level of most fines available for magistrates' courts from a £20,000 fine to an unlimited fine.

For offences committed after March 2015 and charged under the Health and Safety at Work Act Sections 2–6, the maximum penalty in the magistrates' court is an unlimited fine or imprisonment for a term not exceeding 6 months or both. In the Crown Court, the maximum penalty is an unlimited fine or imprisonment not exceeding two years or both.

The situation is Scotland is slightly different. For offences against the Health and Safety at Work Act Sections 2–6 the maximum penalty for a case taken in a Sheriff Court under summary procedure (i.e. without a jury) is £20,000, with an unlimited fine being available if the case is taken under solemn procedure (i.e. on indictment with a jury present).

Cases can be taken under different sections of the Health and Safety at Work Act, and the penalties vary according to which particular section is proved to have been breached. Reference should be made to the Health and Safety (Offences) Act 2008 if more detailed information is required.

What can be said with certainty is that the available penalties for health and safety offences have risen dramatically in recent years, reflecting society's concern about people being injured or killed or suffering ill health at work. Of course, with unlimited fines available to the courts, there has to be some guidance on the size of fines that should be awarded in the event of a guilty verdict. To that end, with effect from 1 February 2016, courts in England and Wales have been given guidance on the appropriate size of fine to apply for particular types of case, as published by the Sentencing Council in a document titled *Health and Safety Offences, Corporate Manslaughter and Food Safety and Hygiene Offences, Definitive Guideline.* This suggests, for example, that a large company with a turnover of £50 million or more that breaches a section of the Health and Safety at Work Act with very high culpability leading to very serious harm, including death, should be fined up to £10M. At the other end of the scale, a company with a turnover of less than £2 million that breaches a section of the Health and Safety at Work Act with low culpability leading to minor injury or harm could be fined £200. Courts in Scotland are not bound by these guidelines and can ignore them, although they may 'take notice' of them when determining sentences.

The increasing size of fines and application of the new sentencing guidelines was recently exemplified in a case brought against Scottish Power Generation Ltd following a non-fatal but serious injury accident at its Longannet generating station in Fife. One of the company's contractors was badly scalded by steam when a valve he was operating failed; he required skin grafts and had to medically retire. The company pled guilty to charges brought under the Health and Safety at Work Act. Whereas 10 or so years ago it might have been fined in the order of £20,000, when the company was sentenced on 31 May 2016 in Dunfermline Sheriff Court the court took notice of the sentencing guidelines and fined the company £2.5M discounted to £1.75M for lodging an early guilty plea. The company is appealing the size of the sentence, but it is clear that the new sentencing guidelines are having an impact.

In similar vein, the Sentencing Council's guideline suggests that companies found guilty of a breach of the Corporate Manslaughter and Corporate Homicide Act 2007 should be fined anything between £180,000 and £20M, depending on the size and turnover of the business and the severity of the offence. These are not trivial sums.

Government reviews of the health and safety regime

When the new UK government came into office in 2010 it quickly launched a review of the health and safety system, driven by a belief that health and safety had become too burdensome on industry and because it believed that a damaging compensation culture had arisen. This had been influenced by a constant stream of media stories that claimed that health and safety was having a stifling impact on society, well-publicised examples being conker games being banned at a school, local authorities banning long-held and traditional public events on the grounds of health and safety, doormats being banned from outside council houses because they were a trip hazard, and police not being allowed to rescue people who had fallen into ponds. These were exacerbated by stories of so-called ambulance chasing solicitors suing companies for compensation for trivial injuries, usually on a 'no win no fee' basis.

To a large extent the stories simply reflected headline-grabbing reporting by sections of the media, but they gained traction with some politicians, the media and sections of public opinion. Organisations like the HSE referred to many of the stories as 'myths'; the HSE even established a 'myth busting' panel to assess and debunk many of the more ridiculous stories, arguing that they were a distraction from the importance of effective health and safety management in industries where the real risks to health and safety exist.

The government cut back the HSE's budget with the consequence that on 1 October 2012 the HSE introduced a charging regime known as Fees For Intervention (FFI) under the terms of the Health and Safety (Fees) Regulations 2012. Essentially, this allows the HSE to recoup its costs from businesses deemed to be in 'material breach' of the law; a 'material breach' occurs when, in the opinion of an HSE inspector, there is or has been a contravention of health and safety law that requires them to issue a notice in writing of that opinion to the duty holder.

The HSE argued that the intention of FFI was to create a level playing field on which those who flouted the law would have to pay, whereas those who complied with it would not have to pay, but the reality is that the HSE needed the income to survive without draconian cuts to its workforce. Its implementation means that many companies have been issued with large bills after a visit by an inspector, particularly after an accident leading to enforcement action, although the average invoice is quite low at about £500. One adverse impact is the change in relationship between the HSE and its 'stakeholders', many of whom are now wary of contacting the HSE for advice in fear of being hit with a bill for a 'material breach'.

In addition to the HSE's budgetary cuts, two significant reviews were carried out. The first report was by Lord Young of Graffam titled *Common Sense Common Safety*, dated 15 October 2010; it focussed on the compensation culture and the role of insurance companies but made several recommendations concerning the legal framework and the work of the HSE and local authorities. The second report was by Professor Ragnar E Löfstedt titled *Reclaiming health and safety for all: An independent review of health and safety legislation*, November 2011; it contained wide-ranging recommendations for change. However, both reports fully supported the principles underpinning the Health and Safety at Work etc Act 1974 and its enforcement by the regulatory bodies, with neither report calling for dramatic or fundamental changes to the regulatory framework.

A full review of these reports is beyond the scope of this book, but the main outcomes of relevance to the book's subject matter are listed here:

- The Reporting of Injuries, Diseases, and Dangerous Occurrence Regulations were changed to reduce the number and types of accidents, ill health, and dangerous occurrences that have to be reported to the enforcing authorities. The 3-day reporting threshold for absence from work caused by work-related injury or ill-health was increased to 7 days.
- The HSE was instructed to focus its efforts on high risk industries and poorly performing businesses, and significantly reduce the number of preventive inspections it conducted in low risk businesses. The criteria it used for selecting which incidents to investigate were changed, effectively reducing the range and scope of its investigation activities.
- The HSE has published simple guidance targeted at lower risk small and medium size enterprises on the management of health and safety and on how to conduct risk assessments; it has published on-line risk assessment tools for low risk premises. It has also withdrawn a number of ACOPs.
- An Occupational Safety and Health Consultants Register has been established, with more than 2400 consultants on the register. The aim is to make sure that companies who engage a consultant get the best advice.
- The HSE published revised guidance on 'portable appliance testing' (INDG236 *Maintaining portable electric equipment in low-risk environments;* Rev 3; Sep 2013) in an attempt to reduce the burden of unnecessary inspection and testing. This was the only outcome directly related to electrical safety.
- From 1 October 2015, self-employed workers whose work activity poses no potential risk to the health and safety of other workers or members of the public were exempted from the requirement to comply with health and safety law. This was achieved through the Health and Safety at Work etc Act 1974 (General Duties of Self-Employed Persons) (Prescribed Undertakings) Regulations 2015. It is almost certainly the case that self-employed people such as electricians carrying out electrical installation and maintenance and similar work would not benefit from this

exemption, because their work has the potential to expose other workers or members of the public to risk.

As for the future, there seems to be little doubt that further reviews and potentially very significant changes will need to be initiated as a consequence of the UK's decision to leave the EU and the outcome of the negotiations with the EU's institutions.

5

The Health and Safety at Work etc Act 1974

Introduction

The Health and Safety at Work etc Act 1974, abbreviated as the HSW Act, does not contain any provisions relating specifically to electrical safety. However, it is an act that has had, and continues to have, an immense impact on the UK's health and safety system, effectively acting as its backbone for in excess of 40 years. It therefore deserves a chapter in this book to outline the main duties imposed on those at work. It is not intended, however, to cover the many other provisions of the act.

The most important features of the act are as follows:

- It is an enabling act, allowing the promulgation of subsidiary regulations.
- It imposes goal-setting, non-prescriptive duties on employers, the self-employed and employees to secure so far as is reasonably practicable the health, safety and welfare of those at work and the health and safety of people, such as members of the public, who may affected by work activity.
- It covers all work activities with the sole exception of those employed as domestic servants in private households.
- It established and defined the general functions of the Health and Safety Executive.
- It sets out the enforcement powers of health and safety inspectors, as discussed in Chapter 4.

The act does not contain any specific requirement for employers and the self-employed to conduct risk assessments, despite the need for such assessments being an implicit consequence of the concept of reasonable practicability. Duty holders had to wait until the arrival in 1992 of the Management of Health and Safety at Work Regulations for an explicit requirement to conduct risk assessments.

Main duties

Section 2

This section of the act sets out the general responsibilities that employers have regarding their employees' health, safety and welfare. The flavour can, perhaps, best be appreciated by considering the duty in Section 2(1) on every employer to ensure, so far as is reasonably practicable, the health, safety and welfare at work of all his employees. The very general nature of this duty reinforces the broad scope of the act and its non-prescriptive character. It also points to the need for considerable discretion by the enforcing authorities when interpreting such a generic duty.

The rest of Section 2 puts a bit more flesh on the general duty expressed in Section 2(1). Important requirements in the context of electrical safety include the provision of plant and systems of work that are safe; the provision of information, instruction, training and supervision as necessary; the maintenance of 'places of work' in a safe condition; and the provision and maintenance of a safe and healthy working environment.

Section 3

This section deals with employers' and self-employed people's duties to persons other than their employees. Essentially, their work activities should not expose persons in their employment, nor others, to risks to their health and safety. They also have a duty to provide others with information about work activities which might affect their health and safety. An example would be a live cable jointing team working in an excavation in a public road: the team would need to ensure that passers-by could not fall into the excavation or otherwise come into contact with the exposed live conductors, perhaps by erecting barriers with suitable warning notices affixed. Alternatively, an electrical contractor who installs a new circuit in a domestic property, but leaves it in a dangerous condition to the risk of the householder, would be in breach of Section 3.

As noted in the preceding chapter, from 1 October 2015, self-employed workers whose work activity poses no potential risk to the health and safety of other workers or members of the public were exempted from the requirement to comply with health and safety law.

Section 4

This section applies to persons in charge of premises where there are workers who are not their employees. They must ensure that the premises, any plant (including the fixed electrical installation) and the means of ingress and egress are safely maintained. An obvious example of the application of this duty is a construction site where the main contractor is the person in charge and in control of the site and who is responsible, therefore, for ensuring that the site is safe for the subcontractors' labour forces.

Section 6

Section 6 applies to manufacturers, designers, importers, and suppliers of articles and substances for use at work. In the context of articles, which would include electrical equipment and apparatus, the principle duties are that they must ensure that the articles are safe when being set, used, cleaned or maintained by a person at work. They must also ensure that operating and maintenance instructions are available and provide revisions of information when it becomes known that something gives rise to a serious risk to health and safety.

The latter requirement is commonly used to ensure that users of equipment are informed when, after an accident investigation has been carried out, it becomes apparent that the equipment needs to be modified in some way to prevent a recurrence of the incident.

The section entitles a supplier to rely on the safety-related research and testing already done by others; he does not have to repeat it. For example, an electrical contractor who installs electrical equipment such as a circuit breaker can rely on the assurances of the device's manufacturer that it has been made and tested to an appropriate British Standard. The contractor, however, retains responsibility for selecting the correct type of circuit breaker for the application, ensuring, for example, that it has adequate breaking capacity for the prospective fault current. However, importers of foreign apparatus must satisfy themselves that the designer's or maker's assurances are valid.

In the event of an incident, it is a defence to prove that the occurrence could not reasonably be foreseen. The argument would be that the scientific and technical knowledge available at the time that the product was marketed was not such as might be expected to reveal the deficiency.

Section 7

Section 7 imposes duties of care on employees for their own and others' safety. They have to cooperate with their employer and anyone else to fulfil their obligations. An example would be when an electrician ignores the instructions and procedures put in place by his employer to work on an electrical installation once it has been isolated and made safe and instead carries out the work live. Such behaviour would probably be a contravention of Section 7, and it is not unusual for cases to be brought before the courts when employees behave in this way and where somebody is harmed as a result.

Section 8

This section prohibits intentional or reckless interference or misuse of anything provided in the interests of health, safety or welfare.

Sections 18–26

These sections of the act deal with enforcement powers. Section 20 is important because it provides inspectors of health and safety with their powers, which were listed in Chapter 4; inspectors are issued with warrants that identify them as inspectors. The next two sections empower inspectors to issue improvement notices and prohibition notices where they believe there is a contravention of a statutory provision and/or there is a risk of serious personal injury. It is an offence not to comply with a notice, but recipients have the right under Section 24 to appeal a notice.

Sections 33 to 42

These sections deal with offences. Section 33 details the types of offences that can be committed under the act; these include but are not limited to

- failing to discharge a duty to which he is subject by virtue of sections 2 to 7;
- contravening any health and safety regulations;
- contravening any requirement imposed by an inspector under section 20 or 25;

- contravening any requirement or prohibition imposed by an improvement notice or a prohibition notice (including any such notice as modified on appeal); or
- intentionally obstructing an inspector in the exercise or performance of his powers.

Section 37 should be of interest to company directors, company secretaries and senior managers because it allows them to be prosecuted as individuals if their company is being prosecuted and if it can be shown that the offence occurred with their consent or connivance, or can be attributed to their neglect. In general, enforcing authorities will always seek to prove the guilt of directors and senior managers if there is sufficient evidence of consent, connivance or neglect.

Recent important case law

Over the years there has been a continuing debate about issues such as the difference between 'safe' as used in Section 2 and the 'absence of risk' as used in Section 3; the extent to which 'foreseeability', or the lack of it, can be used by the prosecution and defence during trials; and whether or not the prosecution needs to prove how an accident was caused in seeking to prove a case under Sections 2 and 3 of the HSW Act. Two recent cases have shed some light on these matters, explained as follows.

R v Chargot Ltd (2008). This case in the House of Lords arose from the death of a dumper truck driver. The important point arising from the judgement was that in the event of an injury occurring at work, when the prosecution can prove that the accident occurred, it can be taken prima facie that the risk was not adequately managed; otherwise the accident would not have occurred. The prosecution does not need to prove what acts or omissions by the accused led to the accident. The burden of proof then transfers to the accused to demonstrate, on the balance of probabilities, that he took all the measures necessary to remove the risk, so far as is reasonably practicable.

R v Tangerine Confectionery Ltd and Veolia ES (UK) Ltd (2011). These two cases, both arising from fatal accidents at work, were considered by the Court of Appeal after the companies had been found guilty of offences under Sections 2 and 3 of the HSW Act. The important outcomes of the judgement were

- The concept of 'safe' in Section 2 of the act is the same as the concept of 'absence of risk' in Section 3 of the act.
- In a trial, once the prosecution has proved that the accused exposed their employees, or those who may be affected by their undertaking, to risk, the burden shifts onto the accused to prove, on the balance of probabilities, that it was not reasonably practicable to do more than was done or that there was no better practicable means to be used to satisfy the relevant duty. The prosecution does not need to establish that the accused caused the injury, just that there was a risk that was not reduced so far as is reasonably practicable.
- In considering risk, the risk must be a material risk and not one that is trivial or fanciful.
- In a prosecution under either Section 2 or 3, the prosecutor is not required to prove that the actual accident which occurred was foreseeable; rather he is required to show that the risk was foreseeable. From the accused's perspective, if the risk is not

foreseeable, it would not be reasonably practicable for them to have taken steps to remove or reduce the risk.

It is generally accepted that the important lesson arising from these cases is that employers must carry out effective risk assessments covering their activities and ensure, so far as is reasonably practicable, that their employees implement the control measures consistently and diligently.

6

The Electricity at Work Regulations 1989

Introduction

The Electricity at Work Regulations 1989 (Statutory Instrument 1989 No 635) are made under the Health and Safety at Work Act. They were promulgated on 1 April 1989 and replaced the long-standing previous regulations on electrical safety, the Electricity (Factories Act) Special Regulations 1908 and 1944. These latter regulations had been drafted by J Scott Ram, who had been appointed in 1902 as the first HM Electrical Inspector of Factories and who, with remarkable prescience, laid down the basic electrical safety principles that persist to this day.

Unlike the older 1908 and 1944 regulations, which only applied to factory premises covered by the Factories Act, the Electricity at Work Regulations 1989 apply to all workplaces covered by the HSW Act. Their publication in 1989 was therefore the first time that detailed electrical safety statutory requirements had been applied to non-factory premises.

The regulations are mostly enforced by the HSE's inspectors of health and safety and by local authority environmental health officers. The HSE's general regulatory inspectors are supported by field-based specialist inspectors who are professionally qualified and experienced electrical engineers. The specialist inspectors of electrical engineering also provide expert support to environmental health officers and, in fatal accident investigations, to the police.

The regulations are supported by the HSE publication HSR25, the *Memorandum of Guidance to the Electricity at Work Regulations 1989*, the latest version of which was published in November 2015. As explained in Chapter 11, there is a range of other guidance publications, many of which are referenced in this chapter.

The regulations

The regulations were originally promulgated in four parts, with Parts I, II and IV having general application and Part III applying only to mines. The regulations in Part III, Regulations 17–28, together with Schedule 1, were revoked in April 2015 by the Mines Regulations 2014, although there is additional guidance for mines in the HSE publication HSG238 *Electrical safety in mines*.

To gain an understanding of the regulations, it is helpful to consider them as having two main blocks: those that deal with electrical hardware and those that deal with working practices. The former are Regulations 4(1), 4(4), 5, 6, 7, 8, 9, 10, 11, 12 and 15. The latter are Regulations 4(2), 4(3), 13, 14 and 16. This distinction is useful because it recognises the fact that high standards of electrical safety will only be achieved if electrical installations and equipment meet appropriate build standards, and if persons at work on electrical systems are competent and adopt safe systems of work.

The following paragraphs give a brief explanation of each of the regulations, with the text of the main block of regulations in Parts II and IV being spelled out.

Part I: Introduction

Regulation 2 – Interpretation

Regulation 2(1) provides interpretations of words used in the rest of the regulations. 'Danger' is defined as the 'risk of injury', where the specific injuries covered by the regulations are listed as death or personal injury from electric shock, electric burn, electrical explosion or arcing, or from fire or explosion initiated by electrical energy; these are the injury mechanisms described in Chapter 1 of this book. Note that other injuries, such as crushing or shearing injuries resulting from an electrical fault in a machine's control system, are not included. The risk of injury arising from control system faults on machinery is more in the purview of the Provision and Use of Work Equipment Regulations than the Electricity at Work Regulations.

'Electrical equipment' includes equipment used to generate, provide, transmit, transform, rectify, convert, conduct, distribute, control, store, measure or use electrical energy. The reach of the regulations is therefore very wide and all-embracing.

Regulation 3 – Persons on whom duties are imposed by these regulations

Regulation 3(1) places the duty to comply with the regulations on employers, the self-employed, and mine and quarry operators, for matters which are within their control. It is most likely that self-employed workers such as electricians and technicians will not be exempted from health and safety legislation by virtue of the Health and Safety at Work etc Act 1974 (General Duties of Self-Employed Persons) (Prescribed Undertakings) Regulations 2015, so this part of the Electricity at Work Regulations will apply to them.

These regulations mean, for example, that if a company were to employ an electrical contracting company to carry out work on their electrical installation, and if they had taken all reasonable steps to ensure the competence of the contractors, if an electrician employed by the contractors were to suffer an electrical injury, the client company would have the defence that they were not in control of the electrical work.

Regulation 3(2) mirrors the duty that the HSW Act, Section 7, places on employees to cooperate with their employers. It also requires that employees must comply with the regulations in relation to matters that are within their control. This is an important duty on individuals and means, for example, that an electrical engineer in charge of a group of people working on, say, an overhead line has duties to ensure compliance with the regulations, in so far as they relate to matters that are within his or her control. This might include, for example, ensuring that the line has been isolated, and that circuit main earths and drain earths have been correctly applied, and that work is restricted to safe areas. The individual may be liable for prosecution under this regulation if something goes wrong because he or she did not exercise adequate control.

Part II: General

Regulation 4 – Systems, work activities and protective equipment

(1) *All systems shall at all times be of such construction as to prevent, so far as is reasonably practicable, danger.*
(2) *As may be necessary to prevent danger, all systems shall be maintained so as to prevent, so far as is reasonably practicable, such danger.*
(3) *Every work activity, including operation, use and maintenance of a system and work near a system, shall be carried out in such a manner as not to give rise, so far as is reasonably practicable, to danger.*
(4) *Any equipment provided under these Regulations for the purpose of protecting persons at work on or near electrical equipment shall be suitable for the use for which it is provided, be maintained in a condition suitable for that use, and be properly maintained.*

Regulation 4 is an 'omnibus' regulation which is fleshed out by subsequent regulations. The four subsidiary regulations are dealt with separately in the following text.

REGULATION 4(1)

This regulation simply says that all systems must be safe, so far as is reasonably practicable. The main means of complying with the requirement is to ensure that electrical equipment, apparatus and installations are designed by competent persons to recognised standards and, as appropriate, built and installed to a good standard.

As far as low voltage distribution systems are concerned, the benchmark standard is BS 7671:2008 *Requirements for electrical installations. The IET Wiring Regulations*, 17th Edition, described in Chapter 12. Indeed, the Memorandum of Guidance, HSR25, specifically comments that installations built and installed in compliance with that standard are likely to achieve compliance with the regulations. Of course, it is axiomatic that the system would need to be installed by a competent person, and inspected and tested to verify its safety before being commissioned and put into use.

In addition to BS 7671, there are very many other standards that offer guidance on the safe design of electrical systems and equipment and which, if followed, provide a presumption of conformity with the regulations. It is impractical to list them all here, and many of them are referenced throughout this book, but two are worthy of mention. BS EN 60204–1 *Safety of machinery. Electrical parts of machinery* is a CEN/CENELEC Type B standard (see Chapter 17) with comprehensive guidance on electrical safety principles relating to machinery. The BS EN 60335 series of standards provides equally comprehensive guidance relating to electrical safety in commercial and domestic equipment such as food mixers, meat slicers, white goods, fly 'electrocutors' and so on.

REGULATION 4(2)

This regulation requires that systems are maintained to prevent danger, but only where maintenance is necessary to prevent such danger; it has the 'so far as is reasonably practicable' caveat. There is no prescription on the form that the maintenance should take, and there is no duty for records of the maintenance to be kept. However, it is generally recommended that records are kept as a means of demonstrating compliance to enforcement authorities and as a means of detecting any deterioration that may occur over time.

There are many different forms of preventive maintenance, examples being on-fault rectification, reliability centred maintenance, periodic calendar-based inspection and testing, operating-hour inspection and testing, and predictive maintenance. The duty holder is free to choose which type of technique to use, so long as it reduces the risk of injury from the electrical system so far as is reasonably practicable.

It is noteworthy that the regulation makes no specific mention of portable appliances, despite the widespread belief that it is only portable appliances that need to be inspected and tested. This belief is reflected in the fact that many organisations have comprehensive and expensive inspection and test programmes for portable appliances (kettles, computers, radios and the like) but ignore the need to maintain their fixed distribution systems and equipment. This is commonly known as the tautological term PAT testing, where PAT means portable appliance testing. Indeed, largely at the behest of companies that offer portable appliance testing services, it has to be said that many organisations over-maintain their portable appliances by carrying out testing too frequently; in some cases, such as Class II appliances used in low risk environments, there may be no need to carry out any testing at all, with preventive maintenance limited to periodic visual inspections. The HSE has published guidance on this topic in HSG107 *Maintaining portable and transportable electrical equipment* and INDG236 *Maintaining portable electric equipment in low-risk environments*. The IET has also published guidance on the matter in its *Code of Practice for In-service Inspection and Testing of Electrical Equipment* (4th edition), which forms the basis of training provided to people who wish to qualify in the inspection and testing of electrical equipment. This topic is covered in greater detail in Chapter 18.

Preventive maintenance of electrical systems most commonly consists of a mixture of visual inspections and tests. Low voltage installations may be maintained following the advice contained in BS 7671:2008, which covers the appropriate visual inspections and periodic tests (see Chapter 16 for a description of the inspections and tests). Advice on the maintenance of high voltage switchgear is published in BS 6626, and the HSE's guidance note HSG230 *Keeping electrical switchgear safe;* 2nd edition, November 2015, provides helpful advice.

As a general comment, it is often best to follow the advice of manufacturers and suppliers on the form of preventive maintenance needed. In most cases the importance of routine visual examinations of equipment, as well as testing where required, cannot be over-emphasised, for the very good reason that most faults (such as damaged insulation on cables which leads to danger) can be detected by a simple visual inspection; things to look for include signs of wear and tear, missing covers and securing bolts or screws, damaged enclosures, signs of overheating and burning, excessive corrosion, and leaking oil or insulating compound. This includes a before-use inspection by the operator of the equipment as well as periodic inspections by a competent person, who does not necessarily need to be an electrically qualified person such as an electrician.

REGULATION 4(3)

This regulation requires that any work activity on or near a system is carried out safely, again with the 'so far as is reasonably practicable' caveat. In this context, the caveat strongly implies that a risk assessment should be done before the work commences. Whereas this is not explicitly spelled out, it is made a specific requirement by the Management of Health and Safety at Work Regulations which embrace all work activity, including electrical work.

The work activities include installation, commissioning, operation, use and maintenance. Those responsible for the work, and those engaged in it, need to be competent to appreciate the electrical hazards and the control measures needed to minimise the risks to an acceptable level; these issues are addressed in Regulations 13, 14, and 16. To this end, the person responsible for managing the work should consider the work activities, the risks that arise, the measures that should be taken to control them and who should be allowed to perform them. If, for example, work has to be done on a low voltage system which necessitates it being made 'dead', the supply to it needs to be isolated, the isolator locked off and a test made to prove that the isolated part is indeed 'dead' and therefore safe to work on. Anyone authorised to carry out these safety precautions has to be familiar with the system, know which isolator or isolators to open and lock off and how to apply the test. The responsible person should also ensure that those who do the work have appropriate technical knowledge to do it properly and, on completion, to test it to prove its safety.

Where work has to be done near a system where there may be danger, the responsible person has to provide for the safety of the workers. For example, painters, who may not be electrically knowledgeable, may have to paint an overhead crane track where there are bare live trolley wires. The painters must be warned about the hazard, and measures must be taken to avoid the danger; some companies, especially in the electricity supply industry, formalise this in a safety document called a Limited Work Certificate or Limitation of Access. The precautionary measures might involve the provision of track stops to limit crane movement, and the painters working in a segregated area where the live conductors are screened to prevent direct contact. Where systems are concealed, however, the danger may be less obvious, but measures still need to be taken. Buried cables are a potential hazard during excavation work, so their location must be ascertained before the work starts and the workers instructed on the precautions to be taken to avoid the cables being damaged in a way that may result in injury.

Where live work is to be carried out, such as live jointing work or fault finding on live equipment, the risk of electrical injury tends to be high, so careful planning is needed. Such work must only be done by competent people who have the appropriate skills and knowledge. The types of precautions that can be taken were explained in Chapter 3 but, in brief, are as follows:

- Ensure that the area of the work activity is barriered off or otherwise delineated and protected.
- Use tools and test equipment suitable for the job.
- Where necessary, shroud off metalwork in the vicinity of the work that may be at other potentials, including earth potential.
- Where it will contribute to risk reduction, arrange for the live worker to be accompanied.
- Use appropriate personal protective equipment.
- If there is a particular method for the work, such as is the case for many types of live cable jointing, ensure that the approved procedure is followed.

An uninstructed person can safely carry out many operation and use activities such as operating a light switch. However, dangerous use often stems from abuse and/or ignorance on the part of the user. The careless operator of a portable angle grinder, for example, might abrade the flexible cord and expose a live conductor, resulting in a potential shock hazard to anyone who might touch it. Again, if the user of a machine exceeds its duty cycle by continuous use or overload, it may overheat, the insulation may fail, and the metal carcass may become live, endangering the user. So it is important that those who are engaged in operation and use activities are trained to avoid danger.

REGULATION 4(4)

Regulation 4(4) deals with protective equipment such as permanent and portable insulating stands or screens, insulating boots and gloves used to prevent direct contact injuries, insulated tools, and antiflash clothing used to protect against flashover burn injuries. In fact, the reach of the regulation is wide and would include, for example, the fencing of an overhead power line crossing a construction site to prevent plant or scaffold poles making contact with the line. It would also include door interlocks on control panels used to prevent access to live apparatus, insulated tools and potential indicators.

Many employers require their electrical staff to provide their own tools for use at work, including insulated tools for use during live work. Whereas this may be standard practice, and the tools may be the personal property of the employees, it does not absolve the employer from ensuring that the tools are suitable for the work being carried out by the electricians and technicians. Employers are well advised to carry out periodic inspections of their employees' tools to ensure that they are suitable and in a safe condition, and are maintained as such.

Note that this regulation does not attract the 'so far as is reasonably practicable' caveat – it is an absolute duty that equipment used for protective purposes has to be suitable, and properly maintained and used.

Regulation 5 – Strength and capability of electrical equipment

No electrical equipment shall be put into use where its strength and capability may be exceeded in such a way as may give rise to danger.

Regulation 5 refers to the ability of equipment, including cables, to withstand the effects of the currents and voltages to which it is likely to be subjected. These include the stresses that will occur when operating at its normal rating, on overload, and under fault conditions and when subjected to mains-borne transients. The key to compliance is the selection of the right equipment by a competent person.

For example, in choosing a direct-to-line motor starter, one which complies with BS EN 60947–1:2007+A2:2014 *Low-voltage switchgear and controlgear. General rules* is preferable because its performance details are ascertainable and because it is tested to the standard. Obviously, it should be rated to match the full load, starting and stalled currents of the induction motor it controls and, also, the duty cycle. Inching requirements and the number of starts per hour need to be known, along with any adverse environmental conditions that may exist.

The next consideration would be fault protection. If a short circuit fault occurred on the cable between the starter and the motor, is the starter capable of safely interrupting the fault current? A loop impedance test may be needed to ascertain the fault level. If the chosen starter is incapable of safely interrupting the fault current, suitable additional protection would be needed on the supply side of the starter. This would ensure that the operation of the combined excess current protective devices would either avoid damage to the starter or limit it so that anyone nearby is not endangered.

Another example is the selection and use of suitably insulated cables. Their voltage and current ratings should not be exceeded where this might cause an insulation failure due to overheating and the consequential exposure of a live conductor that would produce the risk of electric shock and burn injuries to anyone who may touch it.

One element of cable sizing that is sometimes overlooked is the concept of diversity, which considers the probability of a number of electrical loads all being on at the same time. In reality, there are many situations and installations in which it is unlikely that all the loads will be on at the same time. In these circumstances, a diversity factor can be applied to loads such as ovens and lighting, especially in domestic installations, that allows for smaller cables to be specified than would be the case if all the loads were arithmetically summed and the cables sized on the result. A good source of guidance on diversity factors for low voltage installations is the IET's *On Site Guide to BS 7671*.

Yet another example is high voltage switchgear, such as circuit breakers. These devices may be required to clear short circuit faults on circuits where the fault level is extremely high, normally up to 250 MVA on 11 kV distribution circuits, for example. If the circuit breakers' rated fault breaking capacity were to be less than the fault level on the system, short circuit faults would have the potential to cause catastrophic failure of the switchgear, resulting in an explosion or fire. It is not unknown for circuit breakers which are rated at, say, 50 MVA, and which were installed on a network in the 1940s and 1950s when they were adequately rated, to become underrated, because system reinforcements have resulted in increases in the fault level above 50 MVA.

Another cause of devices becoming dangerously underrated stems from increases in the quantity and power rating of rotating plant in the load centres fed from a circuit breaker – the contribution these devices make to fault level is frequently overlooked and should be included in fault level calculations.

Regulation 6 – Adverse or hazardous environments

Electrical equipment which may reasonably foreseeably be exposed to –

(a) mechanical damage;
(b) the effects of the weather, natural hazards, temperature or pressure;
(c) the effects of wet, dirty, dusty or corrosive conditions; or
(d) any flammable or explosive substance, including dusts, vapours or gases, shall be of such construction or as necessary protected as to prevent, so far as is reasonably practicable, danger arising from such exposure.

This regulation has the important stipulation that measures must be taken to prevent dangerous deterioration of equipment exposed to conditions that may adversely affect its performance and integrity. The risk of injury must be prevented so far as is reasonably practicable. The specific influences covered are mechanical damage; the effects of the weather, natural hazards, temperature or pressure; the effects of wet, dirty, dusty, or corrosive conditions; and any flammable or explosive substances, including dusts, vapours or gases.

Compliance requires the selection of appropriate equipment for the prevailing conditions, and those conditions that can be foreseen for the location where the equipment is to be used. Alternatively, it is sometimes the case that the equipment can be otherwise protected to achieve compliance, such as by placing it out of harm's way in the case of potential mechanical damage to cable runs.

The most common protection against mechanical damage is making the construction of the equipment sufficiently robust for its likely applications. For example, portable equipment used on construction sites will need to be able to withstand the very rough treatment commonly meted out by personnel who may not be very aware of the electrical risks. In contrast, electrical equipment designed for domestic use will usually not need to be quite so well protected against mechanical damage. BS EN 62262:2002 *Degrees of protection provided by enclosures for electrical equipment against external mechanical impacts* defines an IK code to deal with the degrees of protection provided by enclosures of electrical equipment against external mechanical impacts. The code ranges from IK00 to IK10, with increasing levels of impact energy.

Providing steel wire armouring on cable is a commonly used method of protecting the cable against mechanical damage, and covering the armouring with an outer plastic over-sheath prevents corrosion and water ingress. Alternatively, non-armoured cable can be placed on cable trays or inside conduit, trunking or ducts to provide protection against damage, including rodent damage.

The IP code associated with protecting enclosures against the ingress of liquids and dusts has been described in Chapter 3. This has important application in securing compliance with Regulation 6.

Corrosive conditions exist in a variety of locations, one example being livestock farms where methane and urea in the environment has been known to cause corrosion in electrical switchgear installed in livestock sheds. RCDs have been known to corrode rapidly in this type of environment, becoming solidly stuck in the 'on' condition and with the failure not being detected until a demand is placed on the RCD to operate or the test button is pressed.

Lightning is a weather hazard that is often overlooked. The 30% of electricity supply failures on overhead line transmission and distribution systems that are due to lightning strikes demonstrate the extent of the risk and the consequential need for the owners to provide protection, so far as is reasonably practicable, to avoid danger from the strikes. Section 3 of the HSW Act also requires employers to conduct their undertakings in such a way as to avoid risks to persons not in their employment. On public supply systems, some attempt is made to protect the high voltage (HV) system, but nothing is usually done to protect the low voltage (LV) overhead distribution system. The consequence is that powerful, high voltage transients can damage the service line within the consumer's premises and, sometimes, his installation. This can result in fire and may result in persons in the premises being injured from electric shock and/or burn injuries from electrical equipment damaged by the lightning impulse, although such events in the UK are thankfully rare.

An extensive range of British Standards covers the prevention of explosions caused by electrical apparatus acting as ignition sources in flammable and dust-laden atmospheres. These give the best route to securing compliance with the requirements of Regulation 6 covering explosive atmospheres. For example, BS EN 60079–10–1:2009 *Explosive atmospheres. Classification of areas. Explosive gas atmospheres* is a harmonised European code of practice which provides guidance on zone classification of explosive gas hazardous areas. BS EN 60079–14:2014 *Explosive atmospheres. Electrical installations design, selection and erection* covers the selection and installation of electrical installations in explosive atmospheres, and BS EN 60079–17 covers the inspection and maintenance of these installations. There are also standards in the BS EN 60079 series which provide guidance on the construction requirements for electrical apparatus for potentially explosive atmospheres; for example, BS EN 60079–1:2014 describes the construction and verification requirements for flameproof Ex d enclosures. The topic is explained in more detail in Chapter 15.

Regulation 7 – Insulation, protection and placing of conductors

All conductors in a system which may give rise to danger shall either:

(a) be suitably covered with insulating material and as necessary protected so as to prevent, so far as is reasonably practicable, danger; or
(b) have such precautions taken in respect of them (including, where appropriate, their being suitably placed) as will prevent, so far as is reasonably practicable, danger.

The intention of Regulation 7 is to prevent electric shock and burn injuries from direct and indirect contact, and fire and explosion consequent on short circuits or leakage currents between circuit conductors or between circuit and other conductors.

Regulation 7(a) requires the insulation to be suitable, so it has to be chosen to suit the environmental conditions and the electrical stresses to which it will be subjected. In hot locations, for example, it has to be heat resistant; in damp places it has to be waterproof; and where there is the possibility of contamination by oils or other chemicals, its insulating properties must not be adversely affected. In addition, it has to withstand the normal voltage and any transient high voltages. The 'as necessary protected' requirement is an additional safeguard, usually for protection against mechanical damage by, for example, running PVC-insulated wiring conductors in conduit or trunking, or it could be for protection against other environmental hazards such as enclosure in a waterproof housing to prevent liquid contamination.

Regulation 7(b) refers to other precautions, including placing out of reach or behind barriers as an alternative to covering with insulation.

Regulation 8 – Earthing or other suitable precautions

Precautions shall be taken, either by earthing or by other suitable means, to prevent danger arising when any conductor (other than a circuit conductor) which may reasonably foreseeably become charged as a result of either the use of a system, or a fault in a system, becomes so charged; and, for the purposes of securing compliance with this Regulation, a conductor shall be regarded as earthed when it is connected to the general mass of earth by conductors of sufficient strength and current-carrying capacity to discharge electrical energy to earth.

As the public supply system employs neutral earthing, the most common method of compliance with this regulation is to connect together and earth exposed and extraneous conductive parts by means of low impedance protective and bonding conductors. This is the EEBADS technique described in Chapter 3. The technique creates an equipotential zone so that when an earth fault occurs these conductive parts are raised to substantially the same potential with respect to the ground (earth). This condition persists until the protective device, be it a fuse or a circuit breaker, interrupts the circuit and clears the fault. Anyone in simultaneous contact with more than one of the conductive parts should not experience a shock because the parts are at about the same potential.

The regulation requires the protective conductors to be effective under fault conditions. To ensure compliance, the prospective fault currents need to be ascertained and conductors selected which will carry these fault currents without damage until the protection operates to clear the fault.

If it is not possible to establish an equipotential zone because, for example, there is a conducting floor (such as a concrete floor) or because equipment is being used outdoors, supplementary or alternative measures must be taken. The use of residual current circuit breakers to provide sensitive earth leakage protection, in addition to the overcurrent protection, is one acceptable option; the RCD would detect earth fault currents, including shock currents flowing to earth, and rapidly interrupt the circuit. Indeed, it is now a requirement of BS 7671 that all socket-outlet circuits, including sockets supplying outdoor equipment, have 30 mA RCD protection, apart from labelled essential service sockets; in workplaces, RCDs may be excluded, for example to prevent nuisance tripping, if a risk assessment indicates that this would not increase risks.

As an alternative, precautions may be taken to prevent earth faults by segregating live conductors from exposed conductive parts by additional insulation. This creates a Class II system for the wiring and connected apparatus to prevent exposed conductive parts becoming charged.

For special applications, compliance may be achieved by other means. For example, the insertion of a 1:1 safety isolating transformer in a mains voltage system would isolate the system on the secondary side from the earthed neutral system on the primary. Earthing one pole of the secondary side through an impedance so as to limit the potential earth fault current to no more than about 5 mA would create a system that would prevent electric shock injuries for a phase-to-earth fault. It is usual to employ a circuit that detects the flow of fault current and trips a circuit breaker controlling the supply. This type of system is often used in test areas.

Similar systems are used where continuity of supply is important and it is undesirable for earth faults to lead to the interruption of the supply, such as in high dependency medical locations. In these systems, the limited fault current flowing causes an audible or visual alarm rather than tripping a circuit breaker.

Fully separated systems, where a safety isolating transformer is inserted but the outgoing supply is not earthed is another option frequently employed in production testing areas where an operator may have to handle or be exposed to contact with uninsulated live conductors. Measures must be taken to prevent one pole of the secondary side of the system becoming inadvertently connected to earth; the separated circuit should be kept as short as possible and the conductors should be well insulated and visible so far as possible. Periodic inspections and insulation resistance tests should be carried out to confirm the continuing earth-free integrity of the system.

Yet another alternative is the employment of a sufficiently low voltage to earth, suitable for the location's environment and usage, to avoid the possibility of a dangerous electric shock. The reduced low voltage system used on construction sites and described in Chapter 3 is an example.

Regulation 9 – Integrity of referenced conductors

If a circuit conductor is connected to earth or to any reference point, nothing which might reasonably be expected to give rise to danger by breaking the electrical continuity or introducing high impedance shall be placed in that conductor unless suitable precautions are taken to prevent that danger.

This regulation aims to prevent open circuits or high impedances arising in the referenced conductors which could cause hazardous potential differences between them and the reference point, and prevent the flow of fault current in systems employing automatic disconnection protection techniques. In most cases, earth is the reference point for supply systems, but there are exceptions such as motor vehicle wiring systems which are not earthed and which use the chassis as both a reference point and a common return. In other non-earthed systems, the protective conductor is connected to one of the supply poles.

As an example, in TN-C-S supplies to domestic premises a break in the combined neutral/earth (CNE) conductor of an overhead service line would cause the metalwork in the premises to become live at or about the supply voltage if any apparatus, such as a dishwasher, were to be connected and switched on. To prevent this type of hazard, the integrity of the CNE conductor, throughout the system, has to be maintained, so all joints have to be properly made and reliable, and fuses and solid state devices are prohibited in these conductors.

Regulation 10 – Connections

Where necessary to prevent danger, every joint and connection in a system shall be mechanically and electrically suitable for use.

Joints and connections are often the circuit locations most vulnerable to failure because they are frequently not entirely suitable for their purpose or because of poor workmanship or because of inadequate maintenance. An example may be the use of terminal blocks to joint cables – the blocks usually do not have strain relief, and it is frequently the case that conductors are left exposed to be touched. This regulation aims to ensure that joints and connections are both mechanically and electrically suitable.

In the main, only competent people should be used for making joints and connections in wiring systems. Skilled cable jointers, for example, should be employed to make joints in paper-insulated lead-sheathed steel wire armoured cables and the like. Trained people, such as electricians or non-electrically qualified but suitably instructed people, should be able to make satisfactory joints and connections in low voltage plastics-insulated wiring.

There is a range of factors that merit consideration when seeking to secure compliance with the regulation. Some of these are

- Environmental
  - (a) *Vibration.* Consider the use of flexible multi-strand rather than single-wire conductors, and a soldered or compression joint rather than pinch screws that may shake loose. Plugs may need to be latched or otherwise secured into their sockets.

(b) *Wet, corrosive or dusty conditions.* Seal or enclose the joint.
(c) *Heat.* A cool wiring chamber may be advisable, and the materials used should not be susceptible to the prevailing temperatures.
(d) *Cold.* Anti-condensation measures may be needed and sealants which harden and crack at low temperatures avoided.
(e) *Dissimilar metals.* Joints and connections between dissimilar metals may require sealing to exclude air and moisture so as to avoid corrosion from electrolytic action.

- Mechanical

(a) *Physical damage.* Protect joints from damage by position or enclosure. The clamping end of pinch screws should be designed to avoid damage to the conductor.
(b) *Strain relief.* Secure conductors to relieve strain from joints and terminations.
(c) *Connections.* Mechanical connections, in joints and terminals, should be designed to clamp conductors firmly and not slacken in service.
(d) *Maintainability.* Connections that have to be periodically inspected and tested, such as earthing connections, should be readily accessible.
(e) *Ventilation.* There should be adequate ventilation, or enclosures should be so designed that overheating does not occur.

- Electrical

(a) *Insulation.* The insulation and air gaps must be suitable for the rated voltage and HV transients to prevent insulation failure and consequential short circuit between conductors.
(b) *Cleanliness.* The conductors to be connected and the components of the joint must be clean. Plug pins and socket receptacles and the contacts of switches, circuit breakers and contactors must be clean and make effective contact.
(c) *Contact resistance.* The contact area of the conductors in the connection must be sufficient to avoid overheating in service.

Regulation 11 – Means for protecting from excess of current

Efficient means, suitably located, shall be provided for protecting from excess of current every part of a system as may be necessary to prevent danger.

Excess current arises from overload, short circuits and earth faults. As the means of protection has to be efficient, it must respond sufficiently rapidly in interrupting the circuit so as to avoid danger. An overload trip, therefore, should be matched to the equipment (especially cables with their maximum current-carrying capacities) it protects so that it will operate before dangerous overheating occurs. The device may also be the means of protection against short circuit so that it must be capable of safely interrupting a fault current that may be orders of magnitude greater than the overload current. This usually entails an inverse time characteristic (i.e. the greater the current the shorter the tripping time so as to limit the damage consequent on a fault and hopefully avoid a fire or explosion).

Every part of a system has to be protected and the 'means' have to be suitably located. In practice, this means that there are a number of devices in series between the power source(s) and the loads. Those nearest the power source(s) usually need to be capable of interrupting higher fault currents than those further away. To ensure discrimination, time lag settings should provide for the device in the circuit nearest to the location of the excess current to operate before those nearer to the power source(s) so as to shut down only a part of the system. BS 7671 provides suitable guidance for locating the devices to protect every part of a low voltage system.

In high voltage systems, the process of ensuring adequate discrimination is known as 'grading', and grading assessments or protection studies are an important aspect of ensuring that such systems are safe; people in charge of high voltage distribution systems should ensure that they are carried out.

Regulation 12 – Means for cutting off the supply and for isolation

(1) Subject to paragraph (3), where necessary to prevent danger, suitable means (including, where appropriate, methods of identifying circuits) shall be available for:

(a) cutting off the supply of electrical energy to any electrical equipment; and
(b) the isolation of any electrical equipment.

(2) In paragraph (1), 'isolation' means the disconnection and separation of the electrical equipment from every source of electrical energy in such a way that the disconnection and separation is secure.

(3) Paragraph (1) shall not apply to electrical equipment which is itself a source of electrical energy but, in such a case as is necessary, precautions shall be taken to prevent, so far as is reasonably practicable, danger.

The main purpose of this regulation is that facilities must be provided for both switching off the power to circuits and for isolating circuits from their power sources. The regulation also includes a requirement to identify circuits, which is usually done by applying labels to switchgear and controlgear to identify the circuits that they are supplying. It is, perhaps, surprising how many circuits are not properly labelled, making it difficult for users and maintainers to know which switchgear controls which circuit, and creating the conditions for errors to be made when isolating circuits for work to be carried out on them.

The difference between the switching required by Regulation 12(1)a and the isolation required by Regulation 12(1)b needs to be understood. In both cases the supply has to be interrupted, but isolation has the additional requirement of securing the point of disconnection. This is usually done by locking a switch in the 'off' position, or by withdrawing a circuit breaker from the busbars, or by removing links or fuse links so that it is safe to work on the isolated part of the system. The means of isolation also serves to prevent inadvertent restoration of the supply while work is in progress.

Push buttons controlling contactors meet the requirements for switching off but do not meet the requirement for isolation because there is a possibility of the contactor coil becoming energised and closing the contacts. Note that isolation normally requires a physical air gap to be present, so electronic circuits such as variable speed drives do not provide for isolation.

A question often arises about where isolators and disconnectors should be located and how many should be provided. There is no straightforward answer to this, because it largely depends upon the circumstances. One school of thought is that each motor in a system should have its own local isolator positioned within a metre or so of the motor, ensuring that anybody working on a motor has direct control over its isolation. Whereas the provision of such local isolation has been standard practice for many years and will usually be the solution, others take the view that in some circumstances it is acceptable for 'group' isolation to be provided. For example, some manufacturers of integrated plant such as automated sawmills provide a single isolator to cover whole sections of the plant, each of which may contain many motors. Their logic is that if maintenance work has to be done on one motor in the section of the plant, the whole section has to be shut down, so one isolator will suffice. In this type of situation it is crucially important that the users of the plant ensure that maintenance personnel are properly trained on isolation procedures (see Regulation 13) and that they will use a group isolator when appropriate, despite the fact that it may be positioned tens of metres from the location of the work activity.

If there are means of both switching off and isolation, the isolator does not have to be capable of interrupting a fault or load current because the supply can be switched off before it is isolated, but if the isolator performs both functions, it would have to be suitable for interrupting those currents safely. Some modern isolators have a set of auxiliary contacts that break before the main power contacts; the auxiliary contacts are used to signal the control system to open a power contactor or similar device so that the power is interrupted before the isolator contacts open.

The last part of the regulation exempts power sources such as generators, primary and secondary batteries and charged capacitors which act as suppliers of energy. However, these power sources need to be protected to avoid danger.

Regulation 13 – Precautions for work on equipment made dead

Adequate precautions shall be taken to prevent electrical equipment, which has been made dead in order to prevent danger while work is carried out on or near that equipment, from becoming charged during that work if danger may arise.

Many accidents have happened, and continue to happen, when somebody working on or near a system that has been made dead experiences an electrical injury when the system is unintentionally switched on. This could be, for example, an electrician working on a system, or painters working in the vicinity of an overhead travelling crane's uninsulated power conductors. The aim of Regulation 13 is that measures should be taken to prevent this happening.

Withdrawing a plug supplying an item of equipment disconnects the equipment. If the plug remains close to a person working on the equipment, and under the person's control, no further action is required to secure the isolation. However, if the plug is remote from the person doing the work and there is a possibility of it being plugged in by somebody else while work is being done on the equipment, measures must be taken to prevent this happening. One means of achieving this is to place the plug inside a lockable box, and there are proprietary lockable boxes available on the market for this purpose.

In many cases isolating a circuit using the facilities provided in accordance with Regulation 12 and applying a unique padlock and warning notices to the point of isolation will be an adequate precaution, as illustrated in Figure 3.15. Figure 6.1 illustrates how a universal locking device can be used to lock a circuit breaker in the 'off' position, thereby isolating the circuit supplied from the circuit breaker. Although universal locking devices

Figure 6.1 Universal locking device used to secure circuit breaker in 'off' position

such as this are readily available, switchgear manufacturers also produce a range of special-to-type locking devices for their particular products.

There are circumstances in which more elaborate and considered precautions may be required. For example, when work is being carried out on dual circuit overhead line equipment, the circuit to be worked on may have been made dead but the other adjacent circuit remains live. Capacitive and inductive coupling between them could energise the isolated line, making it dangerous to touch, so temporary earthing commonly called 'drain earths' would have to be applied to the isolated circuit so as to dissipate any induced energy to earth.

Where it is foreseeable that more than one person may be working on plant or machinery that is isolated, the point of isolation needs to be controlled by all those working on the plant. A common solution is to use a multi-padlock hasp onto which a number of padlocks, typically up to 6–8, can be applied; an example is shown in Figure 6.2. This ensures that the isolator cannot be switched back on until all tradesmen have removed their padlock, signifying that all the work has been completed.

For work on most isolated HV equipment and for work involving, for example, the isolation of more than one supply source, it is advisable to employ a Permit to Work procedure as an additional safety measure. This is essentially a paperwork procedure in which a form is completed to identify the equipment on which work is to be done, how the equipment has been made safe, where the points of isolation and earthing are and

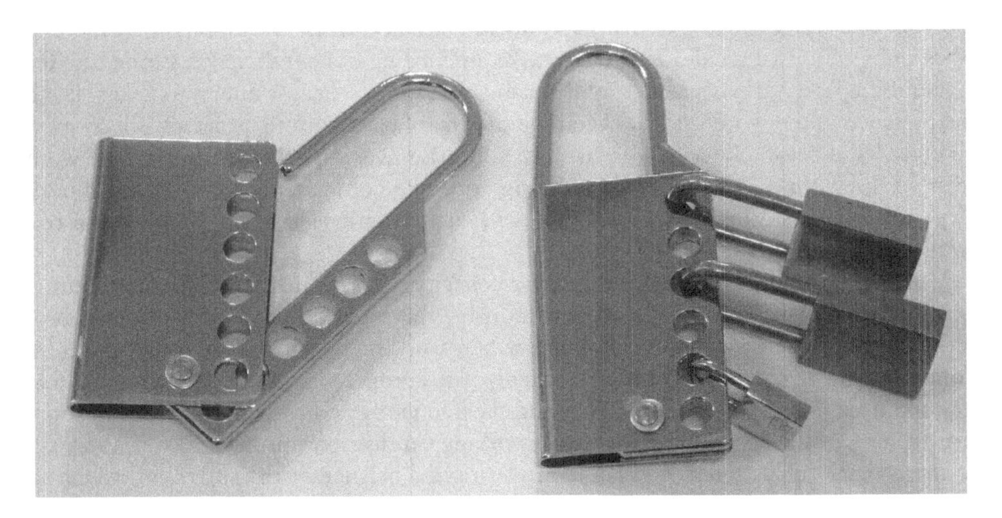

Figure 6.2 Multi-lock hasp

the limits of the work to be done. It is normally signed and issued by the person who made the system safe to a person who is going to carry out the work and who signs the form to indicate that he or she understands its contents. The Permit to Work form is subsequently cancelled once the work has been completed and the system restored to its operational configuration. The advantage of the procedure is that it imposes some structure and rigour onto isolation procedures on systems that have potential for serious injury. Given this, it is important for the procedure to be strictly adhered to and for its continuing implementation to be monitored and audited. Advice on this type of system of work is published by the HSE in guidance note HSG85 *Electricity at work. Safe working practices.*

Where a formal Permit to Work procedure is not used, the responsible person should carry out a risk assessment to devise a procedure that will ensure safety. In the case of the crane painters previously mentioned, for example, they should be told that they should not start work in the vicinity of the bare conductors until the isolator has been locked off, the conductors have been tested to ensure that they are dead, and they have been given permission to proceed. When the work is finished the circuits should not be reenergised until the painters have been withdrawn from the work area.

Regulation 14 – Work on or near live conductors

No person shall be engaged in any work activity on or so near any live conductor (other than one suitably covered with insulating material so as to prevent danger) that danger may arise unless –

(a) it is unreasonable in all the circumstances for it to be dead; and
(b) it is reasonable in all the circumstances for him to be at work on or near it while it is live; and
(c) suitable precautions (including where necessary the provision of suitable protective equipment) are taken to prevent injury.

The main aim of this regulation is to discourage dangerous live working, which over the years has been the cause of many serious and fatal accidents. The emphasis is on doing work on systems that are dead, with live work only being done if there is a strong case for it, and if the risks are acceptable, and if suitable precautions against injury are taken. Note the presence of the word 'and' between the subsidiary parts, meaning that all aspects must be considered before a decision is taken to authorise live work. Note also that the word 'work' is used; this has a very wide meaning with one consequence being that, contrary to popular belief, testing of live parts is a work activity that falls within the scope of this regulation.

Much live working is carried out in the electricity supply industry. This is mostly justified on the basis that it is frequently unreasonable to switch off the supply to consumers or loads such as traffic lights and hospitals when work such as live cable jointing can be done safely by personnel who have the appropriate competences and equipment and follow approved jointing procedures. The industry frequently argues, with some justification, that turning off the power to allow dead working on distribution networks creates risks to the general public and that, when considering the balance of risk, it is safer overall to keep the power on and for their competent staff to work live.

The industry also has a regulatory duty to 'keep the lights on' and has performance criteria set down by OFGEM (the Office for Gas and Electricity Markets) to minimise customer minutes lost (CML) and customer interruptions (CI). Resolving the conflict between, on the one hand, minimising CML/CI figures and, on the other hand, making systems dead to allow safe working has challenged the companies and regulators for some time. However, the courts would be expected to take the view that the safety of the public is more important than companies achieving regulatory business objectives.

This was confirmed in January 2016 at the trial of UK Power Networks following the death of a jogger who ran into a live wire of an 11 kV power line that had fallen from its wooden pole to a height of only 1.5 m above a public footpath. The company, which is the distribution network operator that owned and operated the overhead line, had been contacted about the condition of the line and had sent an engineer to investigate the problem rather than switching the line off, which they could have done but in so doing would have disrupted the supplies to customers. The jogger running on the footpath ran into the line before the engineer arrived and died from burn and electric shock injuries. UK Power Network (Operations) Ltd was fined £1 million for a breach of Section 3(1) of the Health and Safety at Work etc Act 1974. The judge commented that the company's failure was due to its culture, and the HSE said that the death could have been prevented by immediate remote deenergisation of the power network which the circumstances clearly called for.

The enforcement authorities have worked with the electricity supply industry to agree on some measures by which the sometimes conflicting duties of public safety and maintaining supplies can be managed. For example, it has been agreed that, within the constraints of Regulation 14, live cable jointing can be carried out on the low voltage network and live line hands-on working can be carried out on 11 kV and, in some instances, 33 kV overhead lines so as to minimise disruption to supplies. Figure 6.3 illustrates a live line hands-on team working on a live 11 kV overhead line, using an insulated-bucket mobile elevating work platform team and temporary insulation on the live wires.

Strict criteria have been agreed on in relation to these activities, and the authorities will invariably carry out searching enquiries when accidents happen during them.

Figure 6.3 Live working on an 11 kV overhead line (Courtesy of SP Energy Networks)

When an apparent need arises for work to be done on or near potentially dangerous uninsulated live conductors, the responsible person has to be able to justify it. This will involve carrying out a risk assessment that will include consideration of the consequences and the impact of switching off.

Where an electrical fault has halted a production process, for example, the maintenance electrician often needs to work on or near bare live conductors when using a test lamp or voltmeter to locate the fault. This type of live fault finding should only be done where it is not possible or practicable to trace the fault with the supply isolated. Where the work must be done live, the electrician must be competent in live fault finding activity. Test equipment with insulated and fused leads must be used, as well as tools that are insulated to avoid shorting out conductors at different potentials. Consideration should be given to shrouding off adjacent conductors at different potentials, including metalwork of panels and enclosures that may be at earth potential. Temporary barriers may need to be erected to minimise the chances of the electrician being disturbed or distracted. The presence of an accompanying person should also be considered – in the event of something going wrong, the second person's presence would ensure immediate help is available to switch off the supply, drag the electrician clear, summon the rescue services, and perhaps render first aid treatment.

Construction work adjacent to live overhead power lines comes within the remit of Regulation 14. The same considerations apply, and the precautions against injury from contact with the lines are explained in Chapter 14.

The HSE's previously mentioned guidance note, HSG85, includes advice on the decision-making process as to whether work should be done live or dead.

Regulation 15 – Working space, access and lighting

For the purposes of enabling injury to be prevented, adequate working space, adequate means of access, and adequate lighting shall be provided at all electrical equipment on which or near which work is being done in circumstances which may give rise to danger.

This regulation recognises that some electrical work is carried out in dangerous locations where injury should be prevented by, inter alia, having an adequate means of access and adequately and properly lit working space. The regulation should be considered in conjunction with the 'suitable precautions' requirement of Regulation 14(c).

For compliance, it is advisable for the person responsible for the electrical wiring and apparatus such as switchgear and controlgear to consider the requirements at the design stage of a project in conjunction with the architect or other space designer. The aim is to ensure that the requirements are met and thereby save what may be expensive modifications later. An example of the failure to do this arose a number of years ago when there was a vogue for flat-roofed buildings. The planners objected to the roof line being spoilt by lift houses, and so the architects required the lift engineers to install their headgear in a small space on the top floor over the lift shaft between the top of the lift door and ceiling. In a number of cases, this resulted in unsafe working conditions for the maintenance staff and expensive alterations had to be made.

The space requirements will be minimised if equipment can be isolated for work, but even so, where live testing is being carried, or other forms of live work where live conductors are touchable, adequate space and lighting must be available for the work to be done safely. Some guidance on safe clearances may be gleaned from HSE's Memorandum of Guidance on the Regulations, HSR25, which reproduces the clearances in the former switchboard Regulation 17 of the Electricity (Factories Act) Special Regulations 1908 and 1944. The electricity supply industry companies' safety rules also contain information on safety clearances, usually as a function of voltage.

The new Section 729 in BS 7671:2008 offers useful guidance for separation distances in operating and maintenance gangways. It is targeted mainly at areas where there are live overhead busbars, such as internal substations. The precautionary measures are the use of barriers, obstacles and enclosures. Gangways must be at least 700 mm wide, and the distance to live parts placed out of reach must be at least 2.5 m. The dimensions are illustrated in the standard.

The space should also be adequate for safe movement in an emergency. For example, when horizontal drawout circuit breakers are withdrawn from a switchboard, there must be sufficient room for a person to escape from the switchroom in the event of a fire or explosion.

There has to be enough light for safe working. For most purposes, about 150 lux of general lighting should be adequate, supplemented if necessary by local lighting from handlamps or luminaires on pedestals. The lighting should be arranged so as to prevent dazzle and confusing shadows that may cause danger.

Regulation 16 – Persons to be competent to prevent danger and injury

No person shall be engaged in any work activity where technical knowledge or experience is necessary to prevent danger or, where appropriate, injury, unless he possesses such knowledge or experience, or is under such degree of supervision as may be appropriate having regard to the nature of the work.

This regulation recognises that competence is an important prerequisite for 'any work activity' on electrical systems to be undertaken safely and that, where competence is not held, adequate supervision must be provided. The phrase 'any work activity' is obviously very broad, so it would include those who plan the work and give instructions for its execution but do not themselves necessarily participate, as well as those who supervise and do the work on site. There are two areas of risk covered, i.e. the prevention of the risk of injury and the prevention of injury itself. Work on equipment that has been isolated from the supply and earthed is an example of the former because danger has been prevented, but live working entails some risk and is an example of the latter because the safety precautions counter the likelihood of injury occurring. Having said that, the difference between danger and injury in this context is a bit obtuse, and the need for the distinction is not particularly obvious.

The matter of 'competence' is very important and frequently creates confusion and difficulty. For that reason, a chapter has been devoted to the topic, so the interested reader is directed to Chapter 19.

Part IV: Miscellaneous and general

Regulation 29 – Defence

In any proceedings for an offence consisting of a contravention of regulations 4(4), 5, 8, 9, 10, 11, 12, 13, 14, 15, 16 or 25, it shall be a defence for any person to prove that he took all reasonable steps and exercised all due diligence to avoid the commission of that offence.

Regulation 29 provides a due diligence defence in criminal proceedings against alleged infringements of the listed regulations. These are the regulations that have absolute requirements and are not covered by the 'so far as is reasonably practicable' requirement. The intention is to provide some protection for persons who are guilty, for example, of a technical infringement, but who have done their best to comply and who are able to demonstrate this in court.

Regulation 30 – Exemption certificates

(1) Subject to paragraph (2) the Health and Safety Executive may, by a certificate in writing exempt -

(a) any person;
(b) any premises;

(c) any electrical equipment;
(d) any electrical system;
(e) any electrical process;
(f) any activity,

or any class of the above, from any requirement or prohibition imposed by these Regulations and any such exemption may be granted subject to conditions and to a limit of time and may be revoked by a certificate in writing at any time.

(2) The Executive shall not grant any such exemption unless, having regard to the circumstances of the case, and in particular to

(a) the conditions, if any, which it proposes to attach to the exemption; and
(b) any other requirements imposed by or under any enactment which apply to the case,

it is satisfied that the health and safety of persons who are likely to be affected by the exemption will not be prejudiced in consequence of it.

The intention of Regulation 30 is to enable the responsible inspectorate, the HSE, to vary the legal requirements in particular circumstances where they are inappropriate. To date, no exemptions have been issued.

Regulation 31 – Extension outside Great Britain

These Regulations shall apply to and in relation to premises and activities outside Great Britain to which sections 1 to 59 and 80 to 82 of the Health and Safety at Work etc. Act 1974 apply by virtue of Articles 6 and 7 of the Health and Safety at Work etc. Act 1974 (Application outside Great Britain) Order 1977 as they apply within Great Britain.

Regulation 31 extends the geographical scope of the regulations to cover work over, on and under the territorial waters of Great Britain. It includes mines extending under the sea, the construction of oil and gas rigs and pipelines and their operation, diving, the loading and unloading, fuelling and provisioning of any vessel and any work on it except that done by the ship's company, and construction projects such as the Channel Tunnel and the Cardiff Bay barrage.

Regulation 32 – Disapplication of duties

The duties imposed by these Regulations shall not extend to:
(a) the master or crew of a sea-going ship or to the employer of such persons, in relation to the normal ship-board activities of a ship's crew under the direction of the master; or
(b) any person, in relation to any aircraft or hovercraft which is moving under its own power.

Regulation 32 is limited to the exclusion of the work on a sea-going ship by the master and crew and work on aircraft and hovercraft when moving under their own power.

Regulation 33 – Revocations and modifications

(1) *The instruments specified in column 1 of Part I of Schedule 2 are revoked to the extent specified in the corresponding entry in column 3 of that Part.*
(2) *The enactments and instruments specified in Part II of Schedule 2 shall be modified to the extent specified in that Part.*
(3) *In the Mines and Quarries Act 1954, the Mines and Quarries (Tips) Act 1969 and the Mines Management Act 1971, and in regulations made under any of those Acts, or in health and safety regulations, any reference to any of those Acts shall be treated as including a reference to these Regulations.*

This regulation refers to those parts of older law which are either revoked or modified.

Comment

The extension of specific electrical safety legislation beyond the boundaries of the factory gate has been the major contribution of these regulations to the safety system in the UK. This, coupled with the regulations' emphasis on working dead, and the specific provisions relating to maintenance, have had a very positive impact.

It has to be recognised that, even though the regulations have been in force for over 25 years, there is still widespread ignorance of them, especially Regulation 14 concerning live work activities. Perhaps the situation would be improved if undergraduate electrical and electronic engineers at universities were to receive training and instruction on them. At least electricians who are undertaking apprenticeships in the electrical contracting industry receive instruction on them with the result that, as time passes, the electrical workforce is becoming increasingly aware of its legal responsibilities and the number of electrical injuries sustained at work is steadily reducing.

In 2013 the IET produced a document called the *Code of practice for electrical safety management* and has been running a series of associated workshops throughout Great Britain. The main value of the document is that it provides for managers a comprehensive set of pointers to, or checklists on, matters that they need to consider when seeking to comply with their legal duties concerning electrical safety, especially the Electricity at Work Regulations. It doesn't reference the applicable law or offer solutions, but it is a useful tool in the managerial armoury.

7

The Electricity Safety, Quality and Continuity Regulations 2002

Introduction

A brief history of the regulatory regime governing Great Britain's electricity supply industry makes for interesting, if rather convoluted, reading.

Networks for distributing electricity were first developed in Great Britain as localised street lighting systems at the end of the nineteenth century, and they have since evolved to become an interconnected national transmission and distribution network. The first acts governing them were the Electricity Lighting Act 1882 and the Electricity (Supply) Act 1882. This empowered the then Board of Trade to impose safety requirements 'for securing the safety of the public from personal injury or from fire or otherwise'.

The Electricity (Supply) Act 1919 established district electricity boards and Electricity Commissioners who, until their disbandment in 1948, were responsible for developing electricity policy, including policy on public safety. Around this time, only 100 years ago, the Department of Electricity Supply declared that alternating current was the preferred system for electricity supply, although there had yet to be any standardisation of voltage and frequency.

Separate private and municipally owned utilities provided electricity until the 1930s. The Electricity (Supply) Act 1926 aimed to improve efficiency of generation, transmission and distribution by creating an interconnected network, including establishing the Central Electricity Board with the task of building an EHV transmission system. Then the Electricity Act 1947 nationalised the large number of existing private and municipal electricity supply undertakings and formed them into fourteen Area Boards (twelve in England/Wales and two in Scotland) under a central authority, the British Electricity Authority, that eventually morphed into the Central Electricity Generating Board (CEGB).

In 1990, under the Electricity Act 1989, government policy on the privatisation of state assets led to the CEGB being disbanded and private companies being created out of the old boards, although they continued for the most part to combine the functions of generators and suppliers of electricity with that of the distributor in their areas; they were known generically in England and Wales as Regional Electricity Companies (RECs). Whereas some larger commercial customers had a choice of supplier, this option was not extended to all customers until 1998.

On 1 October 2001, under the Utilities Act 2000, the final privatisation step was taken whereby the RECs were required to separate their functions so that the same company could not be both a distributor and a supplier of electricity. This move opened up the energy supply market, allowing new entrants to compete for market share. Meanwhile in

Scotland, two companies, Scottish Power and Scottish Hydro Electric (part of the Scottish and Southern Energy Group), remained vertically integrated with generation, transmission, distribution and supply businesses, albeit with the supply businesses kept at arm's length from the engineering businesses.

The old Electricity Boards were regulated as far as public safety and supply quality were concerned by regulations made by the Electricity Commissioners, particularly the Electricity Supply Regulations 1937 "for securing the safety of the public and for insuring a proper and sufficient supply of electrical energy". These regulations remained in force until, as a prelude to privatisation, they were replaced by the Electricity Supply Regulations 1988 made under the Energy Act 1983. The 1988 regulations were amended on five occasions before they were revoked from 31 January 2003 and replaced by the Electricity Safety, Quality and Continuity Regulations 2002 SI 2002/2665 (ESQCR). These regulations, which are the subject of this chapter, have been amended by the Electricity Safety, Quality and Continuity (Amendment) Regulations 2006.

The Electricity Supply Regulations and their successor ESQCR were enforced by inspectors of the Department of Trade and Industry's (DTI) Engineering Inspectorate. In 2006, as part of the government's 'reducing burdens' and 'regulatory simplification' agenda, enforcement responsibility for the safety elements of the ESQCR transferred to the Health and Safety Executive's specialist inspectors of electrical engineering under the terms of an agency agreement. Responsibility for the content of the regulations and enforcement of the quality and continuity provisions rested with the Department of Energy and Climate Change (DECC), the successor body to the DTI on energy policy matters. In July 2016 the new Prime Minister in the UK government decided to merge DECC with the Department for Business, Innovation and Skills (BIS) to form a new department called the Department for Business, Energy and Industrial Strategy (BEIS), which now has responsibility for the regulations.

The Engineering Inspectorate was disbanded in 2006 as its inspectors transferred to either the HSE or DECC, now BEIS. These new arrangements have taken a considerable period of time to bed in, and it is still not clear precisely how and to what extent they have reduced regulatory burdens on the electricity supply industry.

Scope of the regulations

In similar vein to the 1988 regulations, the ESQCR set out safety requirements aimed at protecting the general public and consumers from danger arising from the generation, transmission, distribution and supply of electricity to consumers. They also specify power quality and continuity standards. In comparison with the previous legislation, the number of duty holders has increased, reflecting the significant changes that took place in the electricity supply industry on privatisation and its consequential increased fragmentation and complexity. The principal duty holders are generators, distributors, suppliers of electricity and meter operators, as well as their respective agents and contractors. Consumers and others also have some duties. Private networks are covered as well as those used for transmitting electricity and distributing it to the general public, particularly in the context of the safety of overhead lines and substations.

Under the terms of the Electricity Act 1989 and Utilities Act 2000, companies that wish to generate, transmit, distribute or supply electricity to the general public must be licensed to do so by the industry's principal economic regulator OFGEM, the Office for Gas and

Electricity Markets. In addition to complying with their statutory duties under ESQCR, the companies must also comply with detailed and extensive standard licence conditions which cover, inter alia, electrical safety requirements. On the other hand, licence holders are granted certain powers by the Electricity Act 1989, such as compulsory purchase of land, and the right to fell trees or cut vegetation that might compromise continuity of supply.

It is important to understand the definitions of the main duty holders because there is considerable scope for confusion.

Generators

A generator is an organisation that generates electricity at high voltage for the purpose of supplying consumer's installations via a network, where the term 'network' means an electrical system comprising all the conductors and other equipment used to conduct electricity from the source or sources of voltage to one or more consumer's installations, street electrical fixtures or other networks. Offshore installations were originally excluded from the definition of 'network', although the 2006 amendment extended the scope of the regulations to explicitly include the territorial sea of Great Britain and beyond into a region known as the Renewable Energy Zone. This means that the regulations addressing public safety and supply interruption apply equally to generating stations that are installed either offshore or onshore.

It is important to note that generating companies that generate electricity at low voltage are excluded from the definition and, therefore, the duties under these regulations. This is interesting from the point of view of the number of low voltage units that are increasingly being connected to the distribution networks as the country's generation capacity becomes more and more distributed, particularly as environmental policies lead to whole-scale closures of base-load generating fossil fuel plants. There may be a need in the future for low voltage generators to come within scope of the regulations.

Distributor

A 'distributor' means a person who owns or operates a network; operators of railway networks, trams, trolley vehicles, and other forms of guided transport are excluded from the definition. This means that transmission and distribution companies in the electricity supply industry are 'distributors'.

There are currently three Transmission Operators (TOs) permitted to develop, operate and maintain a high voltage system within their own distinct onshore transmission areas. These are National Grid Electricity Transmission plc (NGET) for England and Wales, Scottish Power Transmission Ltd for southern Scotland and Scottish Hydro Electric Transmission plc for northern Scotland and the Scottish island groups.

The term also embraces the Distribution Network Operators (DNOs), the companies that stem from the old area boards and of which there are fourteen, as well as Independent DNOs (IDNOs). IDNOs are organisations that develop, operate and maintain local electricity distribution networks located within the areas covered by the DNOs and which are directly connected to the DNO networks or indirectly to the DNO network via another IDNO. IDNO networks are mainly extensions to the DNO networks serving new housing and commercial developments. They were created as part of the government's drive to introduce competition into the business of distributing power to end users.

Private distributors would also be captured by this definition, examples being the owner of a caravan park who distributes electricity to individual park homes or caravan pitches, and a petrochemical company that generates electricity on site and distributes it to production modules. These are known as unlicenced distributors, and the extent to which they ought to be captured by the provisions of ESQCR has been a controversial subject since the publication of ESQCR. One body of opinion is that the intention of parliament was that the Electricity Act 1989, and hence its subsidiary legislation including ESQCR, was targeted at licenced businesses operating in the public electricity supply industry, which would exclude these unlicenced distributors from the scope of the regulations other than the small number of regulations dealing with matters such as the height of overhead lines and their proximity to buildings and structures. Another body of opinion argues that the unlicenced distributors are distributors as defined in the regulations and must therefore satisfy all the distributor duties under the regulations. It can only be hoped that this matter is tested in the courts before too long so as to remove this uncertainty.

Suppliers

The term 'supplier' refers to organisations that contract to supply electricity to consumers. The supplier does not own the wires that connect to a consumer's installation but operates in the electricity market to purchase electricity under the trading rules of the British Electricity Trading and Transmission Arrangements and then sell that electricity on to consumers.

Meter operators

A meter operator is an organisation that installs, maintains or removes metering equipment at or near the supply terminals on consumers' premises. In general, meter operators are contracted by suppliers, but some larger consumers arrange for their own meter operators. The Meter Operation Code of Practice Agreement (MOCOPA) is a voluntary agreement between a distributor and a meter operator, enabling the meter operator to install and connect meters to the distribution system and covers safety, technical and other interface issues.

The regulations

Structure

The regulations are set out in 8 parts and 5 schedules, as follows:

Part I Introductory

1 Citation, commencement and interpretation
2 Application of regulations
3 General adequacy of electrical equipment
4 Duty of cooperation
5 Inspection of networks

Part II Protection and earthing

6 Electrical protection
7 Continuity of the supply neutral conductor and earthing connections
8 General requirements for connection with earth
9 Protective multiple earthing
10 Earthing of metalwork

Part III Substations

11 Substation safety

Part IV Underground cables and equipment

12 General restriction on the use of underground cables
13 Protective screens
14 Excavations and depth of underground cables
15 Maps of underground networks

Part V Overhead lines

16 General restriction on the use of overhead lines
17 Minimum height of overhead lines, wires and cables
18 Position, insulation and protection of overhead lines
19 Precautions against access and warnings of dangers
20 Fitting of insulators to stay wires

Part VI Generation

21 Switched alternative sources of energy
22 Parallel operation

Part VII Supplies to installations and to other networks

23 Precautions against supply failure
24 Equipment on a consumer's premises
25 Connections to installations or to other networks
26 Disconnection of supply, refusal to connect and resolution of disagreements
27 Declaration of phases, frequency and voltage at supply terminals
28 Information to be provided on request
29 Discontinuation of supplies

Part VIII Miscellaneous

30 Inspections, etc for the Secretary of State
31 Notification of specified events
32 Notification of certain interruptions of supply

Schedules

The following sections describe the regulations, with most emphasis placed on those covering safety matters as opposed to those covering quality and continuity requirements.

Part I: Introductory

Regulation 3 – General adequacy of electrical equipment

Regulation 3(1) sets out the top-level general duties on generators, distributors and meter operators concerning their equipment, which must be

(a) sufficient for the purposes for and the circumstances in which it is used; and
(b) so constructed, installed, protected (both electrically and mechanically), used and maintained as to prevent danger, interference with or interruption of supply, so far as is reasonably practicable.

The use of the word 'sufficient' in the first part of the regulation creates some problems because it is not clear what it means in practice and because the regulation is absolute with no specific defence of either reasonable practicability or due diligence. Duty holders would be advised to interpret it as meaning that their equipment should be designed to appropriate national standards so as to be in compliance with the more detailed regulations. The second part is more akin to other safety legislation in that it places a duty on the organisations to prevent danger so far as is reasonably practicable – a very generic and broad duty.

The duty to maintain is a challenging one, simply by virtue of the extremely large inventory of equipment that has to be maintained; it includes, for example, the service connections and metering equipment in every property in Great Britain. There is some expansion on the generic duty in the standard licence conditions; for example, the SLC for distributors requires them to inspect their equipment – cables, cable-heads, and cut-outs, in properties once every 2 years.

Regulation 3(2) applies only to generators and distributors and concerns the safety of their overhead lines and substations which, as evidenced by accident history, are locations which present the greatest risk to members of the public. It requires them to

- Conduct risk assessments of their assets covering the risks from interference, vandalism and unauthorised access. In addition to the commonplace events such as youths

vandalising assets or people flying kites or fishing near overhead lines, an obvious activity to be covered by this assessment would be criminal activity relating to metal theft, a problem that has blighted this industry, and others, for the past 10 years or so.
- Record the risk assessment and keep it up to date. This is important because, should an accident occur, the investigating police or inspectors will ask to see the risk assessments – if they cannot be produced there is a likelihood of a charge for breach of this regulation.
- Implement the control measures identified by the risk assessment.

Regulation 3(3) requires generators and distributors to take reasonable steps to ensure that the public are made aware of dangers which may arise from activities carried out in proximity to overhead lines, and to indicate the means by which those dangers may be avoided. This requires the companies to take proactive steps to warn the public; examples of such activities include

- visiting schools and leisure clubs to give presentations on the dangers from overhead lines;
- creating leaflets that provide advice on the matter and distributing them to organisations at risk, especially farmers and construction companies; and
- posting warning notices on river banks crossed by overhead power lines where fishing is known to take place, and on recreational areas crossed by overhead lines where sports are played or kites are flown, and so on.

In the unfortunate event of a serious injury or death of a member of the public from contact with an overhead line, the investigating authority will examine what proactive measures the owner of the line takes to warn the public about the danger from overhead lines. Doing nothing, although an option adopted by some distributors, is not a good option when under regulatory scrutiny following an incident.

Regulation 4 – Duty of cooperation

This regulation requires all four principal duty holders to cooperate with each other to ensure compliance with the regulations. The importance of this requirement should not be underestimated. For example, the smart meter replacement programme, which will see 53 million new smart electricity and gas meters installed in thirty million British properties by 2020, will require very close cooperation and coordination between distributors, suppliers, meter operators and their contractors. The scale of the programme, coupled with the inherent risks of specifically recruited and trained staff changing so many electricity meters, in properties that may have unsafe features such as old VIR cables and damaged cut-outs, means that the safety and business risks need to be very carefully managed. The government's judgement is that the benefits in terms of better and more accurate billing (no more estimated bills), and the putative benefit of greater energy efficiency, outweigh the risks and considerable costs.

Regulation 5 – Inspection of networks

This regulation requires generators and distributors to inspect their networks with sufficient frequency so as to ensure, so far as is reasonably practicable, that they comply with the provisions of the regulations. Essentially, this means periodic inspection of overhead lines, substations, and associated equipment and security arrangements. The companies must keep records of the inspections of overhead lines and substations for a period of at least 10 years.

Overhead line inspections are carried out to

- ensure that the lines meet the statutory height requirements of Regulation 17;
- check that poles and pylons have Schedule 1 safety signs (see later for description) and any required anticlimbing precautions attached and that they are in good condition;
- detect and rectify any damage or deterioration that may lead to danger;
- detect any change in land usage that might lead to a change in risk categorisation with a consequential change in preventive measures if necessary; and
- ensure that the lines remain clear of trees and vegetation.

The various DNOs and TOs have different policies on the nature and frequencies of their line inspections, some of which will be conducted by inspectors walking the routes of the lines and others by helicopter or, of increasing interest, drones. Typically, DNOs carry out safety inspections of their high voltage lines every 4 years, with full safety and engineering inspections conducted every 12 years. Where lines pass through areas of woodland or fast-growing vegetation, tree-clearance patrols may be required annually or biennially.

Substations need to be inspected periodically to ensure that security features such as fences, walls and barbed wire are in good condition; that no vegetation or other material has grown or been placed near the fencing or walls to allow easy access; and that safety and other signage is in place.

Part II: Protection and earthing

Regulations 6 to 10 place specific duties on generators, distributors and, in the case of Regulation 10, meter operators in relation to providing protection against excess and earth leakage currents, the continuity of the supply neutral conductor, general requirements for connection with earth, multiple earthing of the neutral, protective multiple earthing (PME) supplies on consumers' premises and metalwork earthing.

The main duties are to

- Install devices (fuses and circuit breakers) to prevent danger, so far as is reasonably practicable, from excess current and earth leakage current.
- Take all reasonable precautions to ensure continuity of the supply neutral conductor.
- Connect electricity systems to earth as close as possible to the source of voltage and meet other general earthing requirements set out in Regulation 8.
- Where a combined neutral and protective conductor is used to supply a consumer's premises, the supplier must meet the protective multiple earthing (PME) requirements of Regulation 9. The neutral conductor of PME supplies must not be connected through to metalwork in caravans or boats. (Refer to Part 7 of BS 7671 and Chapter 12 of this book for clarification.)
- Ensure that exposed metalwork that is not acting as a phase conductor is earthed, where necessary to prevent danger.

Regulation 7(2) requires that no generator or distributor shall introduce or retain any protective device in any supply neutral conductor or any earthing connection of a low

Figure 7.1 Illegal fused cut-out with fuse in the neutral

voltage network which he owns or operates. Regulation 2(4) indicates that this requirement does not apply to any distributor's fusible cut-out brought into use on or before 31st December 1936, until 10 years after the coming into force of these regulations. In other words, all of the pre-1937 fusible cut-outs with fuses in the neutral circuit installed in consumers' premises in Great Britain, of which there were many, should have been removed by distributors by 2013. An example of such a now-illegal cut-out is shown in Figure 7.1.

Part III: Substations

Regulation 11 is concerned with the safety of substations, an important topic in the context of the unfortunately high number of fatal and major accidents in recent years that have occurred during attempts to steal metal from substations. In the south of Scotland, for example, in the 4-year period from 2012 to 2016 there were 1200 break-ins to substations, with 3 deaths and 20 arrests. In one typical incident in Chesterton in England in 2011, a 46 year old man suffered fatal shock and burn injuries while cutting copper earthing strip off a concrete pillar supporting exposed 33 kV busbars in a primary substation. He had used a ladder both to get over the fence and to gain access to the copper at the higher levels of the pillar, at which point he had come too close to the live busbars.

Another long-standing problem has been so-called child trespassing, where children and youths attempt to gain entry into substations for a variety of reasons such as retrieving balls, vandalism, or simply the challenge and excitement.

The regulation requires generators and distributors to do the following:

- Enclose substations where necessary to prevent danger or unauthorised access, so far as is reasonably practicable. If the substation contains external uninsulated live conductors, such as busbars and wires, the area must be enclosed by a fence or wall not fewer than 2.4 metres in height. Most substations will need to be enclosed with a fence or wall, but there are circumstances where the risk is sufficiently low so that no enclosure is necessary.
- Ensure that, so far as is reasonably practicable, Schedule I safety signs are displayed on the fence or walls; there is a property notice placed in a conspicuous position which gives the location or identification of the substation, and the name and contact telephone number of each generator or distributor who owns or operates the substation equipment; and any other danger signs, such as 'Danger – Keep Out' notices, that may contribute to safety and security are displayed.
- Take all reasonable precautions to minimise the risk of fire associated with the equipment.

There are a variety of engineering standards published by the network operators' industry body, the Energy Networks Association (ENA), that provide guidance on the design requirements for security enclosures in substations. A good source of advice is the publication *Child trespassers in electricity substations*, published by the Electricity Council in 1987. In essence, the main requirement is for fences, walls and gates to be constructed to make it very difficult for them to be climbed without climbing aids and for signs to be attached to make the danger very clear to anyone who may be thinking about illegally entering the substation.

Part IV: Underground cables and equipment

Regulations 13 and 14 place duties on generators and distributors in relation to their underground cables and equipment, apart from those in generating stations and substations where health and safety legislation would be applied. The term 'equipment' will include items such as link boxes.

Regulation 13 requires that cables and associated equipment have an earthed, continuous metallic screen enclosing the conductors, which may be used as the neutral. The intention is to protect a careless excavator from injury, should the screen be penetrated with a tool and make contact with a phase conductor. The regulation also requires that joints and cable terminations in low voltage systems have some form of mechanical protection.

Regulation 14 requires that cables are buried deep enough to avoid damage from land use, as explained in greater detail in Chapter 14. They must also be protected against mechanical damage; although not specified in the legislation, the regulator's preferred hierarchy of protection methods is

(i) Cable installed in a duct with marker tape above
(ii) Cable installed in a duct only

(iii) Cable laid direct and covered with a protective tile
(iv) Cable laid direct and covered with marker tape
(v) Some other method of mark or indication

Although this protection is not proof against a carelessly wielded pneumatic drill, it is otherwise fairly effective. Note that this requirement was extended to low voltage cables by the ESQCR in 2003, with the predecessor Electricity Supply Regulations requiring only high voltage cables to be so protected. There is no requirement for the companies retrospectively to apply protection to their buried low voltage cables, but it does mean that there are thousands of miles of low voltage cables currently buried below ground without this protection. This is a situation that will last for very many years.

Underground cable plans are dealt with under Regulation 15. This is an important provision in the context of reducing the number of occurrences of underground cables being struck, with tens of thousands of such events occurring each year. Generation and distribution companies must maintain maps indicating the position and depth of cables and other elements of their distribution systems and make the maps available to anyone who can show reasonable cause. This includes local planning authorities and construction companies and utilities such as gas, water and telecommunication companies.

Most of the companies now hold their maps in electronic form and make them available via the internet, although they may charge for the service, as authorised to do so by Regulation 15(4). Since the requirement for the production of such plans came into force only in 1988, the companies are exempted from recording pre-1988 cables on their plans unless it is reasonably practicable for them to be included. In reality, most companies should have complete records, but this provision recognises that some legacy information may have been lost over time. For this reason, and others, the plans should not be treated as an accurate or compete representation of the situation below ground; it is for this reason that safe excavation techniques are of fundamental importance for avoiding contact with underground services, as explained in greater detail in Chapter 14.

Part V: Overhead lines

Regulations 16 to 20 are about the safety of overhead lines, which Regulation 16 stipulates must not be energised at a nominal voltage greater than 400 kV. The most important provisions concern safety clearances from the ground or from buildings and structures. The regulations apply to all owners of overhead supply lines, so they apply not just to licenced TOs, DNOs and IDNOs but also to unlicenced owners of overhead lines.

Regulation 17 and its associated Schedule 2 set out the minimum heights above ground of the phase conductors of uninsulated overhead lines, as shown in Table 7.1.

An exception to these heights are conductors that connect support-mounted equipment such as transformers, circuit breakers, fuses and switches to the lines supported by the pole or pylon. These conductors, if not insulated, must be at least 4.3 metres above ground. However, there would be a requirement for anticlimbing measures to be installed on these types of poles or pylons to prevent any non-authorised person climbing them; an example of this is seen in Figure 7.2 which shows an H-pole carrying a transformer on its cross-arm. There are anticlimbing brackets with barbed wire attached to each of the two poles. This requirement is covered in Regulation 19 and designs are based on the

Table 7.1 Minimum heights above ground of overhead lines

Voltage of overhead line	*Minimum height above road accessible to vehicular traffic (m)*	*Minimum height above ground other than road accessible to vehicular traffic (m)*
Not exceeding 33,000 volts	5.8	5.2
Greater than 33 kV and less than or equal to 66 kV	6.0	6.0
Greater than 66 kV and less than or equal to 132 kV	6.7	6.7
Greater than 132 kV and less than or equal to 275 kV	7	7
Greater than 275 kV and less than or equal to 400 kV	7.3	7.3

industry standard ENA Technical Standard 43–90 *Anticlimbing measures and safety signs for overhead lines.*

Regulation 18 refers to the safety of overhead lines that are 'ordinarily accessible', meaning that the conductors could be reached by hand if any scaffolding, ladder or other construction was erected or placed on, in, against or near to a building or structure. Such lines must be made dead, or insulated, or otherwise adequately protected to prevent danger. This is reinforced in Regulation 18(5) which requires that "No overhead line shall, so far as is reasonably practicable, come so close to any building, tree or structure as to cause danger." As explained in Chapter 14, the industry interpretation of these requirements is contained in a benchmark technical standard produced by the ENA – ENA TS 43–8 *Overhead Line Clearances.*

Regulation 19 deals with the aforementioned anticlimbing measures and, in addition, the requirement to fit safety signs to all supports carrying high voltage overhead lines and uninsulated low voltage lines. The design of the safety sign is specified in Schedule 1 to the regulations and is reproduced in Figure 7.3. Two safety signs can be seen in Figure 7.2 attached to both poles just above the anticlimbing brackets – the sign has black lettering and symbology on a yellow background.

The final regulation in this part, Regulation 20, requires that insulators are fitted in stay wires at a height greater than 3 metres above ground. It applies to uninsulated overhead lines carried on non-metallic supports such as wooden and concrete poles. Stay wires are fitted to some supports to stabilise them, perhaps because the support is positioned in soft ground or because the line turns through an angle at the particular support and is therefore applying uneven tension that may pull the support over. The purpose of the insulator is to prevent the lower section of the stay wire from becoming live under fault conditions and creating a danger to livestock or people who may touch it.

Part VI: Generation

Regulations 21 and 22 cover the situations where someone wishes to connect a generator to a distribution network either as a switched alternative supply (Regulation 21) or to run in parallel (i.e. permanently connected) with the distribution network (Regulation 22).

Figure 7.2 Anticlimbing brackets with barbed wire installed on H-pole supporting transformer and 33 kV overhead line

Figure 7.3 ESQCR Schedule 1 safety sign

There is a large number of emergency standby generators in places such as hospitals and large shopping centres, and distributed generation such as photovoltaic cells in domestic and commercial premises is increasingly common.

In the case of the switched alternative supply, the requirement is that the generator cannot operate in parallel with the network, usually achieved by having a break-before-make changeover switch to switch between the two alternative types of supply.

The main thrust of Regulation 22 is that a generator, which includes sources such as photovoltaics and wind turbines, may be connected and operated in parallel with a distribution network only if it has the necessary and appropriate equipment to prevent danger or interference with that network or with the supply to consumers, so far as is reasonably practicable, and, where the source of energy is part of a low voltage consumer's installation, the installation complies with BS 7671 *Requirements for electrical installations*. Moreover, the generator must be configured to disconnect in the event of the loss of the DNO's supply; and the DNO must be advised of the intention to connect the generator in parallel with the network before, or at the time of, commissioning it.

If the generator has an output of greater than 16 A per phase, the person responsible for the installation must also ensure that the personnel and procedures associated with the installation can prevent danger, so far as is reasonably practicable, and that the connection of the generator satisfies specific standards agreed with the relevant DNO. The national standards for connections are set out in ENA Engineering Recommendations G59, *Recommendations for the Connection of Generating Plant to the Distribution Systems of Licenced Distribution Network Operators*, and G83, *Recommendations for the Connection of Type Tested Small-scale Embedded Generators (Up to 16 A per Phase) in Parallel with Low-Voltage Distribution Systems*. Every DNO and IDNO in Great Britain has procedures for securing compliance with G83 and G59, and there are detailed guides on their application published on the ENA and the companies' websites.

Part VII: Supplies to installations and to other networks

Regulations 23 to 29 contain requirements relating to the supply of electricity to consumers and other networks. The main safety-related provisions are as follows.

Regulation 24 requires distributors or meter operators to ensure that all their equipment in a consumer's premise is suitable and safe, so far as is reasonably practicable, and is protected by a fuse or circuit breaker. The fuse or circuit breaker, which belongs to the distributor and is commonly known as the cut-out, must be in a sealed enclosure.

There has been a considerable amount of discussion over the years about whether or not electricians not employed by a DNO are allowed to break the cut-out's security seal to allow the fuse to be removed or circuit breaker to be switched off, commonly to allow the supply to be isolated so that the consumer unit can be upgraded or replaced safely. Most DNOs do not allow this to be done, preferring their own staff or contractors to remove the seal and replace it once the work is complete, usually at a cost to the owner of the premises. A small number of DNOs do allow it to be done, so long as the electricians are members of accredited trade associations such as the National Inspection Council for Electrical Installation Contractors (NICEIC); the electricians must inform them of their actions and apply approved temporary labels once the work is complete. Those electricians who do remove the seal without the express permission of the DNO put themselves at risk not just of injury but also of civil action by the DNO.

Regulation 24(4) requires distributors to provide an earth connection to consumers, unless the distributor has good reason to believe that it would be unsafe to do so. In some cases, for example in remote rural locations, the distributor may not be able to ensure that the external earth loop impedance will be sufficiently low, in which case the consumer would need to establish a TT supply using their own earth electrodes.

Regulation 25 requires distributors to refuse to provide a connection to a consumer if they believe that the installation does not comply with BS 7671 or if they believe that the connection to their network is not safe.

Regulation 26 empowers distributors to disconnect consumers or other networks or street furniture if they believe that the installation or its use may be dangerous or interfere with their or other networks. The distributor must give the consumer notice in writing of required remedial works, but if these are not carried out, he may then disconnect or refuse to connect the consumer. If that happens, the distributor must again give the consumer written notification of why he has taken the action. The distributor is also empowered to disconnect consumers without prior notification when justified on the grounds of safety, but he must give the consumer a written explanation of why this has been done and the remedial action required before he can be reconnected.

Both Regulations 25 and 26 provide an appeals procedure though the Department for Business, Energy and Industrial Strategy (BEIS) in the event of a dispute.

Regulation 27 requires the supplier to provide its customers with information on the supply voltage, number of phases and frequency. Unless otherwise agreed between the consumer, supplier and distributor, the frequency must be 50 Hz +/- 1% and the voltage, on low voltage supplies, must be 230 V +10%/-6% between phase and earth. On HV supplies the tolerance is +/-6% on supplies up to 132 kV and +/-10% on supplies at and above 132 kV.

Regulation 28 requires distributors to provide consumers with the following information on request:

(a) the maximum prospective short circuit current at the supply terminals;
(b) for low voltage connections, the maximum earth loop impedance of the earth fault path outside the installation;
(c) the type and rating of the distributor's protective device or devices nearest to the supply terminals;
(d) the type of earthing system applicable to the connection; and
(e) the information specified in Regulation 27(1), which apply, or will apply, to that installation.

Part VIII: Miscellaneous

Regulation 30 requires generators and distributors to facilitate inspections by appointed inspectors. These will either be inspectors appointed by BEIS who have an interest in quality and continuity issues, or specialist inspectors appointed by the HSE who have an interest in safety issues.

Regulations 31 and 32 require generators, distributors and meter operators to report a range of specified events to the HSE with respect to safety incidents, and to BEIS with respect to loss of supply incidents. Reporting criteria are covered in Schedules 3 and 4. The list is extensive and results in excess of 12,000 events, including 'near misses', reported to the enforcing authorities each year. In these days of reducing burdens and minimising state interference on safety-related matters, there is scope for the reporting requirements to be reduced and refocussed on priority issues that really matter.

Requests for exemptions are dealt with in Regulation 33, and enforcement procedures are addressed in Regulation 34, which covers the issuing by inspectors of enforcement notices on generators, distributors, meter operators or consumers.

Enforcement

According to the HSE's prosecutions database, available at www.hse.gov.uk/prosecutions/, there have been no prosecutions for breaches of ESQCR since enforcement responsibility for public safety matters transferred to HSE. In fact ESQCR is not a recognised statutory regulation in the HSE's prosecutions or enforcement databases despite the HSE's inspectors having issued a number of Regulation 34 notices. However, there have been prosecutions under the HSW Act Section 3 arising from events that were in reality breaches of ESQCR, such as low overhead lines leading to injury. It is not possible from publicly available information to determine how many such prosecutions have taken place.

Relationship between the ESQC Regulations 2002 and the Electricity at Work Regulations 1989

It might seem strange that there should be two sets of electrical safety legislation with overlapping requirements on matters such as safety of equipment, earthing, excess current and earth leakage protection, and maintenance. The anomaly arises from the historical development of the safety legislation, which includes the electricity supply legislation described at the start of this chapter and the Factories Act and its subsidiary electrical safety regulations, the Electricity (Factories Act) Special Regulations 1908 and 1944.

The duty holders under ESCQR are primarily businesses operating within the electricity supply industry, whereas those under the EAWR are employers, the self-employed and employees in all workplaces. This means that the ESQCR's principal duty holders – generators, distributors, meter operators and suppliers – all have additional duties under the EAWR, whereas the vast majority of EAWR duty holders have very limited duties under ESQCR and principally only in their capacity as consumers of electricity.

The complexity is enhanced by the reporting requirements. All businesses have duties under the Reporting of Injuries, Diseases and Dangerous Occurrences Regulations 2013 to report accidents, ill health and dangerous occurrences. The principal ESQCR duty holders have additional requirements under Regulation 31 to report a range of incidents, considerably broader in scope than RIDDOR, to the same regulator, the HSE.

There is no doubt that the split in responsibilities between the two sets of regulations has the capacity to create confusion and adds an unnecessary level of complexity to the regulatory regime. The obvious solution is for the safety-related regulations in ESQCR to be transferred to, and amalgamated with, the EAWR to create a single coherent and readily understandable set of electrical safety regulations. At the same time, the incident reporting criteria could, with considerable benefit, be rationalised and modernised to reduce the reporting burden on the companies operating in the electricity supply industry.

8

European-derived law with electrical and electrotechnical control system safety requirements

Introduction

Legislation enacted in recent years to transpose into UK law the provisions of European directives has had a major impact on Great Britain and Northern Ireland's health and safety systems. Whilst only the Electrical Equipment (Safety) Regulations are targeted specifically and uniquely at electrical safety, many of the directives have requirements that have an influence on electrical safety, including the safety of electrotechnical control systems and the prevention of electrical ignition of explosive atmospheres.

As observed in Chapter 4, some of the regulations stem from directives aimed at ensuring the free movement of goods throughout the European Union, the legal basis for which is Articles 26 and 28–37 of the Treaty on the Functioning of the European Union, known as the Treaty of Rome; many of these directives have a subsidiary but very important safety content. Other regulations stem from directives targeted at social provisions, including occupational health and safety, perhaps the most important one being the so-called Framework Directive (Directive 89/391/EEC – Council Directive of 12 June 1989 on the introduction of measures to encourage improvements in the safety and health of workers at work).

The first tranche of European-derived legislation covering worker health and safety that had a significant impact in the UK was introduced in 1992 and was known as the 'six pack'; it comprised:

- The Management of Health and Safety at Work Regulations
- The Display Screen Equipment Regulations
- The Manual Handling Operations Regulations
- The Personal Protective Equipment at Work Regulations
- The Provision and Use of Work Equipment Regulations
- The Workplace Health, Safety and Welfare Regulations

In addition to directives that set out objectives that all EU countries must transpose into their national law by a given date, the EC also enacts regulations that are binding legislative acts to be applied in their entirety across the EU without the need for separate national legislation. At present, there are no such regulations relating directly to electrical or control system safety, but any that do exist for other hazards will immediately cease to apply on the day the UK leaves the EU.

The transposition of European law into British and Northern Irish law is a complex legal subject which attracts a considerable amount of debate amongst the legal profession;

this debate is beyond the scope of this book. However, this chapter describes the main legislation of relevance to electrical and control system safety and is split into two sections: the first section deals with single market legislation and the second section deals with occupational health and safety legislation.

This legislation will continue to apply until such time as the UK leaves the EU in response to the momentous referendum decision made on 23 June 2016. The early expectation is that worker health and safety legislation represented by the six pack and other regulations could be retained as domestic legislation, albeit amended to take account of the UK's independence from the EU's governing order and, perhaps, to reduce to some extent the perceived burden of implementing EU-inspired legislation. The product safety legislation relating to the free movement of goods will require significant changes unless the UK negotiates continuing membership of the common market, in which case all the duties relating to compliance with essential health and safety requirements and conformity procedures would continue to apply. At the time of writing this text in late July 2016, there is considerable uncertainty about the future of European-derived legislation, but what can be said with certainty is that the scale of the challenge in deciding what to do, what to repeal, what to replace and what to amend will be enormous.

Legislation governing the free movement of goods

As already noted, this legislation stems from the desire to establish a single market within the EU. In reality, it extends to the full European Economic Area (EEA) so that goods can be moved across national boundaries within the EEA without undue hindrance on the basis that they are safe for use in all of the member countries. The EEA includes the twenty-eight EU member states (twenty-seven once the UK leaves the EU), plus the four member states of the European Free Trade Association: Iceland, Liechtenstein, Switzerland and Norway. EFTA may also include the UK in the future if the exit negotiations take that particular route.

The directives considered in this section are known as 'New Approach' directives and are characterised by containing essential safety-related technical requirements which must be satisfied before the equipment can be placed on the market within the EEA. The usual route to complying with these essential requirements is to comply with a European harmonised standard, which provides a presumption of conformity with the requirements; harmonised standards that may be used for this purpose are listed in the Official Journal of the European Union.

The CE mark, when attached to a product, signifies that the manufacturer of the particular product, or his representative in the EU, affirms that the product complies with the essential requirements of relevant directives; in the main, there will also be paperwork in the form of declarations or certificates of conformity on which the manufacturer details which directives and standards apply, and a technical file that contains prescribed information.

In 2008 the European Commission initiated a New Legislative Framework (NLF) aimed at improving market surveillance and the quality of conformity assessments, and clarifying the use of CE marking. It also created a new set of measures for use in product legislation and defined new terms and responsibilities to be used in this legislation. One of the new terms is 'economic operator', which includes manufacturers, authorised representatives, importers and distributors, all of whom have duties under the new legislation.

One of the outcomes of this was revisions to a number of directives that member states had to adopt by 20 April 2016; three of these amended directives and their associated UK legislation are covered in this chapter: the Low Voltage, EMC and ATEX Directives. At the time of writing (July 2016), the government was in the process of finalising the most effective and proportionate way to implement the directives into UK legislation and expected the transposition process to be completed during 2016, although it is highly questionable whether this change will proceed in the light of the Brexit vote. If it does proceed to cover the small number of years that the UK has left as a member of the EU, the process would not change the essential safety requirements with which the products must comply, and so products conforming to the current requirements should in any event be compliant with the new requirements in terms of their safety. However, the government advised that economic operators should, from 20 April 2016, fulfil all their obligations as set out in the relevant directives, including a declaration of conformity against the relevant new directives. Notified bodies are in the process of being appointed by the Secretary of State under the new directives. Once a notified body has been appointed, they will be able to carry out conformity assessments and issue conformity certificates under the new regime, regardless of when the new UK regulations come into force.

The Electrical Equipment (Safety) Regulations 1994

Scope

These regulations were made on 15 December 1994 by the Secretary of State for Trade and Industry under the Consumer Protection Act 1987 to implement the amended Low Voltage Directive 93/68/EC of 22 July 1993. They extended and replaced the Low Voltage Electrical Equipment (Safety) Regulations 1989 and came into force on 9 January 1995 to apply to electrical equipment placed on the market within the EU, but they were not applicable to equipment placed on the market before 1 January 1997. After this date, all new equipment had to comply.

The regulations are affected by the changes made under the New Legislative Framework and should therefore be updated during 2016 to reflect the requirements of the new Low Voltage Directive 2014/35/EU of 26 February 2014, albeit with the Brexit caveat. This update, if it proceeds, would not affect the essential safety content of the regulations but would change some of the procedural aspects of placing equipment on the market within the EU.

The regulations apply to electrical equipment designed or adapted and supplied to operate between 50 V and 1000 V a.c. or between 75 V and 1500 V d.c.; they do not apply to components that cannot function on their own as equipment. The term 'supply' has a broad interpretation and would include, for example, electrical equipment that forms part of the rental agreement for rented domestic accommodation and equipment that is sold over the internet. Duty holders in terms of the NLF are the economic operators. The exclusions are equipment for use in explosive atmospheres, radiological and medical apparatus, parts for goods and passenger lifts, electricity meters, domestic plugs and sockets, electric fence controllers, and equipment in aircraft, railways and ships.

Compliance

In order to be compliant, electrical equipment must be

(a) safe;
(b) constructed in accordance with good engineering practice in relation to safety matters and in particular designed and constructed to ensure that it is safe by providing protection against electric shock which relies on a combination of insulation and earthing or which achieves that level of protection by other means; and
(c) in conformity with the principal elements of the safety objectives for electrical equipment set out in Schedule 1 to the regulations.

The safety objectives in Schedule 1, which must be complied with if a CE mark is to be applied to the apparatus, amount to the following:

(a) The provision of a manufacturer's mark, typically in the form of a rating plate, or a written notice providing data on essential characteristics that will ensure the equipment is used safely. In practice, the data to be marked will include but not be limited to the manufacturer's brand name or trademark, the supply voltage, power, frequency and construction class (Class I, Class II or Class III).
(b) The equipment is designed and manufactured so as to ensure persons and domestic animals are protected against electric shock and burn hazards and non-electrical dangers. Examples of the latter would include the provision of guards on chain saws, circular saws and grinders.
(c) That dangerous temperatures, arcs or radiation are not produced. On the face of it, this clause would prohibit electric arc welding where dangerous electric arcs are necessarily produced for the process. The wording could, with advantage, be amended to make it clear that the production of such arcs under controlled conditions by instructed persons so as to minimise the hazard is acceptable.
(d) The equipment has to be robust enough to withstand rough handling and be suitable for use in the environmental conditions for which it is designed.
(e) It has to be suitably protected against overload.

Equipment made to a relevant harmonised standard or, failing that, an international standard or, failing that, a national standard, is regarded as being sufficiently safe for compliance. If there are no applicable standards, the equipment is still acceptable if the manufacturer can demonstrate compliance with the safety objectives. The manufacturer may test the equipment himself to prove its safety or have it examined and tested, possibly by an approved testing laboratory. Any test results must be included in the technical documentation to be drawn up by the manufacturer.

The manufacturer must produce an EU declaration of conformity which refers to the relevant harmonised standards. He has to maintain technical documentation containing prescribed information and from 1 January 1997 apply the CE mark to the product, or its packaging, or its instruction sheet or its guarantee certificate.

The regulations apply to equipment as supplied and exclude any defects caused by improper installation, maintenance or use, but the maker must provide any necessary instructions.

Guidance and enforcement

For domestic-type appliances, enforcement is by local authorities and usually by their trading standards officers. For equipment used by persons at work, enforcement is by the HSE.

The Department of Trade and Industry, now the Department for Business, Energy and Industrial Strategy (BEIS), issued a guidance document, URN07/616 dated March 2007, which provided comprehensive guidance on the regulations. It is assumed that a new one will be produced for the latest version of the regulations once they are finalised in 2016.

The penalties for infringement are an unlimited fine, or imprisonment not exceeding 6 months where there is a risk of death or injury to a person, or 3 months otherwise, or both a fine and imprisonment.

The Supply of Machinery (Safety) Regulations 2008

Scope

The Supply of Machinery (Safety) Regulations 2008, as amended by the Supply of Machinery (Safety) (Amendment) Regulations 2011, transpose into UK law the Machinery Directive (2006/42/EC) and came into effect on 29 December 2009. These regulations replaced the Supply of Machinery (Safety) Regulations 1992, amended in 1994 and 2005, which had transposed earlier versions of the Machinery Directive.

The Supply of Machinery (Safety) Regulations 2008 are made under the European Communities Act 1972 and promulgated by the Secretary of State for Business, Innovation and Skills. Whereas this means they are not regulations made under the HSW Act, they are enforced by the HSE for machinery used at work.

The Machinery Directive is a New Approach Directive and, as such, requires the following approach to conformity for machinery being placed on the market within the EEA:

(a) the machinery must meet all of the relevant essential health and safety requirements (EHSRs) for the product, to the state of the art, and undergo conformity assessment as required by the directive;
(b) be CE marked;
(c) have a declaration of conformity;
(d) have user instructions in the language of the end user; and
(e) the manufacturer or his representative in the EU must compile a technical file to demonstrate compliance with the Directive's requirements. This must be kept securely for at least 10 years from the last dates of manufacture and be made available on request to market surveillance authorities, such as the HSE.

The regulations are concerned with the safety of machinery and apply to all machinery placed on the market in the European Economic Area. It includes machinery imported from non-EU countries. They cover all the hazards, including electrical hazards, associated with the installation, use, cleaning, maintenance and dismantling of machinery, where the term 'machinery' includes in addition to the standard meaning of the term,

(a) interchangeable equipment;
(b) safety components;
(c) lifting accessories;
(d) chains, ropes and webbing;
(e) removable mechanical transmission devices; and
(f) partly completed machinery.

There is a long list of exclusions, set out in Schedule 3 of the regulations. These include fairground machinery; machinery intended to move performers during artistic performances; means of transport by air, on water and on rail networks; and other machinery covered by other regulations such as electrical equipment subject to the Electrical Equipment (Safety) Regulations 1994 where the risks are predominantly electrical. The regulations are not, therefore, predominantly concerned with electrical safety, but where machines are electrically powered or have electrotechnical control systems, the integrity and safety of these systems need consideration to secure compliance.

Compliance

The principle duty holder under the regulations is the 'responsible person' who is, in relation to machinery or partly completed machinery, the manufacturer of that machinery or partly completed machinery, or the manufacturer's authorised representative in the EU.

The responsible person must comply with requirements relating to the application of a CE mark signalling compliance of the machinery with the relevant EHSRs; the generation and retention of a technical construction file containing prescribed information; and the drawing up of declarations of conformity or declarations of incorporation (depending upon whether the product is a complete machine that can operate on its own or is destined for incorporation into other machinery). The responsible person must follow the appropriate conformity assessment procedures which may include, for certain types of more dangerous machinery listed in Annex IV to the regulations (so-called Annex IV machines), submitting the product or its technical file to an approved body; examples of Annex IV machines are

- woodworking machines such as circular saws, hand-held planers, and thicknessers;
- presses, including press-brakes, for the cold working of metals, with manual loading and/or unloading, whose movable working parts may have a travel exceeding 6 mm and a speed exceeding 30 mm/s;
- injection or compression plastics-moulding machinery with manual loading or unloading;
- certain machinery for underground working;
- vehicle servicing lifts; and
- power-operated interlocking movable guards designed to be used as safeguards in machinery.

Annex I of the regulations contains an extensive list of EHSRs which must be complied with by manufacturers or suppliers; they have a similar status to the safety objectives of the Electrical Equipment (Safety) Regulations. Amongst them are quite generic safety requirements covering electrical systems, including

(i) Protection against hazards from the electricity supply and static electricity (EHSRs 1.5.1 and 1.5.2)
(ii) The provision of means for isolation from energy sources (EHSR 1.6.3)
(iii) Safety and reliability of control systems (EHSR 1.2.1)
(iv) Stopping devices (EHSR 1.2.4)
(v) Failure of the power supply (EHSR 1.2.6)

The most commonly used route to compliance with these EHSRs is designing and building machines, devices or components in conformity with the provisions of a harmonised standard describing the particular machine, device or component. The standards are produced by Technical Committees and their working groups of experts within CEN, the European standards-making body. The working groups are formed from experts from nations within the EU and will generally represent manufacturers and users of machines, and the regulatory bodies such as the HSE. The standards may be used once their details have been published in the Official Journal of the European Union. These standards play such an important and fundamental role in machinery safety that it is difficult to see how they will not continue to be used by British manufacturers after the UK's exit from the EU.

There are 3 generic types of harmonised standards covering the safety of machines: Types A, B and C.

(i) Type A standards are basic safety standards that have broad application to all machinery and safety components. Perhaps the most important of these is BS EN ISO 12100:2010 *Safety of machinery. General principles for design. Risk assessment and risk reduction*.
(ii) Type B standards are subdivided into 2 sub-types: Type B1 and Type B2. Type B1 standards are application standards, an example being EN 60204–1 *Safety of machinery. Electrical equipment of machines. General requirements.* This standard provides guidance on the means of providing protection against electrical injury on machinery as well as on the provision of safe control systems; it can therefore be considered as a basic standard covering electrical safety principles. Type B2 standards cover particular safety components and devices; an example is EN 61496–1 *Safety of machinery. Electro-sensitive protective equipment. General requirements and tests.*
(iii) Type C standards relate to specific types or groups of machines, examples being palletisers, guillotines, form-fill-seal machines and press-brakes. Manufacturers who build machinery in conformity with a Type C standard have a 'presumption of conformity' with the EHSRs.

Manufacturers will clearly be keen to have access to Type C standards for their particular type of machinery. Unfortunately, progress with the production of Type C standards, in particular, has been quite slow, and this has made it difficult for some manufacturers who are aiming to comply with the EHSRs and the regulations. Where an appropriate Type C standard does not exist, manufacturers may use conformity with Type A and Type B standards as a route to compliance.

Guidance

The Department for Business Innovation and Skills (now BEIS) produced comprehensive guidance on the regulations titled *Machinery, Guidance notes on the UK Regulations*, dated September 2009, which can be obtained free of charge from the gov.uk website.

Enforcement

Enforcement is by the HSE for machinery and safety components for use at work and by trading standards officers for items for domestic use.

The Electromagnetic Compatibility (EMC) Regulations 2006

Scope

These regulations were made by the Secretary of State of the Department of Trade and Industry to implement the Electromagnetic Compatibility Directive (2004/108/EC), coming into force on 20 July 2007 and applying throughout the UK. The regulations repealed, replaced, and to some extent simplified, previous versions.

The regulations are affected by the changes made under the New Legislative Framework and should therefore be updated during 2016 to reflect the requirements of the new EMC Directive 2014/30/EU of 26 February 2014, albeit with the Brexit caveat. This update, should it proceed, would not affect the essential safety content of the regulations but would change some of the procedural aspects of placing equipment on the market within the EU.

The regulations apply to electrical and electronic equipment in the form of either apparatus or a fixed installation and have the purpose of ensuring that any electromagnetic disturbance generated by the equipment does not exceed a level above which radio and telecommunications equipment and other equipment cannot operate as intended, and that the equipment itself has an adequate level of immunity to electromagnetic disturbance.

The regulations refer to "electromagnetic disturbance", meaning any electromagnetic phenomenon which is liable to degrade the performance of relevant apparatus. Quoted examples of phenomenon are

(a) electromagnetic noise;
(b) unwanted signals; and
(c) changes in the propagation medium.

In reality, these will include conducted phenomena, such as harmonics, signalling voltages, oscillatory transients and voltage fluctuations, and radiated phenomena in the electromagnetic spectrum.

A signal or emission which is a necessary function, or consequence of the operation, of relevant apparatus is not taken to be electromagnetic disturbance if, in relation to that apparatus, the signal or emission is permitted, and does not exceed specified limits.

Exclusions

The following equipment is excluded from the scope of the regulations:

(i) electromagnetically benign apparatus;
(ii) equipment presented at trade fairs or similar events, if a sign displayed visibly on or near the equipment clearly indicates that it is not compliant with these regulations and cannot be placed on the market or put into service, or both, until it is made compliant with those requirements; and
(iii) equipment covered by other directives and legislation, in whole or in part, or otherwise excluded from the EMC Regulations 2006.

Requirements

The general requirements are that equipment must be designed and manufactured, having regard to the state of the art, so as to ensure that

(a) the electromagnetic disturbance it generates does not exceed a level above which radio and telecommunications equipment or other equipment cannot operate as intended; and
(b) it has a level of immunity to the electromagnetic disturbance to be expected in its intended use which allows it to operate without unacceptable degradation of its intended use.

The requirement specific to fixed installations is that, with a view to meeting the essential requirements set out in the regulations, they must be installed

(a) applying good engineering practices, which must be documented with the documents held by the responsible person for inspection by the enforcement authority for as long as the fixed installation is in operation; and
(b) respecting the information on the intended use of its components.

Compliance

Compliance with the regulations relating to apparatus can be secured through one of two routes:

(i) The manufacturer may self-certify that the product complies with an appropriate European harmonised standard or with the essential requirements. There is a significant number of standards that can be used, and these are referenced in the government's guidance document referred to below in the section titled 'Guidance'. The manufacturer must draw up documentation that provides evidence of conformity and produce a declaration of conformity.
(ii) The manufacturer may employ a notified body to assess compliance and produce a statement of compliance, thereby allowing the manufacturer to produce a declaration of conformity.

Compliance of fixed installations which are intended to be used permanently can be secured by meeting the essential requirements in respect of emissions and immunity, but they need not be subjected to conformity assessment, or have a declaration of conformity, and they are not required to be CE marked. In general terms, a fixed low voltage installation that complies with BS 7671 *Requirements for electrical installations* is likely to meet the requirement for installation to good engineering practice.

There must always be a person, known as the responsible person, with responsibility for ensuring that, when used, a fixed installation complies with the essential requirements. The responsible person does not have to be an EMC expert and may seek appropriate advice in fulfilling their obligations. It has to be said that very few electrical installations have a nominated responsible person and there is widespread ignorance of this requirement, possibly because the regulatory authorities adopt a very light touch on its enforcement.

Guidance

The Department for Business Enterprise and Regulatory Reform, which has been replaced by BEIS, issued a guidance document titled *Guide to the Electromagnetic Compatibility Regulations 2006* dated August 2008 (Reference URN 08/1192), which can be obtained free of charge from the gov.uk website.

Enforcement

Enforcement responsibility rests primarily with trading standards departments of local authorities, or the Department of Enterprise, Trade and Investment in Northern Ireland, for most industrial and consumer products and fixed installations. Trading standards officers will generally only react to complaints or incidents, rather than taking proactive action in respect of the regulations.

OFCOM may enforce the regulations in relation to the protection and management of the radio spectrum.

The Equipment and Protective Systems Intended for Use in Potentially Explosive Atmospheres Regulations 1996

Scope

The Equipment and Protective Systems Intended for Use in Potentially Explosive Atmospheres Regulations 1996 (SI 1996/192) were made by the Secretary of State for Trade and Industry, coming into force on 1 March 1996. They were amended by the Equipment and Protective Systems Intended for Use in Potentially Explosive Atmospheres (Amendment) Regulations 2001 (SI 2001/3766). In Northern Ireland, the regulations are the Equipment and Protective Systems Intended for Use in Potentially Explosive Atmospheres Regulations (Northern Ireland) 1996 (SR 1996/247). They implement the provisions of Directive 94/9/EC, the so-called ATEX Directive. From here on in, they will be referred to as the EPS Regulations.

The EPS Regulations are affected by the changes made under the New Legislative Framework and should therefore be updated during 2016 to reflect the requirements of the new ATEX Directive 2014/34/EU of 26 February 2014, albeit with the Brexit caveat. This update, should it proceed, would not affect the essential safety content of the regulations but would change some of the procedural aspects of placing equipment on the market within the EU.

The EPS Regulations apply to manufacturers and suppliers of equipment intended for use in potentially explosive atmospheres, as well as those who put such equipment into service within the EEA for the first time. As in the case of the Supply of Machinery (Safety) Regulations, the main duties are placed on the shoulders of a 'responsible person', who is usually the manufacturer of the product.

A very important factor is that the term 'equipment' extends beyond electrical equipment and includes machines, apparatus, fixed or mobile devices, control components, instrumentation, and protective systems which, separately or jointly, are intended for the generation, transfer, storage, measurement, control and conversion of energy or the processing of material and which are capable of causing an explosion through their own potential sources of ignition. The regulations cover such equipment used below ground, on the surface and on fixed offshore installations, but not floating production platforms (FPPs) and floating production storage and offloading vessels (FPSOs). This means, for example, that an internal combustion engine falls within the remit of the regulations as much as an electrical switch or luminaire.

The EPS Regulations extend to all types of potentially explosive atmospheres, including those consisting of combustible dust. Indeed, an explosive atmosphere is defined as mixtures with air, under atmospheric conditions, of flammable substances in the form of gases, vapours, mists or dusts in which, after ignition has occurred, combustion spreads to the entire unburned mixture, i.e. there is an explosion. This definition has the effect of excluding explosions as a result of increases in pressure, or as a result of chemical interactions.

The main intent of the legislation is to remove the need for documentation and testing for each individual European market. Since the ATEX Directive is a New Approach Directive, manufacturers only have to CE mark their products once to show compliance with the directive, with the product then being able to be traded freely throughout the EEA.

Exclusions

The following products are excluded from the regulations:

(a) medical devices intended for use in medical environments;
(b) equipment and protective systems where the explosion hazard results solely from the presence of explosive or chemically unstable substances;
(c) equipment intended for use in domestic and non-commercial environments where potentially explosive atmospheres may only rarely be created, solely as a result of the accidental leakage of fuel gas;
(d) Personal Protective Equipment covered by Directive 89/686/EEC as amended by Directives 93/95/EEC, 93/68/EEC and 96/58/EC;
(e) sea-going vessels and mobile offshore units together with equipment and protective systems on board such vessels or units;
(f) means of transport (i.e. vehicles and their trailers intended for transporting passengers and/or goods by air or on road, rail or water networks), while vehicles intended for use in a potentially explosive atmosphere are within scope of the regulations; and
(g) equipment specifically designed for military purposes.

Duties

The main duty of the responsible person (economic operator once the 2016 regulations are enacted) is to ensure that equipment destined for use in potentially explosive atmospheres satisfies the relevant Essential Health and Safety Requirements (EHSRs), which are listed in Schedule 3 of the regulations. The EHSRs have three sections: common requirements, supplementary requirements for equipment and supplementary requirements for protective systems. Protective systems are defined as design units which are intended to halve incipient explosions immediately and/or to limit the effective range of explosion flames and explosion pressures. Protective systems may be integrated into equipment or separately placed on the market for use as autonomous systems.

There are additional duties relating to the information and instructions to be provided, as well as the markings that must be applied to the equipment. All equipment and protective systems must be CE marked once their conformity to the EHSRs has been assessed.

Categorisation

The regulations define two main groups of equipment: Group I equipment is for use in underground mines and other high hazard areas where very high levels of protection would be justified, and Group II equipment is for use in areas where explosive atmospheres from gases, vapours, mists and/or combustible dusts could exist for varying lengths of times.

The groups are then subdivided into a range of different categories. Group I equipment is subdivided into categories M1 and M2. For Group II equipment, there are 3 categories (1, 2 and 3) that are broadly equivalent to the well-known hazardous area Zones 0, 1 and 2 (Zones 20, 21 and 22 in the case of dust atmospheres) – see Chapter 15 for information on zoning.

Conformity and marking

Compliance with the EHSRs can be achieved through conformity with harmonised standards, of which there are very many covering the construction requirements for equipment; for electrical equipment, most harmonised standards are in the EN 60079 series, such as BS EN 60079–1:2014 *Explosive atmospheres. Equipment protection by flameproof enclosures "d"*. In the absence of a harmonised standard, appropriate and relevant national standards or specifications may be used or, alternatively, the manufacture may demonstrate compliance with the EHSRs.

The regulations provide a range of options for assessing conformity depending on the equipment's group and sub-group. However, as far as electrical equipment is concerned, conformity assessment through a notified body will be the norm, following EC type examination processes and procedures. There are additional requirements for product quality control and third-party audits. Conformity assessment will ultimately allow the generation of a declaration of conformity and the application of the CE mark to the equipment.

In addition to the CE mark, the equipment must be marked to enable full identification of its protection category. The marking must at least contain the following:

(a) The specific explosion protection mark, ⟨Ex⟩, together with the mark indicating the equipment group and category, and, relating to equipment group II, letter 'G' (concerning explosive atmospheres caused by gases, vapours or mists) and/or 'D' (concerning explosive atmospheres caused by dust), together where necessary with all information essential for safe use, e.g. gas group, temperature class, and Ex protection concept. An example is shown below:

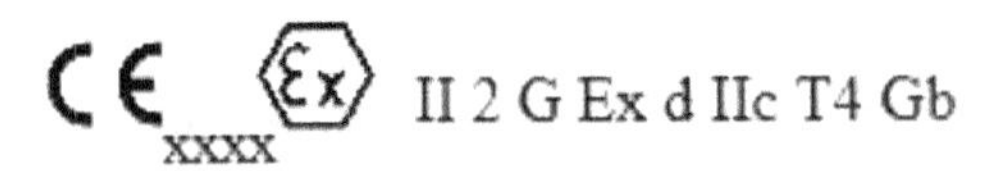

Where:
CE is the CE mark
xxxx is the Notified Body's registration number
Ex is the European explosion protection mark
II is the equipment group
2 is the equipment category (2 = equipment for use in Zone 1 or Zone 2 areas)
G means Gas ('D' for dust)
Ex means explosion protection (EEx if compliant with a harmonised standard)
d is the protection type code ('d' = flameproof)
IIC is the gas group (IIC = acetylene or hydrogen as representative gases)
T4 is the temperature code (T4 = 135°C)
Gb is the equipment protection level

(b) The name and address of the manufacturer
(c) The designation of series or type and serial number
(d) The year of production
(e) Restricted or other safety-related conditions of use

In the UK, there are eight notified bodies providing conformity assessment services, basing their assessments on the harmonised standards that exist for explosion-protected electrical equipment; the main standards are identified in Chapter 15.

Guidance

The Department of Trade and Industry (now BEIS) published a guidance document, titled *Equipment and protective systems intended for use in potentially explosive atmospheres; Guidance Notes on the UK Regulations* (Second edition, February 2002, reference URN 02/609).

Enforcement

The HSE is the sole enforcing authority for regulations in the UK.

Legislation from occupational safety and health directives

As already noted, this legislation stems from the social policy elements of the treaty, with particular emphasis on worker health and safety. The following text covers the main legislation falling within this category.

The Management of Health and Safety at Work Regulations 1999

These regulations transpose into British law the Framework Directive (89/391/EC), although they also enact directives relating to pregnant workers, young people and temporary workers at work. They used to have an associated Approved Code of Practice, but this was withdrawn in 2013 as part of the government's agenda to reduce regulatory burdens on industry. In Northern Ireland, they are promulgated as the Management of Health and Safety at Work Regulations (Northern Ireland) 2000.

The regulations have wide-ranging provisions relating to the management of health and safety (including specific requirements relating to young persons and new and expectant mothers). The most important provisions in the context of this book are explained as follows.

Regulation 3 – Risk assessment

Regulation 3 sets out the legal requirement for employers and the self-employed to conduct risk assessments, a crucially important tool in the management of health and safety and one that many organisations fail to get right.

The duty on employers is to carry out a suitable and sufficient assessment of

(a) the risks to the health and safety of his employees to which they are exposed whilst they are at work; and
(b) the risks to the health and safety of persons not in his employment arising out of or in connection with the conduct by him of his undertaking.

This is for the purpose of identifying the measures an employer needs to take to comply with the requirements and prohibitions imposed upon him by or under the relevant statutory provisions.

In the context of this book, this would include work activities such as live working and the isolation of high voltage systems.

In similar vein, unless excluded from the duty by virtue of The Health and Safety at Work etc Act 1974 (General Duties of Self-Employed Persons) (Prescribed Undertakings) Regulations 2015, a self-employed person must carry out a suitable and sufficient assessment of

(a) the risks to his own health and safety to which he is exposed whilst he is at work; and
(b) the risks to the health and safety of persons not in his employment arising out of or in connection with the conduct by him of his undertaking

for the purpose of identifying the measures he needs to take to comply with the requirements and prohibitions imposed upon him by or under the relevant statutory provisions.

Where an employer has five or more employees, he must record the significant findings of the assessment and any group of his employees identified by it as being especially at risk.

So, whereas none of the specific electrical safety regulations make any mention of risk assessment, there is a duty under these regulations to conduct such assessments and record them if five or more people are employed. As previously observed, it is difficult to see how duty holders can make judgements about the reasonable practicability of measures without carrying out risk assessments.

The essential processes in conducting a suitable and sufficient risk assessment are as follows:

- identifying the hazards in the workplace, such as electricity;
- identifying who might be harmed and how, such as an electrician carrying out live fault finding work on production machinery and potentially experiencing an electric shock or burn injuries;

- evaluating the probability of the harm occurring and the severity of any consequential injury, i.e. the risks, and deciding on the appropriate controls that need to be put in place to eliminate or reduce the risk, over and above any controls already in place;
- recording the risk assessment (principally what the risks are, what is already being done to control them, what further action is needed and when the action has to be completed and by whom) and details of any particular groups of employees who have been identified as being especially at risk; and
- reviewing and updating the risk assessment to take account of any changes that may occur in the workplace.

The level of detail and complexity in a risk assessment should be matched to the level of risk – the higher the risk, the greater should be the level of detail. So, for example, an engineer working inside a high voltage compound in which there are live uninsulated busbars would expect to have a more detailed risk assessment than an electrician working on the installation of a domestic low voltage electrical distribution system. However, it is important to try to keep risk assessments as brief but as informative as possible – an assessment the thickness of *War and Peace* will not be read, and whereas it may be 'sufficient', it would probably not be regarded as being 'suitable'.

There is no prescribed format for risk assessments, and there is a wide variety of techniques and formats in common use. Many organisations use risk graph or risk table techniques that link the probability of an event to its severity in graphical form. At the higher risk end of the spectrum, companies in high hazard industries will carry out quantified risk assessments using numerical techniques to demonstrate that risks have been reduced to a tolerable level; such numerical techniques are not needed for the type of work normally conducted in the electrotechnical and similar industries, where qualitative assessments are perfectly acceptable.

Of particular importance is that the risk assessment should be treated as a 'live' management document that is not simply filed away and forgotten. As required by Regulation 10, the workers who are at risk need to be instructed on the hazards, risks and the associated control measures, although they do not necessarily need to see the risk assessment document itself. They could be instructed on the assessments' contents at events such as project briefings and toolbox talks.

It is quite common for the control measures to be set out in method statements and other written procedures or safe systems of work. Examples of this include written safe isolation procedures for maintenance activity and written live working procedures during, for example, commissioning activity on a motor control centre. These documents together with their associated risk assessments make up the safety documentation package for a particular work activity.

A good source of advice on carrying out risk assessments is the HSE's publication *5 Steps to Risk Assessment*. In addition to this, the HSE has generated useful risk assessment tools and example assessments, available on its website: www.hse.gov.uk.

Regulation 4 – Principles of prevention to be applied

This regulation sets out the fundamental principles of health and safety management in Europe, including for the timebeing the UK. It requires employers to implement the following hierarchy of preventive or protective measures and, as such, expands the general duties in the Health and Safety at Work Act Section 2:

(a) avoiding risks;
(b) evaluating the risks which cannot be avoided;
(c) combating the risks at source;
(d) adapting the work to the individual, especially as regards the design of workplaces, the choice of work equipment and the choice of working and production methods, with a view, in particular, to alleviating monotonous work and work at a predetermined work-rate and to reducing their effect on health;
(e) adapting to technical progress;
(f) replacing the dangerous by the non-dangerous or the less dangerous;
(g) developing a coherent overall prevention policy which covers technology, organisation of work, working conditions, social relationships and the influence of factors relating to the working environment;
(h) giving collective protective measures priority over individual protective measures; and
(i) giving appropriate instructions to employees.

Perhaps the most important thing to pick out of this list is that the law requires employers to eliminate risks or, failing that, to take other precautions such as guarding machinery before instructing employees on how to avoid harm.

Regulation 5 – Health and safety arrangements

This regulation requires employers to put in place effective health and safety management arrangements covering the planning, organisation, control, monitoring and review of the preventive and protective measures. Where the employer employs five or more employees, the management arrangements must be written down.

The effect of this is that employers of five or more people must have a written health and safety policy document. In the electrotechnical industries, this would include the need to have a set of electrical safety rules or similar procedures where justified by the nature of the work being carried out. For example, a company with a high voltage distribution network on its premises would be expected to have a set of safety rules or procedures covering matters such as switching operations, safe isolation procedures, use of safety documents such as Permits to Work and Sanctions for Test and the authorisation processes for people working on the system.

Regulation 11 – Cooperation and coordination

This regulation requires employers who share a workplace (whether on a temporary or a permanent basis) to cooperate with each other regarding health and safety arrangements and to take all reasonable steps to coordinate the measures taken to comply with the legislation. They must also take all reasonable steps to inform the other employers concerned of the risks to their employees' health and safety arising out of or in connection with the conduct their undertakings.

This has quite wide application in the electrotechnical sectors because it is often the case that electrical workers find themselves working in the same locations as other companies' workers, good examples being construction sites and joint user substations.

Enforcement

Enforcement of these regulations is by HSE regulatory inspectors and local authority environmental health officers.

Provision and Use of Work Equipment Regulations (PUWER) 1998

These regulations, which came into force on 5 December 1998, replaced the original Provision and Use of Work Equipment Regulations (PUWER) 1992. They transpose into GB law the Work Equipment Directive (89/655/EC) as amended by the Amending Directive to the Use of Work Equipment Directive, which extended the original directive to include mobile work equipment, lifting equipment (although lifting equipment is not actually covered in PUWER 98) and the inspection of work equipment. In Northern Ireland, they are enacted as the Provision and Use of Work Equipment Regulations (Northern Ireland) 1999. They have the main objective of ensuring that work equipment does not cause risks to health and safety.

The term 'work equipment' has very broad meaning and includes equipment, machinery, appliances, apparatus, tools or installations for use at work (whether exclusively or not); it therefore includes all electrical and electronic apparatus and equipment used at work. The regulations are supported by an Approved Code of Practice and guidance material published in the HSE's document L22.

As far as electrical and electronic systems are concerned, there are a number of relevant regulations that are of interest.

Regulation 6 – Inspection

Regulation 6, on inspection, requires that work equipment is inspected when it is first put into use and then at routine intervals thereafter if the equipment is likely to deteriorate in its conditions of use. The ACOP specifies that the inspections should be done by a competent person, without giving very much guidance on the meaning of the term 'competent person'; however, see Chapter 19 for information on competence as far as electrical work is concerned. The regulation also requires the inspection to be recorded; although the recording method is not specified, or what the contents of the records should be, there is information on this in the guidance material.

There is frequently a question raised when referring to inspection of electrical apparatus and equipment about the extent to which tests should be carried out. The stated intention of the regulation is to ensure that the equipment is safe to operate, in which case it would be reasonable to argue that properties such as insulation resistance and earth continuity should be tested as part of the 'inspection' activity. However, it would also be reasonable for the competent person carrying out the inspection to determine what maintenance has been done on the equipment to comply with the Electricity at Work Regulations, Regulation 4(2). If he is able to confirm that appropriate inspections and tests have been carried out to a satisfactory standard, he may be able to dispense with this type of test on the basis that they have already been carried out. In that case, he would restrict himself to a visual examination of the insulation, terminal connections and so on. He would, however, be well advised to carry out such tests himself if there is any doubt about the integrity of the electrical insulation and earthing.

Functional tests of safety-related control systems, such as emergency stops and guard interlocks, should always be carried out as part of an inspection carried out under this regulation.

Regulation 11 – Dangerous parts of machinery

This regulation is of interest because it sets out the hierarchy of safeguarding techniques on machinery that underpins all the safety requirements set out in harmonised standards for machinery safeguarding, as well as the Essential Health and Safety Requirements of the Supply of Machinery (Safety) Regulations. The hierarchy is as follows:

(a) the provision of fixed guards enclosing every dangerous part or rotating stock-bar where and to the extent that it is practicable to do so, but where or to the extent that it is not, then
(b) the provision of other guards or protection devices where and to the extent that it is practicable to do so, but where or to the extent that it is not, then
(c) the provision of jigs, holders, push-sticks or similar protection appliances used in conjunction with the machinery where and to the extent that it is practicable to do so, but where or to the extent that it is not, then
(d) the provision of information, instruction, training and supervision.

The important point to note is that relying on people to follow safe working procedures in order to prevent injury is at the bottom of the hierarchy. The law requires that employers take engineering measures to eliminate risk before relying on human beings to follow instructions. This is entirely consistent with the principles in Regulation 4 of the Management of Health and Safety at Work Regulations, as previously described.

Regulations 14 to 18 – Control system provisions

These regulations contain particular requirements relating to control systems, including those that carry out safety functions. This topic is covered in more detail in Chapter 17, but the following summarises the requirements.

Regulation 14 covers the requirement for equipment to have some form of start control to ensure that it can only be started as a result of a deliberate action. It would not be safe, for example, for a hand-held electric drill to start as soon as the power to the drill is switched on; there must be a start button to ensure that it will only start when the user wants it to start.

A similar argument holds true when the machine can change its operating characteristics, such as speed, when that might lead to a dangerous condition for the operator. An example of this would be a robotic palletising machine that moves at crawl speed when it is in 'teaching' mode; it would not be safe if the machine were to suddenly move at full operational speed unless and until the operator has given a specific command to the machine's control system to enter 'operational' mode.

It is accepted, however, in Regulation 14(3), that many automated machines will start, stop and change mode as part of their operating cycles; this is acceptable so long as the machine is adequately safeguarded to prevent injury to the operators.

Regulation 15 stipulates that equipment must be provided with some means to bring it to a safe condition in a safe manner. This will normally be some form of stop control that, when activated, brings those parts that have dangerous motion or other dangerous feature such as high temperature or pressure to a safe state. This does not need to be an instantaneous stop; there are many examples of machines in which the control system will need to go through a sequence of control actions to bring the equipment to a safe condition in a controlled manner. The technical requirements for stop devices and functions can be found in BS EN 60204–1:2006+A1:2009 *Safety of machinery. Electrical equipment of machines. General requirements*, in which the stop function is categorised in one of three categories: 0, 1 and 2. A category 0 stop is one in which power is removed immediately, usually by electromechanical means such as a contactor; a category 1 stop is one in which the machine is brought to a safe stop under control means and then power is removed, usually by electromechanical means; a category 2 stop is one in which the machine is brought to a controlled stop and power is retained on the system once the safe condition has been achieved.

Regulation 16 requires that work equipment must be provided with one or more emergency stop controls, but only where they can and will contribute to risk reduction. The decision on whether or not to provide emergency stop devices and, if so, how many and where they should be located should form part of the risk assessment carried out on the machine. The requirements for emergency stop actuators are published in BS EN ISO 13850:2006 *Safety of machinery. Emergency stop. Principles for design*. Note that category 2 stops must not be used for the emergency stop function, although either category 0 or 1 stops may be used.

Regulation 17 covers the suitability of control devices on work equipment. This addresses the positioning, visibility, identification and other characteristics of control buttons, switches and other actuators.

Regulation 18 covers the safety integrity of control systems that carry out safety functions. Such control systems need to have sufficient integrity, in terms of fault tolerance and reliability, when compared to the amount of risk reduction they are aiming to achieve. This subject is treated in detail in Chapter 17.

Regulation 19 covers the means for isolating sources of energy and, as far as electrical energy is concerned, has the same effect as the Electricity at Work Regulations, Regulations 12 and 13.

Enforcement

These regulations are enforced by HSE regulatory inspectors and local authority environmental health officers.

The Personal Protective Equipment at Work (PPE) Regulations 1992

These regulations transposed into British law the Use of Personal Protective Equipment Directive, 89/656/EEC. They were made on 25 November 1992 and came into force on 1 January 1993. In Northern Ireland, the directive was transposed as the Personal Protective Equipment at Work Regulations (Northern Ireland) 1993.

They should not be confused with the Personal Protective Equipment Regulations 2002, which address the responsibilities of manufacturers and suppliers of PPE. These regulations are undergoing a 2-year transition in effect from February 2016 to implement requirements set out in new EU regulations called the PPE Regulation (EU) 2016/425.

The onus is on the employer or self-employed to assess the risks in the workplace and to provide and maintain appropriate personal protective equipment which is worn by the worker. In the electrical risk context this would comprise, for example, insulating boots, gloves and possibly arc flash PPE for live working, and semi-conducting footwear and clothing where it is necessary to avoid static discharges, such as in operating theatres or during the electrostatic spraying of flammable paints and powders. For arc welding, gloves, aprons and footwear to resist hot metal spatter are required and a helmet with a visor and suitable lenses is required to protect the face, reduce the glare and cut out the UV radiation. Where noxious fumes are produced, and/or in confined conductive spaces, the helmet may need connecting to a fresh air supply and the operator should be provided with insulating clothing and footwear to counter the electric shock hazard.

The regulations require that PPE is used only as a last resort, after other methods have been taken to control risks. Employers must ensure that suitable PPE is provided and that

- it offers adequate protection for its intended use;
- those using it are informed of the hazards it is designed to avoid and are adequately trained in its safe use;
- it is properly maintained and any defects are reported; and
- it is returned to its proper storage after use.

The HSE has published guidance in HSG53 *Respiratory protective equipment at work. A practical guide* and INDG174 *Personal protective equipment (PPE) at work. A brief guide.*

The Construction (Design and Management) Regulations 2015

Made on 22 January 2015 by the Secretary of State of the Department of Work and Pensions to transpose in Great Britain the requirements of Directive 92/57/EEC on the implementation of minimum safety and health requirements at temporary or mobile construction sites, the Construction (Design and Management) Regulations 2015, commonly known as the 'CDM Regulations', came into force on 6 April 2015. They replaced the CDM Regulations of 2007, which in turn replaced the CDM Regulations of 1994.

In Northern Ireland, as of early 2016 the 2015 regulations that apply in GB have not been adopted, so the Construction (Design and Management) Regulations (Northern Ireland) 2007 still apply.

The regulations extend the general requirements of the Health and Safety at Work Act to specific duties relating to construction work, with the intention of countering the somewhat cavalier attitude to health and safety that has been endemic in the construction industry, resulting in it having an appalling accident record.

The regulations place duties on clients, mainly commercial clients because domestic clients' duties will normally be taken by other duty holders such as a contractor or principal contractor; principal contractors; contractors; principal designers; and designers relating to health and safety in construction activity. These new regulations place much more emphasis on the role of the clients than previous versions. This is because clients have a crucial influence over how projects are run, including the management of health and safety risks. Whatever the project size, the commercial client has contractual control, appoints designers and contractors, and determines the money, time and other resources for the project.

In general terms, the regulations apply to 'construction work', which is defined as the carrying out of any building, civil engineering or engineering construction work and includes

(a) the construction, alteration, conversion, fitting out, commissioning, renovation, repair, upkeep, redecoration or other maintenance (including cleaning which involves the use of water or an abrasive at high pressure, or the use of corrosive or toxic substances), decommissioning, demolition or dismantling of a structure;
(b) the preparation for an intended structure, including site clearance, exploration, investigation (but not site survey) and excavation (but not pre-construction archaeological investigations), and the clearance or preparation of the site or structure for use or occupation at its conclusion;
(c) the assembly on site of prefabricated elements to form a structure or the disassembly on site of the prefabricated elements which, immediately before such disassembly, formed a structure;
(d) the removal of a structure, or of any product or waste resulting from demolition or dismantling of a structure, or from disassembly of prefabricated elements which immediately before such disassembly formed such a structure; and
(e) the installation, commissioning, maintenance, repair or removal of mechanical, electrical, gas, compressed air, hydraulic, telecommunications, computer or similar services which are normally fixed within or to a structure.

The definition does not include the exploration for, or extraction of, mineral resources, or preparatory activities carried out at a place where such exploration or extraction is carried out (including demolition and dismantling).

From an electrical safety perspective, it should be noted that construction work includes the installation, commissioning, maintenance, repair and removal of fixed electrical installations in properties.

A construction project must be notified to the HSE if it will last longer than 30 working days and have more than 20 workers working simultaneously at any point in the project or if it will exceed 500 person days. This means, for example, that most electrical maintenance work on fixed installations will not be notifiable.

The commercial client's duties include making suitable arrangements for managing a project, including making sure that other duty holders are appointed as appropriate, that sufficient time and resources are allocated, that relevant information is prepared and provided to other duty holders, that the principal designer and principal contractor carry out their duties and that welfare facilities are provided.

Principal designers must plan, manage, monitor and coordinate health and safety in the pre-construction phase of a project. This includes identifying, eliminating or controlling foreseeable risks, ensuring designers carry out their duties, and preparing and providing relevant information to other duty holders. They must also liaise with the principal contractor to help in the planning, management, monitoring and coordination of the construction phase.

Designers must, when preparing or modifying designs, eliminate, reduce or control foreseeable risks that may arise during construction and the maintenance and use of a building once it is built. They must also provide information to other members of the project team to help them fulfil their duties.

Principal contractors must plan, manage, monitor and coordinate health and safety in the construction phase of a project. This includes liaising with the client and principal designer, preparing the construction phase plan, organising cooperation between contractors and coordinating their work. They must also ensure that suitable site inductions are provided, reasonable steps are taken to prevent unauthorised access, workers are consulted and engaged in securing their health and safety, and welfare facilities are provided.

Contractors must plan, manage and monitor construction work under their control so it is carried out without risks to health and safety. For projects involving more than one contractor, they must coordinate their activities with others in the project team; in particular, they must comply with directions given to them by the principal designer or principal contractor. For single contractor projects, they must prepare a construction phase plan.

Workers on construction sites must be consulted about matters which affect their health, safety and welfare. However, they also have duties under the regulations. In particular they must take care of their own health and safety, and of others who might be affected by their actions, report anything they see which is likely to endanger either their own or others' health and safety, and cooperate with their employer, fellow workers, contractors and other duty holders.

So far as the electrical subcontractor is concerned, if the electrical installation is designed to comply with the requirements of the Electricity at Work Regulations 1989, and in such a manner as to minimise the risks entailed in its construction and subsequent maintenance, the contractor will be likely to be in compliance with the CDM Regulations. It is important, however, that the contractor cooperates fully with the other CDM duty holders and employs staff who are competent both in terms of the Electricity at Work Regulations and the CDM Regulations. In this respect he has to cooperate with the client, other designers and the principal contractor. One obvious requirement is to ensure safe access to those parts of the installation that require periodic attention. In his role as a subcontractor, he has to provide information for incorporation in the construction phase plan, cooperate with the principal contractor and ensure that his employees are properly briefed on health and safety relevant to the site. Where the electrical contractor is the only contractor, such as may occur for a rewiring project, he becomes the principal contractor and has to shoulder the relevant responsibilities.

The HSE has published guidance on the regulations in the form of guidance note L153 *Managing health and safety in construction. Construction (Design and Management) Regulations 2015. Guidance on Regulations.* There is also a considerable amount of industry guidance available on the topic. There is no longer an Approved Code of Practice, although it is possible that one will be published in due course.

There is no doubt that, since their inception, the CDM Regulations have had a very positive impact on the management of construction activity and the way in which health and safety arrangements are planned and recorded. Some people argue that they have increased the amount of paperwork and bureaucracy involved in construction projects, although one aim of the 2015 version is to reduce this part of the burden. However, it is a moot point whether or not these changes have worked their way down to the 'coal face' and had an impact on the way in which the tradesmen themselves conduct their activities and take account of health and safety arrangements.

The Dangerous Substances and Explosive Atmospheres Regulations 2002 (DSEAR)

In Great Britain, these regulations transpose into law two directives: the safety aspects of the Chemical Agents Directive 98/24/EC (CAD) and the Explosive Atmospheres Directive 99/92/EC (ATEX). The regulations place duties on employers and the self-employed to protect against risks from fire, explosion and similar events arising from dangerous substances used or present in the workplace. In order to transpose requirements of the Chemical Agents Directive, with effect from June 2015, they also cover gases under pressure and substances that are corrosive to metals. In Northern Ireland, the regulations are the Dangerous Substances and Explosive Atmospheres Regulations (Northern Ireland) 2003.

The dangerous substances covered by the regulations include solvents, paints, varnishes, flammable gases, liquefied petroleum gas (LPG), dusts from machining and sanding operations and dusts from foodstuffs. These all have the potential to create an explosive atmosphere when mixed with air, which might catch fire or explode in the presence of an ignition source. The types of work activities covered include, but are not limited to,

- storage and use of gases such as oxygen, acetylene, and LPG;
- storage and use of flammable liquids such as solvents, paints, inks, and petroleum;
- storage, transport and use of flammable powders; and
- chemical and petrochemical manufacturing, processing and warehousing.

The regulations require duty holders, in decreasing order of preference, to take measures to protect the health and safety of their employees by preventing the formation of explosive atmospheres, by avoiding the ignition of explosive atmospheres and by mitigating the effects of an explosion such as by providing explosion relief and suitable PPE for personnel. Duty holders must also have appropriate emergency procedures in place should a fire or explosion occur.

In order to achieve this, the employer must assess the risks and classify the workplace into zones (Zones 0, 1 and 2), the nature of which follow the well-known hazardous area classification scheme explained in Chapter 15 relating to the probability of an explosive atmosphere being present. There are three zones (Zones 20, 21 and 22) relating specifically to the presence of combustible dusts.

The main implication from an electrical safety perspective is that ignition sources, including those from electrical equipment and electrostatic discharges, must be excluded from the hazardous areas. Only equipment and protective systems that meet the requirements of the Equipment and Protective Systems Intended for Use in Potentially Explosive Atmospheres Regulations 1996 should be used and installed. As explained earlier in this chapter, such equipment will be CE marked and, in Europe, carry the Ex symbol. Equipment and protective systems in use before July 2003 can continue to be used, provided the risk assessment shows it is safe to do so.

Guidance

There is an ACOP associated with the regulations which, together with guidance material, is published by the HSE in its publication L138 *Dangerous substances and explosive atmospheres. Dangerous substances and explosive atmospheres regulations 2002 approved code of practice and guidance.*

9

Other legislation covering electrical safety

Building regulations

England and Wales

The Building Regulations 2010 (as amended) are made under the Building Act 1984 and apply to buildings in England and Wales. They impose requirements on people carrying out "building work", which includes work on the electrical installation in the property. Building work must be carried out so that it complies with the applicable requirements set out in Parts A to P of Schedule 1 of the regulations, with Part P covering the subject of electrical safety.

Part P is a very general statement of requirement, stating that "reasonable provision shall be made in the design and installation of electrical installations in order to protect persons operating, maintaining or altering the installations from fire or injury". This applies only to low voltage and extra-low voltage installations that are in dwellings, in the common parts of a building serving one or more dwellings but excluding power supplies to lifts, and in a garden (including sheds and outhouses), or in or on land associated with a building where the electricity is from a source located within or shared with a dwelling.

Under the terms of the Building Act 1984, the Department for Communities and Local Government (DCLG) has published a set of approved documents containing practical guidance on achieving compliance with the regulations. Approved Document P provides guidance on satisfying the electrical safety requirements of Part P. The document has the status of guidance, meaning that duty holders can chose to adopt other routes to compliance with the regulations but they would need to be able to demonstrate that they achieve the same outcome as far as safety is concerned. In that sense, it is easiest to follow the approved documents.

In general terms, Approved Document P advises that compliance with Part P of the regulations can be achieved by installing, and inspecting and testing, electrical installations that satisfy the technical requirements of BS 7671 *Requirements for electrical installations*. This is the route to compliance adopted in the vast majority of building works. There are one or two requirements in the approved documents that are additional to BS 7671, such as the requirement that in new dwellings switches and socket-outlets must be between 450 mm and 1200 mm above finished floor level, a requirement derived from Part M of the regulations.

Only a limited set of electrical installation work needs to be notified to local building control officers; it includes

- installation of a new circuit;
- replacement of a consumer unit; and
- any addition or alteration to an existing circuit in a special location, meaning locations within specified zones in a bathroom or shower room or in a room containing a swimming pool or sauna.

All other electrical work does not need to be notified to the local authority although it will still need to meet the general requirement of Part P to be safe.

There are three procedures that may be used in England and Wales to certify that notifiable work satisfies the requirements of the regulations:

- the work can be self-certified by a registered competent person;
- the work can be certified by a registered third-party certifier, i.e. a registered competent person who does not do the installation work but inspects and tests it to check for compliance; or
- the work can be certified by a local authority building control officer.

There are a number of 'competent person schemes' in England, authorised by DCLG, and in Wales, authorised by the Welsh Government. The electrical competent person schemes are run by organisations such as Benchmark, BSI, BESCA, Blue Flame Certification, Certsure, NAPIT, OFTEC and Stroma. People who train under the schemes become registered 'domestic installers' and thereby should have the set of competences set out in DCLG's minimum technical competence (MTC) requirements for competent person schemes, as expanded in the Electrotechnical Assessment Specification (EAS) published by the IET (see also Chapter 19).

There are two levels of domestic installer: defined scope and full scope. Full scope installers' work includes the design, installation and verification of full domestic electrical installations, including installing new consumer units or ring main circuits. Defined scope installers carry out electrical work as part of their main trade work activity, e.g. plumbers, heating installers, kitchen and bathroom fitters.

Domestic installers are not full scope electricians with the competences to install and maintain, for example, three-phase installations in commercial and industrial premises.

Scotland

Scotland has a different set of building regulations and certification procedures than England and Wales. The primary legislation is the Building (Scotland) Act 2003, which came into force on 1 May 2005. The Building (Scotland) Regulations 2004, made under the 2003 Act, apply to the design, construction or demolition of a building, the provision of services, fittings or equipment in or in connection with a building, and the conversion of a building.

Schedule 5 of the regulations sets out functional standards that must be achieved; the functional standard for electrical safety is very general and states that every building must be designed and constructed in such a way that the electrical installation does not (a) threaten the health and safety of the people in, and around, the building; and (b) become a source of fire. The standard does not apply to an electrical installation (a) serving a building or any part of a building to which the Mines and Quarries Act 1954 or the Factories Act 1961 applies; or (b) forming part of the works of an undertaking to which regulations for the supply and distribution of electricity made under the Electricity Act 1989 apply.

There is an additional functional standard for 'electrical fixtures' in domestic buildings with an electricity supply which states that every building must be designed and constructed in such a way that electric lighting points and socket-outlets are provided.

The Scottish Government's Building Standards Division publishes comprehensive technical handbooks providing advice on how the functional standards may be achieved; there is one volume for domestic buildings and another for non-domestic properties. Both technical handbooks advise that, for low voltage installations, "an electrical installation should be designed, constructed, installed and tested such that it is in accordance with the recommendations of BS 7671:2008".

The technical handbook for domestic properties provides additional requirements for fixtures, over and above those required by BS 7671; in particular, it stipulates that a dwelling should be provided with at least the following number of 13A socket-outlets:

- four within each apartment;
- six within the kitchen, at least three of which should be situated above worktop level in addition to any outlets provided for floor-standing white goods or built-in appliances; and
- an additional four anywhere in the dwelling, including at least one within each circulation area on a level or storey.

The regulations stipulate that certain electrical work may only be carried out after a building warrant has been issued by the local authority. Whether or not electrical work requires a building warrant is quite complicated and depends upon the type of property (flat, house up to two storeys or house three storeys or higher) and the nature of the work. For example, wiring in new socket-outlets in a flat or a house of three storeys or higher does require a building warrant, whereas the same job in a single or two storey house does not. Information on the requirements for building warrants for electrical work is published by the Scottish Government's Building Standards Division on its website.

If a building warrant is required, the work may be self-certified as being compliant with BS 7671 if it is carried out or inspected by a person registered and approved under the Certification of Construction (Electrical Installations to BS 7671) Scheme; otherwise it would need to be certified by a local authority verifier. The Scottish Government has approved two scheme providers: SELECT (the Electrical Contractors Association of Scotland), and the NICEIC.

Northern Ireland

The Building Regulations (Northern Ireland) 2012 are made by the Department of Finance and Personnel and administered by Northern Ireland's 26 District Councils. Some forms of building works require applications for approval to be made to the appropriate District Council who then administers the application and check compliance. There are no specific requirements in the regulations covering electrical safety in the same way as in the other British regulations, but there would no doubt be an expectation that electrical installations would be installed in compliance with BS 7671.

Legislation dealing with safety of rented dwellings

It is important for landlords who rent properties for occupation by households to ensure that the power distribution systems and electrical equipment in those properties are safe and properly maintained. There is a range of legislation placing duties to that end on landlords in the social housing and private sectors. The law varies between the individual nations in Great Britain and is fully described in Chapter 20 of this book.

The Plugs and Sockets etc (Safety) Regulations 1994

These regulations are made under the Consumer Protection Act 1987 and apply to electrical appliances intended for domestic use, although many of these will also be used in workplaces.

Part I of the regulations applies to plugs, sockets and adaptors rated at not less than 200 volts and their associated fuses. All standard plugs designed to engage a socket made to the dimensions of BS 1363:1984 *Specification for 13A fused plugs and switched and unswitched socket outlets* must be of a type that has been approved by a notified body. Other devices such as sockets, fuse links, plugs, adaptors etc must conform to the appropriate British standard listed in Schedule 2 of the regulations. Schedule 1 lists some types of apparatus that are excluded, an example being a moulded-on Europlug connected to a shaver supply unit conforming to BS 3535: Part 1 (now BS EN 61558–1).

Part II of the regulations applies to appliances ordinarily intended for domestic use which are rated at not less than 200 volts and a maximum rated input of not more than 13 amps. There is a small number of exceptions, as follows:

- any fixed luminaire, being a luminaire which cannot easily be moved from one place to another, either because it can only be removed with the use of a tool, or because it is intended for use out of easy reach;
- any ceiling-rose connector (that is to say a connector designed to hold up overhead electric lighting fittings);
- any electric light designed and intended to be located within a recess in a wall or ceiling;
- any appliance which is fitted with an RCD plug;
- any appliance which is fitted with a plug transformer;
- any appliance which is fitted with a plug other than a standard plug which is designed to engage with a compatible portable multiple socket-outlet;
- any appliance which is intended to be permanently connected to the fixed wiring of the mains system other than by means of a plug and socket.

Apart from these excepted items, all appliances must be fitted by the supplier with a standard plug of a type approved by a notified body and which is fitted with an appropriately rated fuse conforming to British Standard 1362. The requirement does not apply to appliances fitted with a non-UK plug complying with the safety provisions of the standard IEC 60884–1 *Plugs and Socket-Outlets for Household and Similar Purposes, General Requirements,* and fitted with a type-approved conversion plug that can only be removed by the use of a tool. This requirement for the supplier to fit plugs, most of which are moulded plugs, has had a considerable benefit in reducing the need for consumers to fit their own plugs and possibly make mistakes while doing so. The downside is that there is an increasing proportion of the population growing up without learning how to change a plug, but the safety benefits outweigh this disadvantage.

Part III of the regulations stipulates the information that must accompany all standard plugs and conversion plugs. Such devices must be legibly marked or bear a label that signifies that the device has been approved and the identity of the body that carried out the approval. Where it is necessary for the safe operation of any standard plug or conversion plug that the user should be aware of any particular characteristics, the necessary information must be given by markings on the plug or, where this is not practicable, in a notice accompanying the plug. Any such instructions must be given in English.

The regulations are enforced by local authority Trading Standards Officers.

Explosive Regulations 2014

These regulations govern the safety of explosives operations involving duty holders such as employers, private individuals and other people manufacturing and storing explosives. The regulations are the outcome of an extensive review carried out by all the relevant regulators to simplify and rationalise the previous legislation covering explosives safety.

Regulation 26 requires measures to be taken to exclude sources of ignition from explosives, including electrical and electrostatic sources. Appendix 3 to the HSE's guidance note on the regulations, publication L150, contains specific guidance on electrical safety in explosives operations and is essential reading for everyone with responsibility in this area. Its content on electrostatics is detailed, helpful and prescriptive, unlike the content on electrical installations which is very generic and does not add very much value.

Cinemas

The Cinematograph (Safety) Regulations 1955 and subsequent amendments, made under the Cinematographic Act 1952, lay down the requirements for the electrical installations in cinemas. Where both the general and safety lighting are electric, there need to be two sources of supply to ensure maintenance of the emergency lighting in the event of mains failure. These regulations are enforced by the local authorities.

10

The Corporate Manslaughter and Corporate Homicide Act 2007

Introduction

The Corporate Manslaughter and Corporate Homicide Act came into force on 6 April 2008. The offences covered by the act are called corporate manslaughter in England, Wales and Northern Ireland, and corporate homicide in Scotland.

Before this act was introduced companies could be prosecuted for gross negligence manslaughter but only if a senior manager or director of the company, known as a 'directing mind', could also be prosecuted for the offence. This concept of the 'directing mind' and the need to prove its involvement made it difficult for corporate manslaughter prosecutions to succeed. One of the most well-known examples arose from the Herald of Free Enterprise tragedy in 1987 when the car ferry operated by Townsend Thoresen, which later became part of P&O European Ferries, capsized with the loss of 193 lives as it departed the port of Zeebrugge. The root cause was the ferry leaving port with its bow doors open and the car deck becoming flooded. The corporate manslaughter case against the company failed because the various acts of negligence could not be attributed to any senior individual who was a 'directing mind', leading to a public outcry about the ineffectiveness of the legislation.

The new act was introduced to make it easier for organisations, not individuals, to be prosecuted for gross negligence manslaughter where a senior management failing has caused a fatality. However, individuals are still liable to prosecution for gross negligence manslaughter should the police find evidence to support a charge. In addition, senior managers and directors remain liable to charges under Section 37 of the Health and Safety at Work Act if the enforcing authority can show that an offence occurred with their consent or connivance, or can be attributed to their neglect.

The offence

An offence is committed if the way in which an organisation's activities are managed or organised both causes a person's death and amounts to a gross breach of a relevant duty of care owed by the organisation to the deceased. An organisation is guilty of an offence only if the way in which its activities are managed or organised by its senior management is a substantial element in the breach. Cases are always heard before a jury and, on conviction, a penalty of an unlimited fine may be imposed, as well as the organisation being instructed to remedy the breach and/or publicise its conviction and failures.

Corporate manslaughter and homicide investigations of work-related deaths are conducted by the police, usually under the terms of the Work-Related Death Protocol for liaison between the police and other enforcing authorities. However, where there is a prosecution under the act, there will also be a prosecution brought under health and safety legislation by the appropriate enforcing authority such as the HSE. So if the corporate manslaughter case fails, there is still a possibility of conviction under health and safety legislation.

To some extent this is because the prosecution's burden of proof is different. Under the corporate manslaughter legislation, the prosecution has to prove that the way in which an organisation's activities were managed or organised caused a person's death and amounted to a gross breach of the relevant duty of care. In contrast, under health and safety legislation, the prosecution merely needs to prove the presence of a risk to health and safety, and it is then for the employer to show on the balance of probabilities that it was not reasonably practical to do more than was done to comply with the duty.

The duality of prosecutions in the case of work-related deaths is exemplified by the case of CAV Aerospace Ltd. The company was convicted in July 2015 of corporate manslaughter following the death of its employee Paul Bowers and fined £600,000. It was also fined £400,000 for contravening the Health and Safety at Work etc Act 1974, and the company was required to pay £125,000 in prosecution costs.

Gross breach of a relevant duty of care

A relevant duty of care is one of several duties of care owed by an organisation under the law of negligence. Section 2(1) of the act defines it in the following terms:

(a) a duty owed to its employees or to other persons working for the organisation or performing services for it;
(b) a duty owed as occupier of premises;
(c) a duty owed in connection with:

   (i) the supply by the organisation of goods or services (whether for consideration or not),
   (ii) the carrying on by the organisation of any construction or maintenance operations,
   (iii) the carrying on by the organisation of any other activity on a commercial basis, or
   (iv) the use or keeping by the organisation of any plant, vehicle or other thing;

(d) a duty owed to a person who, by reason of being a person within subsection (2), is someone for whose safety the organisation is responsible.

The matter of whether or not there has been a gross breach is for the jury to decide, although a fundamental principle is that the organisation's conduct must have fallen far below what could have been reasonably expected. Section 8 of the act advises that the jury must consider whether the evidence shows that the organisation failed to comply with any health and safety legislation that relates to the alleged breach and, if so, how serious that failure was and how much of a risk of death it posed. The jury may also consider the extent to which the evidence shows that there were attitudes, policies, systems or accepted practices within the organisation that were likely to have encouraged any such failure, or to have produced tolerance of it; it should also consider any health and safety guidance that relates to the alleged breach.

The term 'senior management' refers to a manager who plays a significant and influential role in the management of the whole, or a substantial part, of the organisation's activities. Middle managers, supervisors, project managers and the like would not be classed as 'senior management' in this context – it is people above that level who would potentially be caught by the legislation.

Exempted organisations

A number of classes of organisations are outside the scope of the act. The following are wholly excluded:

- Public policy decisions such as strategic funding decisions and other matters involving competing public interests. However, decisions about how resources are managed are not exempted.
- Military activities, including potentially violent peacekeeping operations and those dealing with terrorism and violent disorder. Related support and preparatory activities and hazardous training are also exempt.
- Police operations, dealing with terrorism and violent disorder. This also extends to support and preparatory activities and hazardous training.

Some organisations are partially excluded, an example being emergency response personnel in certain circumstances.

Convictions

As of February 2016 there had been eighteen successful convictions under the act, all of them in England, Wales and Northern Ireland, with none in Scotland. The average fine is around £250,000, with the largest being £700,000 and the lowest £8000. However, in the latter case of the smallest fine, the company had to pay £4000 in prosecution costs and the company's sole director was fined £183,000 with £8000 in costs and banned from holding the position of director for 5 years; so the headline figure tends to hide the true costs.

In another case that exemplifies the mix of penalties, in July 2015 Huntley Mount Engineering was fined £150,000 for corporate manslaughter after the death of an apprentice who had recently joined the company. A company director was jailed for 8 months for offences under Section 2 and Section 37 of the Health and Safety at Work Act 1974. His son received a 4-month jail sentence, was suspended for a year, and was fined £3000.

It is notable that all the organisations convicted have been small to medium size enterprises, with no convictions of the large corporations at whom the act was meant to be targeted.

Example electrical safety and control system cases

There has been only one corporate manslaughter case involving an electrical accident. This was taken against PS & JE Ward, a small company that runs a flower nursery in Norfolk. It was charged after the death of an employee, Grzegorz Pieton, in 2010 from electric shock and burn injuries when a trailer he had been towing came into contact with an 11 kV overhead power line at the edge of a field at the nursery. He had raised the tipper trailer to offload some soil and, while it was being pulled by the tractor he was driving, it struck the overhead line. He most probably suffered his fatal injuries when he climbed out of the tractor cab while the trailer was still in contact with the live wire.

The company faced both corporate manslaughter charges and charges brought under health and safety legislation. It was acquitted of the corporate manslaughter charge, mostly because it was successfully argued that the deceased was working under his own initiative and had received training on the dangers from overhead lines; this was the first successful defence of a corporate manslaughter charge. However, the company was fined £50,000 plus costs of £47,932 for breaching Section 2(1) of the Health and Safety at Work Act, largely for failures relating to the suitability of its risk assessments.

In another case, this time involving a machinery control system, Pyranha Mouldings Ltd was fined £200,000 following the death of an employee, Alan Catterall. A company director was sentenced to 9 months in prison, suspended for 2 years, and fined £25,000 and both the company and the director were required to pay costs of £90,000 between them. The employee had become trapped inside an industrial oven which was switched on by another employee while he was still inside it following some maintenance activity. The oven doors closed automatically and trapped him inside the oven but he had no means of escape or raising the alarm, and he died from his injuries. The company had designed and manufactured the oven, including its safeguarding systems, and had devised the work procedures that the deceased had been following.

11

Standards and guidance

Introduction

The UK's health and safety legislation is non-prescriptive and goal-setting, placing the responsibility for controlling risks to health and safety on those organisations or people who create the risks in the first place. In general, these organisations must manage the risks so as to reduce them so far as is reasonably practicable.

In such a system it is essential that duty holders are able to obtain information, advice and guidance on how to reduce the risks created in their workplaces so far as is reasonably practicable. This is not just to allow them to develop and implement effective safety management practices but also to allow them to demonstrate that they are legally compliant if challenged by an enforcement authority or during court proceedings.

The other side of the coin is that the availability of such guidance material provides the enforcement authorities with ammunition to show that a duty holder is non-compliant, should that be the case, in the event of enforcement action being taken.

This chapter describes the status of the different types of guidance material and identifies the main sources of guidance in the field of electrical safety.

Types and status of guidance material

The law

The law itself should be treated as providing guidance, with some items of legislation being more helpful than others. For example, whereas the Health and Safety at Work Act's strictures that risks must be reduced so far as is reasonably practicable may not be very helpful in deciding what actually needs to be done, the Electricity at Work Regulations' (Regulation 7) requirement that conductors must be insulated or placed out of harm's way leaves little room for doubt on what needs to be done. The question of how it is to be done is covered by other guidance material, but the fundamental legal duty is pretty clear and obvious.

The status of this type of guidance is also quite clear: it must be complied with.

Approved codes of practice

Approved Codes of Practice (ACOP) are published by the HSE, usually after extensive consultation with the public. They are meant to offer quite detailed advice on how to comply with the law covered by the ACOP, with the principle that if an employer follows the advice there will be a presumption of conformity with the relevant legislation.

There is no duty to comply with an ACOP, because employers are free to adopt alternative solutions in order to comply with the law. However, an ACOP has a special legal status in that if a company is prosecuted for a breach of health and safety law, and it is proved that it did not follow the relevant provisions of the ACOP, it will need to show that it has complied with the law in some other way or a court will find it in breach of the law. For this reason, companies are well advised to find out if there is an ACOP that is applicable to their risks and to make sure that they comply with its advice and guidance; for the removal of any doubt, the Electricity at Work Regulations 1989 do not have an associated ACOP.

In his 2011 review of the UK's health and safety system, Professor Ragnar Löfstedt noted that there were fifty-three ACOPs published by the HSE and recommended that they should all be reviewed. The HSE subsequently conducted a public consultation on its proposals for the revision, consolidation or withdrawal of fifteen ACOPs and on proposals for minor revisions, or no changes, to a further fifteen ACOPs. As far as this book is concerned, the most important outcome of the consultation was the withdrawal of the ACOP on the Management of Health and Safety at Work Regulations 1999, the publication of a new consolidated ACOP L138 on the Dangerous Substances and Explosive Atmospheres Regulations 2002, and the publication of a revised ACOP L22 on the Provision and Use of Work Equipment Regulations 1998.

Guidance

Guidance comes in many forms. It is published by enforcing authorities such as the HSE, professional bodies such as the Institution of Engineering and Technology (IET), trade bodies such as the Electrical Contractors' Association and the Energy Networks Association, manufacturers such as Siemens and Pilz, charitable organisations such as Electrical Safety First (ESF) and individuals such as the author of this book.

The guidance issued by the HSE is perhaps the most important because the enforcing authority declares that, by following it, organisations will normally be doing enough to comply with the law and that health and safety inspectors may refer to this guidance when seeking to secure compliance with the law. However, there is no duty to comply with the guidance, and inspectors may not use it in the same way as an ACOP in the case of a prosecution. Nonetheless, the HSE's guidance has more legal authority than the guidance issued by other non-regulatory bodies. On the other hand, it has the disadvantage of being quite generic in nature and not targeted at any one particular industrial or commercial sector. In recent years, the HSE's guidance has lost much of the technical content that once made it such a valuable source of information for engineering professionals; it now tends to focus on safety management issues rather than giving helpful and targeted detailed technical advice.

Guidance produced by bodies such as the IET and ESF are more focussed and technically authoritative than the HSE's guidance. For example, the IET's suite of guidance notes on BS 7671 (see Chapter 12) sets down benchmark guidance on matters such as inspection and testing of electrical installations. Again, there is no duty to comply with this guidance but the courts generally accept it as being the principal source of advice on how to secure compliance with the law.

Standards

The originator of standards in the UK is the British Standards Institution (BSI). It publishes not just British-derived standards but also harmonised European standards produced by CEN, CENELEC and ETSI, and international standards published by the International Electrotechnical Commission (IEC) and the International Standards Organisation (ISO). UK versions of international standards are often double- or treble-prefixed, such as BS EN or BS EN ISO.

Standards usually have the same status as guidance. They offer a route to compliance, but there is no duty to comply with them unless they are specifically mentioned in legislation, such as in the ESQCR's requirement for compliance with BS 7671 in some circumstances (see Chapter 8).

European harmonised standards produced under the terms of, for example, the Low Voltage and Machinery Directives, have a special status. This is because compliance with them provides a presumption of conformity with the essential health and safety requirements, or safety objectives, of the relevant directive. Again, there is no duty to comply with a harmonised standard because the legal duty is to comply with the essential requirements, but the presumption of conformity offers a very helpful and smooth route to securing legal compliance.

It's going to be interesting to see what happens to the standards-making process once the UK leaves the EU. Over the past 40 years or so the UK has been deeply engaged with EU bodies in the development of standards covering a huge range of products – machines, appliances, food products, medical equipment, toys, cosmetics, and so forth. It will be a very significant step if the UK disengages from this activity and returns to producing its own standards, which may happen if the nation does not negotiate some form of free trade agreement with the EU.

Guidance on electrical safety

This section identifies some of the main sources of guidance on electrical safety. Given the large number of documents covering the subject, the list by necessity has to be selective and is based on the author's experience of accident investigation and witness appearances in court.

HSE guidance

HSR25 Memorandum of guidance on the Electricity at Work Regulations 1989

This guidance note, last updated in November 2015, sets out the Electricity at Work Regulations 1989 (EAWR) and provides top-level guidance on their meaning and routes to compliance. It is, to all intents and purposes, the foundation guidance note on the regulations.

HSG85 Electricity at work, safe working practices

This guidance note, last updated in 2013, covers the important topic of safe working practices on electrical systems. Its main focus is safe isolation procedures and safety during live work, with an explanation of the decision-making process that underpins compliance with EAWR Regulation 14 and the types of precautions that need to be taken during any authorised live work. Unfortunately the document is silent on the subject of arc flash protection, the creeping introduction into the UK of arc flash surveys and the insistence of some employers that their staff wear arc flash PPE whenever they work on or near electrical systems.

HSG107 and INDG236 Maintaining electrical appliances

These two publications cover the maintenance of electrical equipment, a topic commonly but often mistakenly known as Portable Appliance Testing (PAT). HSG107 is the main guidance on the topic, covering the maintenance of all forms of electrical equipment in all environments. INDG236 covers the routine inspection and testing of appliances in low risk environments such as offices, shops and hotels, and represents the HSE's latest attempt to bring some common sense into the topic of which types of equipment need to be inspected and tested and how often it needs to be done, if at all (see Chapter 18).

GS6 Avoiding danger from overhead power lines

Since contact with overhead power lines has historically accounted for roughly half of all work-related deaths from electrical causes, the importance of this guidance note and its recommended precautionary measures should not be underestimated (see Chapter 14). It is the benchmark guidance on the topic and is widely used by the DNOs whose lines are being struck and by the industries who tend to strike them most – construction and agriculture. The latest version was published in 2013.

HSG47 Avoiding danger from underground services

A failure to follow the precautionary procedures set out in this guidance is a common cause of underground cables being struck during excavation works (see Chapter 14). The document, last updated in 2014, is essential reading for anyone managing excavation work where there is a risk of underground utilities being struck.

L138 Dangerous Substances and Explosive Atmospheres Regulations 2002. Approved Code of Practice and guidance

This ACOP, published in 2013, covers, amongst other things, the selection, installation and maintenance of electrical equipment that could potentially act as an ignition source in explosive atmospheres.

Other HSE guidance documents

Some other HSE guidance documents addressing electrical safety and worthy of note are

- GS38 Electrical test equipment for use by electricians (updated 2015);
- INDG354 Safety in electrical testing at work (updated 2013);
- HSG230 Keeping electrical switchgear safe (updated 2015);
- GS50 Electrical safety in places of entertainment (updated 2014); and
- INDG139 Using electric storage batteries safely (2011).

The Institution of Engineering and Technology

Code of practice for in-service inspection and testing of electrical equipment (4th edition)

This publication forms the basis of training provided to people who wish to qualify in, and work on, the in-service inspection and testing of electrical equipment. It builds on the HSE's publications HSG107 and INDG236, providing more comprehensive guidance on the inspection and testing procedures and practices.

Guidance notes on the application of BS 7671

As explained in Chapter 12, BS 7671 is a considerably long and complicated British Standard that is at the heart of achieving safe low voltage power distribution systems. The IET has for many years produced informative and helpful explanatory guidance notes on the standard, the main ones being

- Guidance Note 1 – Selection and erection
- Guidance Note 2 – Isolation and switching
- Guidance Note 3 – Inspection and testing
- Guidance Note 4 – Protection against fire
- Guidance Note 5 – Protection against electric shock
- Guidance Note 6 – Protection against overcurrent
- Guidance Note 7 – Special locations
- Guidance Note 8 – Earthing and bonding
- The On-Site Guide – An abbreviated guidance document for electricians
- Electrical installation design guide

Code of practice for electrical safety management

This book provides guidance to duty holders on the topics that they need to be addressing if they are to implement effective safety management practices and procedures. It does not refer to the legal requirements or offer solutions for compliance, but it is a helpful pointer towards the full range of safety management issues that have to be addressed.

Electrical Safety First

This charitable organisation, which is linked to the National Inspection Council for Electrical Installation Contractors (NICEIC), publishes a variety of Best Practice Guides. From a work-related electrical safety perspective the one that has had most impact is Best Practice Guide 2 *Guidance on the management of electrical safety and safe isolation procedures for low voltage installations*. It was produced in conjunction with the HSE in an attempt to codify safe isolation procedures to be adopted in the construction and allied industries as part of a national effort to drive down the number of serious and fatal accidents arising from the failure to isolate systems securely. The principles set out in the guide form the basis of training provided to contracting industry electricians throughout the country.

The Energy Networks Association (ENA)

The ENA is the trade association for the transmission and distribution companies in the electrical supply industry and is the custodian and originator of much of the guidance covering safety in the industry. The main documents of interest fall into the categories of Technical Specifications (TS) and Engineering Recommendations (ER), of which there are hundreds. Some important examples are

- TS 09–22: Protection of cable installations against the effects of fire
- TS 43–8: Overhead line clearances
- TS 43–90: Anti-climbing measures and safety signs for high voltage overhead lines
- TS 43–103: Low voltage overhead line shrouding materials
- ER G55: Safe tree working in proximity to overhead lines
- ER G59/2: Recommendations for the connection of generating plant to the distribution systems of licensed DNOs
- ER S5/1: Earthing installations in substations

Engineering Equipment Materials Users Association (EEMUA)

EEMUA produces a range of guidance publications. For example, its Publication 186 *A Practitioner's HANDBOOK for POTENTIALLY EXPLOSIVE ATMOSPHERES* provides helpful guidance on the law and standards for preventing electrical ignition of potentially explosive atmospheres.

British Electrotechnical and Allied Manufacturers' Association (BEAMA)

BEAMA produces a range of guidance publications, an example of which is its informative RCD Handbook, *BEAMA Guide to the Selection and Application of Residual Current Devices.*

Standards covering electrical safety

The following list in Table 11.1 is intended to identify the main standards containing requirements for electrical safety. It is by no means exhaustive and excludes, for example, the many standards for electric cables; indeed, a complete and exhaustive list would occupy very many pages and would probably be out of date as soon as it is finished, given the pace of technological change. A comprehensive list of standards for electrical installations, including cables and cable containment, can be found at Appendix 1 of BS 7671.

Table 11.1 List of main standards covering electrical safety

Effects of electric current		
Number	*Title*	*Comment*
DD IEC/TS 60479–1:2005	Effects of current on human beings and livestock. General aspects.	Although there are five parts to EC/TS 60479, these two parts provide core guidance on the effects of electric current on humans.
IEC/TS 60479–2:2007	Effects of current passing through the human body. Part 2: Special aspects.	

Low voltage electrical installations and equipment		
Number	*Title*	*Comments*
BS 88	Low voltage fuses	Multi-part standard for low voltage fuses of the type used in cut-outs and similar applications
BS 1363	13 A plugs, socket-outlets, connection units and adaptors	Multi-part standard for three-pin plugs etc for use in the UK; fuses for use in plugs are to be to BS 1362.
BS 6423:2014	Code of practice for maintenance of electrical switchgear and controlgear for voltages up to and including 1 kV	
BS 6626:2010	Code of practice for maintenance of electrical switchgear and controlgear for voltages above 1 kV and up to and including 36 kV	Sets out safe systems of work for HV systems, including use of Permits to Work
BS 7375:2010	Distribution of electricity on construction and demolition sites. Code of practice.	See Chapter 13 for details
BS 7430:2011+A1	Code of practice for protective earthing of electrical installations	Describes system earthing requirements, including for mobile generators
BS 7671:2008+A3	Requirements for electrical installations (IET Wiring Regulations 17th edition)	Benchmark standard for safety of low voltage electrical installations; see Chapter 12 for details.
BS 7909:2011	Code of practice for temporary electrical systems for entertainment and related purposes	Describes how to apply BS 7671 principles to temporary installations at events
BS EN 50110–1:2013	Operation of electrical installations. General requirements.	Part 2 of the standard; contains national requirements
BS EN 60309	Plugs, socket-outlets and couplers for industrial purposes. General requirements.	Multi-part standard describing pin configuration and colours of industrial plugs etc of the CEEForm variety
BS EN 60335	Household and similar electrical appliances	Standard with in excess of a hundred parts covering safety of a large variety of electrical appliances
BS EN 60529:1992+A2	Specification for degrees of protection provided by enclosures (IP code)	Code for construction of enclosures to prevent ingress of moisture and objects

(Continued)

Table 11.1 (Continued)

Low voltage electrical installations and equipment		
Number	*Title*	*Comments*
BS EN 60898	Specification for circuit breakers for overcurrent protection for household and similar installations	Three-part standard for circuit breakers commonly known as miniature circuit breakers (MCB)
BS EN 60947	Specification for low voltage switchgear and controlgear	Multi-part standard for circuit breakers, contactors, switches, disconnectors, etc
BS EN 61008–1:2012 +A11:2015	Residual current operated circuit breakers without integral overcurrent protection for household and similar uses (RCCBs). General rules.	Standard for RCDs
BS EN 61009–1:2012 +A11:2015	Residual current operated circuit breakers with integral overcurrent protection for household and similar uses (RCBOs). General rules.	Standard for RCBOs
BS EN 61140:2002 +A1:2006	Protection against electric shock. Common aspects for installation and equipment.	Sets out the principles of protection against electric shock
BS EN 61439	Low voltage switchgear and controlgear assemblies	Multi-part standard for distribution boards and consumer units
BS EN 61558	Safety of power transformers, power supplies, reactors and similar products	Multi-part standard covering transformers, isolating transformers, shaver supply units and the like
BS EN 62305	Protection against lightning	Multi-part standard on the requirements for lightning protection

Electrical equipment in explosive atmospheres		
Number	*Title*	*Comments*
PD CLC/TR 50427:2004	Assessment of inadvertent ignition of flammable atmospheres by radio frequency radiation. Guide.	
BS EN 60079	Explosive atmospheres	Multi-part standard describing design requirement for explosion protection of electrical equipment, hazardous area classification, gas detectors and so on

Machinery safety – Electrical safety, control systems and interlocking		
Number	*Title*	*Comments*
BS EN 1037:1995 +A1:2008	Safety of machinery. Prevention of unexpected start-up.	Type B standard for prevention of unexpected start-up of moving parts, such as when interlocked guards have been opened and persons are in the danger zone
BS EN ISO 12100:2010	Safety of machinery. General principles for design. Risk assessment and risk reduction.	Type A standard setting out principles for risk control on machinery
BS EN ISO 13849–1:2015	Safety of machinery. Safety-related parts of control systems. General principles for design.	Type B1 standard covering safety requirements design and integration of safety-related parts of control systems (SRP/CS), including the design of software
BS EN ISO 13850:2015	Safety of machinery. Emergency stop function. Principles for design.	Type B standard for emergency actuators, colours etc
BS EN ISO 14119:2013	Safety of machinery. Interlocking devices associated with guards. Principles for design and selection.	Type B standard for the design of interlocking devices used on guards
BS EN 60204	Safety of machinery. Electrical equipment of machines.	Type B multi-part standard covering low and high voltage systems on machinery. Part 1 sets out the core principles of electrical and control systems safety on machinery.
BS EN 61496	Safety of machinery. Electrosensitive protective equipment.	Multi-part Type B standard covering devices such as photoelectric guards and laser scanning devices
BS EN 62061:2005+A2:2015	Safety of machinery. Functional safety of safety-related electrical, electronic and programmable electronic control systems.	Type B standard – machinery sector implementation of IEC 61508; alternative to 13849–1; see Chapter 17.

(*Continued*)

Table 11.1 (Continued)

Electromagnetic compatibility		
Number	*Title*	*Comments*
BS EN 61000	Electromagnetic compatibility	Multi-part standard covering emissions, immunity, and testing and measurements requirements
BS EN 60801–2:1993	Electromagnetic compatibility for industrial-process measurement and control equipment. Electrostatic discharge requirements.	

Electrostatics		
Number	*Title*	*Comments*
BS 5958–1:1991	Code of practice for control of undesirable static electricity. General considerations.	Comprehensive guidance on the generation of static and the means of prevention and control
BS EN 61340	Electrostatics	Multi-part standard covering principles, measurement techniques and control
PD CLC/TR 60079–32–1:2015	Explosive atmospheres. Electrostatic hazards, guidance.	

BS 7671:2008 *Requirements for Electrical Installations* – IET Wiring Regulations, 17th edition

Introduction

The first edition of what are now called the Institution of Engineering and Technology (IET) Wiring Regulations for electrical installations was published in 1882, under the title 'Rules and Regulations for the Prevention of Fire Risks Arising from Electric Lighting'. Since then the document's successor editions have acted as the benchmark design standard for ensuring that low voltage installations do not jeopardise the safety of persons and livestock.

The document has traditionally been known as 'the Wiring Regulations' or simply 'the Regs', although since 1992 it has been published as a British Standard, BS 7671 *Requirements for Electrical Installations*.

The 17th edition of the Wiring Regulations was published in January 2008 as BS 7671:2008 and came into effect on 1 July 2008, superseding the 16th edition which had first been published in 1992 and amended twice thereafter. It consists of 464 pages which, when compared to the four pages in the first edition, indicates the extent to which it has become a very lengthy and complex standard. However, it has a very important place in the UK's electrical safety system and, as such, justifies a separate chapter describing its status, scope, structure and some of its main provisions. Not all of the very many requirements are described, since that would necessitate a considerably larger book, so the chapter should not be treated as a substitute for checking the detailed content of the standard.

As well as being structurally different to the 16th edition, and having a new numbering system, the 17th edition and its subsequent amendments contain some significant new technical requirements, the most notable being

- an increased requirement for RCD protection, notably on socket-outlet circuits and circuits in so-called thin walls;
- changes to disconnection times in systems using automatic disconnection as a means of protecting against electric shock;
- a requirement for consumer units in dwellings to be non-combustible (i.e. made of metal) or located in a non-combustible enclosure;
- changes to the zoning areas in rooms containing baths and showers, with an opportunity both to omit supplementary equipotential bonding if the circuits supplying the location are RCD-protected, and to install socket-outlets in the room (subject to a distance requirement);
- new support requirements for cables passing over emergency exit routes;

- the introduction of a minimum voltage factor (C_{min}) to take account of the impact of low voltages, at the bottom end of the permissible range, and concomitant reductions in the maximum values of earth fault loop impedance needed to achieve the required disconnection times;
- a raft of new 'special locations'.

The standard is the British version of IEC Publication 60364 *Electrical Installations of Buildings*, which is also a CENELEC harmonised document. It incorporates the provisions of CENELEC harmonised standards, as considered and agreed by the joint IET/BSI Technical Committee JPEL/64, but also includes special national conditions. It has been amended on three occasions since 2008; the narrative in this chapter includes the substance of the amendments.

Scope

The standard establishes the accepted safety parameters for the designers, erectors and testers of low voltage installations not exceeding 1000 Va.c. and 1500 Vd.c. There is a list of instances, such as installations in explosive atmospheres, where the standard needs to be supplemented by the requirements of more specific British Standards. There is also a list of exclusions, as follows:

- systems for the distribution of electricity to the public – this is mainly DNOs' and IDNOs' supplies under the terms of the ESQC Regulations;
- railway traction equipment, rolling stock and signalling equipment;
- motor vehicles (although there are requirements relating to caravans and caravan parks);
- equipment on aircraft, offshore platforms and ships;
- equipment in mines where covered by other Statutory Regulations;
- radio interference suppression equipment (unless it affects the safety of electrical installations), lightning protection equipment (refer to BS EN 62305), and those aspects of lift installations covered by BS 5655 and BS EN 81;
- electrical equipment of machines where covered by BS EN 60204;
- electric fences, which are covered by BS EN 60335–2–6; and
- the d.c. side of cathodic protection systems.

Status

BS 7671 has the status of a British Standard code of practice and therefore has the equivalent status of guidance as described in Chapter 11. The term 'Wiring Regulations' should not be misinterpreted as meaning that the standard is a statutory instrument.

The standard is recognised as the benchmark for the design of low voltage electrical installations in GB and some other countries. In each British country, it is referenced in technical standards and guidance supporting the building regulations and property rental legislation. The Memorandum of Guidance to the Electricity at Work Regulations refers to it and notes that installations that are designed in conformity with it are likely to achieve compliance with the Electricity at Work Regulations 1989. However, there is no specific requirement or suggestion that new editions of the standard should be applied retrospectively; given that the standard is amended every 2–3 years, a requirement to apply each amendment retrospectively to existing installations would represent a considerable and unwarranted financial burden.

Having made that point, employers with duties under the Electricity at Work Regulations 1989 need to assess whether amendments that increase safety would be reasonably practicable to implement. For example, the considerable safety benefits arising from the installation of RCDs in final circuits will in many circumstances mean that introducing RCDs into existing circuits would be a reasonably practicable precaution against electrical injury. The absence of RCDs should be brought to the attention of duty holders by persons carrying out periodic inspection and tests of installations, as described later in this chapter, giving the duty holder the opportunity to consider the matter.

The standard is also commonly used in civil litigation concerning the safety of consumer's electrical installations.

The standard contains requirements that go beyond what would be required for compliance with health and safety legislation. For example, there are some requirements that are specific to protecting livestock from electrical injury in agricultural premises – health and safety law does not cover the safety of livestock. This means that care must be taken when using BS 7671 to check compliance with the law.

Structure

The standard comprises seven parts and fourteen appendices. The parts are

Part 1: Scope, object and fundamental principles
Part 2: Definitions
Part 3: Assessment of general characteristics
Part 4: Protection for safety
Part 5: Selection and erection of equipment
Part 6: Inspection and testing
Part 7: Special installations or locations

Each part comprises a number of chapters. The chapters are subdivided into sections or clauses that contain individual 'regulations', which are numbered using a 100s numbering system, the same format as the CENELEC harmonised document. The only exceptions to this are those regulations introduced in Amendment 3 that are special UK conditions and which are numbered using 200s, an example being regulation 421.1.201 which is a UK requirement for consumer units to be either made of non-combustible material or to be housed in a cabinet or enclosure made of non-combustible material.

Enforcement

Based on the content of Approved Document P in England and Wales, and the technical handbooks in Scotland, building control officers in local authorities throughout Great Britain will use BS 7671 to ensure compliance with the electrical safety requirements of the buildings regulations. They will not issue completion certificates for new buildings and other notifiable projects unless the electrical installation conforms to BS 7671.

It is common for local authority licensing bodies to stipulate compliance with BS 7671 as a license condition. This includes licensing, for example, places of entertainment, restaurants and public houses, public events, and houses of multiple occupancy in the private and social housing rental sectors. The local authorities enforce these license conditions as well as the legal requirements covering the safety of electrical installations in dwellings (see Chapter 20).

The HSE's Electrical Inspectors will generally use BS 7671 as a main guide when checking the safety of electrical distribution systems and compliance with the Electricity at Work Regulations. They will also take account of the references to BS 7671 in the Electricity Safety Quality and Continuity Regulations (see Chapter 7).

Many electrical contracting companies who install electrical distribution systems in conformity with BS 7671 are registered with the National Inspection Council for Electrical Installation Contracting (NICEIC), a United Kingdom Accreditation Service (UKAS) accredited organisation concerned with ensuring the safety of electrical installations. Some are also members of the industry's trade associations: the Electrical Contractors' Association (in England and Wales) and SELECT (in Scotland). All three organisations periodically inspect the work of their member companies against the standards laid out in BS 7671. The ultimate penalty for non-compliance is deregistration.

Guidance

The standard can be quite difficult to follow and to interpret, so it is helpful that the IET has produced a series of explanatory Guidance Notes to support it. The following Guidance Notes are published as separate documents:

Guidance Note 1: Selection and Erection
Guidance Note 2: Isolation and Switching
Guidance Note 3: Inspection and Testing
Guidance Note 4: Protection against Fire
Guidance Note 5: Protection against Electric Shock
Guidance Note 6: Protection against Overcurrent
Guidance Note 7: Special Locations

The On-site Guide

In addition to the IET's guidance material, there are a few textbooks devoted specifically to the standard and which are good reference material.

Part 1: Scope, object and fundamental principles

Chapter 11: Scope

The regulations in this chapter lay out the broad scope of the standard, as already explained previously. Regulation 114.1 advises that adherence to the standard may be used in a court of law to claim compliance with a statutory requirement. There is a list of the relevant statutory regulations in Appendix 2 of the standard.

Regulation 115.1 refers to licensing and other authorities, many of which require compliance with BS 7671 as a license condition, as already noted. Designers and erectors need to be aware of, and comply with, these license conditions where they apply.

Chapter 12: Object and effects

This chapter simply states that the standard aims to provide systems that are safe and which function properly. Regulation 120.3 advises that any departure from the standard must not compromise the safety of the installation and should be noted in the Electrical Installation Certificate.

Chapter 13: Fundamental principles

Section 131 – Protection for safety

Section 131.1 lists the injury mechanisms that the standard aims to prevent. In addition to the electrical injuries explained in Chapter 1 of this book, the standard covers under-voltages and overvoltages and electromagnetic disturbances likely to cause injury or damage; injury caused by mechanical movement of electrically powered equipment (impact, shearing, crushing etc injuries) and injuries caused by the interruption of power supplies and safety services.

The rest of Section 131.1 briefly describes the main protection requirements that are expanded in the rest of the standard:

- Protection against electric shock, both by direct contact and indirect contact. Direct contact requires 'basic protection' and indirect contact requires 'fault protection', both terms having been introduced in the 17th edition.
- Protection against thermal effects, particularly with regard to minimising the risk of ignition of flammable materials from high temperatures and arcing, and preventing people being burned from material that becomes excessively hot. In the latter context, it is generally accepted that 60°C is the maximum temperature that external surfaces should be able to reach without the risk of causing burn injuries.
- Protection against overcurrent – overload current and fault current – that may cause excessive temperatures and electromechanical stresses. Overcurrent may be caused by a system being overloaded, for example where the current drawn by the loads exceeds the current-carrying capacity of the cables, whereas fault current is caused by a fault such as a phase to phase or phase to earth short circuit.
- Protection against voltage disturbances, such as overvoltages caused by lightning or switching operations, and undervoltages including loss of supply and dips in the supply voltage caused by disturbances on a DNO's distribution network. In addition, the installation must be immune to electromagnetic emissions created by itself or the equipment connected to it.

Sections 132 and 133 give advice on the general parameters that must be taken into account when designing an installation and selecting equipment. This does not cover simply the electrical characteristics of the installation (voltage, frequency, nature of demand,

and so on) but also the very important factor of the environmental conditions in which the installation is to be used. For example, the highly corrosive atmosphere of an electroplating factory will demand very different design solutions when compared with the installation in a cinema where large numbers of members of the public congregate.

Section 134 addresses, among other things, the need for installations to be verified by using inspection and testing to ensure that the provisions of BS 7671 have been met.

Part 2: Definitions

Part 2 provides an extensive list of definitions of terms used throughout the standard. The fact that definitions now include electric vehicles, electric vehicle charging points, and photovoltaic sub-definitions is a good illustration of the way in which the standard has to evolve to keep up with emerging and developing technologies.

There is also a list of symbols and abbreviations used throughout the standard.

Part 3: Assessment of general characteristics

Part 3, which consists of six chapters, identifies the initial factors that the designer should consider when planning an installation to ensure its viability and safety.

It is usually essential to contact the local DNO early and advise the total load and possible future increases so that an adequate supply will be available when required. Discussions will also need to take place about the type of supply, be it three-phase or single phase, the number and configuration of conductors, the type of system earthing (TN-S, TN-C-S, TT etc – see Chapter 3), and any connected generation that may be being installed such as standby diesel generators, photovoltaics or wind turbines. The DNO or IDNO should provide information on the system voltage, including its frequency and harmonic content, the prospective short circuit current and external earth fault loop impedance, and the type and rating of the overcurrent protective fuse or circuit breaker at the origin of the installation.

Spaces for substations, standby facilities and switchrooms, and any cable duct and tray requirements should be discussed with the distributor and whoever is responsible for the site and buildings as soon as possible so that provision can be made in the plans.

Regulations 314.1 to 314.4 require the installation to be divided into circuits with a separate way for each final circuit in a distribution board or consumer unit, and for separate circuits where required by the regulation or to prevent danger. In this respect, the user should be consulted to identify the hazards. Where a lighting failure could cause danger, for example, there should be more than one lighting circuit, or emergency lighting may be appropriate. Although not mentioned in the regulation, it is advisable to provide spare circuits for future use.

The environmental conditions and other external influences, listed in Appendix 5 of the standard, should be considered to see if any apply and what should be done to counter them. For example, in rural areas, the electricity supply might be an overhead line in a lightning-prone zone where it may be prudent to install voltage surge suppressors (see new section 443) and perhaps fireproof the supply intake. In nurseries and schools, the installation should be childproof and electric heaters located out of reach. Other factors affecting the selection of equipment are detailed in Part 5.

Connected equipment with harmful characteristics is referred to in Chapter 33 – Compatibility. The characteristics include transient overvoltages which may emanate from the supply or from switching inductive loads, harmonic currents, d.c. feedback, high frequency oscillations, large starting currents and large fluctuating loads. Appropriate remedial action is required to prevent adverse effects particularly to sensitive electronic apparatus such as computers and programmable electronic safety-related systems. Where there are no practicable remedies to counter large starting currents and the fluctuating loads of arc furnaces, for example, the DNO has to be advised so that it can make suitable alterations to its network and to avoid the risk of being denied a supply or being disconnected under the provisions of the ESQC Regulations.

The maintainability requirement of Chapter 34 entails the selection of durable equipment which can be readily opened for servicing, and its location in an accessible position. This is a requirement frequently overlooked by architects who often have to shoehorn electrical equipment into modern space-limited buildings, adversely affecting the ability to maintain the equipment safely.

Chapter 35 highlights the need to consider the provision of safety services such as emergency lighting and firefighting equipment. The types of supplies to these services, such as batteries and generators, need to be planned and factored into the overall design.

Part 4: Protection for safety

Part 4 lists the methods of providing protection against direct and indirect electric shock; thermal effects (fire, burns and overheating); excess of current (overcurrent and fault current); overvoltage, undervoltage and EMC in Chapters 41, 42, 43 and 44 respectively.

In the following text, a direct shock is from contact with a live part which is intentionally live. An indirect shock is from contact with an exposed conductive part or an extraneous conductive part made live from a fault.

Chapter 41: Protection against electric shock

The protective measures in this chapter are founded on the principle that parts that are live at a hazardous voltage must not be accessible, so-called basic protection, and all conductive parts that are ordinarily accessible must not become live at a hazardous voltage either in normal use or under single fault conditions, so-called fault protection. The protective techniques described in the chapter are identical to those explained in Chapter 3 of this book and the requirements in the standard are summarised below.

Section 411 – Automatic disconnection of supply

Earthed equipotential bonding and automatic disconnection of supply (EEBADS) is the standard protection method in buildings in the UK. The relevant requirements are in Subsections 411–3 to 411–6 for installations with TN, TT and IT system earthing arrangements. As explained in Chapter 3 of this book, the method consists of the creation of an equipotential zone by bonding simultaneously accessible extraneous and exposed conductive parts directly to the main earthing terminal or indirectly via the protective conductors, thus minimising the potential difference between them in the event of an earth fault. This, combined with the use of protective devices such as fuses, circuit breakers and RCDs, ensures that fault current is interrupted within the prescribed time and that danger cannot occur for the duration of the fault.

The maximum disconnection times under earth fault conditions are matched to the risk and are set out in section 411.3.2, including Table 41.1; e.g. in TN systems it is 5 seconds for distribution circuits and, in final circuits not exceeding 32 A, 0.4 seconds for 230 V circuits and 0.2 seconds for 3-phase 400 V circuits. In the unusual event that these disconnection times cannot be achieved, supplementary equipotential bonding conductors must be installed so as to connect together exposed and extraneous conductive parts and circuit protective conductors, thereby minimising the potential for electric shock until the fault is cleared.

These regulations refer only to metallic parts and ignore the risk from contact with other extraneous conductive parts such as walls and floors made of, for example, brick, stone, quarry tiles or concrete which are conductive particularly when damp and which are often in ground contact and may therefore remain at earth potential. Metal reinforced concrete floors could perhaps be made part of the equipotential zone by a bonding connection to the reinforcement, which should be effective if the metal mesh is electrically continuous. Where this is impracticable and there is a substantial shock risk, an insulating floor covering could be provided.

The disconnection time for earth fault current in any circuit is determined by the circuit voltage, the earth fault loop impedance, Z_s, and the time-current characteristic of the protective device in the circuit, be it a fuse, a circuit breaker, or an RCD. The maximum values of Z_s in TN circuits with nominal voltages of 230 V and a required disconnection time of 0.4 seconds are set out in Table 41.2 for circuits protected by fuses, and in Table 41.3 for circuits protected by circuit breakers or RCBOs. Table 41.3 also covers the Z_s values for 5 second disconnection time in distribution circuits protected by circuit breakers and RCBOs; Table 41.4 provides similar information for distribution circuits protected by fuses.

Where the disconnection times cannot be achieved using a fuse or circuit breaker, most commonly because the external loop impedance is too high, a non-delayed RCD may be used so long as the Z_s values in Table 41.5 are satisfied; e.g. a maximum value of 1667 Ω for an RCD with a rated residual operating current of 30 mA. However, the RCD must be additional to a fuse or circuit breaker provided for overload protection.

RCDs must also be installed to provide additional protection for socket-outlets with a rated current not exceeding 20 A – that is to say, all the socket-outlets in dwellings and places such as offices and shops. The only exceptions to this are

(i) a socket-outlet supplying a particular item of equipment such as a freezer, where economic loss may occur as a result of nuisance tripping, but only if the socket-outlet is labelled to indicate that it is not protected by an RCD; and
(ii) socket-outlets in workplaces where a documented risk assessment concludes that RCD protection is not necessary for risk reduction purposes.

Although the electricity distribution companies have converted a substantial number of TT systems to TN-C-S systems, some TT systems are still in use, mostly in rural areas where it is often difficult or impracticable to obtain and maintain the low earth fault loop impedance values specified in Tables 41.2, 41.3 and 41.4. In these cases, the overcurrent protection should be supplemented by RCD protection as required in Section 411.5.

The particular requirements for IT systems, which are not commonly used in fixed installations, are covered in Section 411.6. Perhaps the most important provision is that the commonly used generator supplies with IT earthing, such as the small unreferenced generators widely used to supply mobile food outlets, should be protected by an RCD that may operate in the event of the first earth fault.

Sections 411.7 and 411.8 address the automatic disconnection requirements for Functional Extra-Low Voltage (FELV) and reduced low voltage, respectively; these types of systems were described in Chapter 3 of this book. In reduced low voltage systems, for fault protection there must be a fuse or circuit breaker in each phase conductor, or an RCD, and the value of Z_s has to be such that a disconnection time of 5 seconds is achieved. Table 41.6 provides appropriate values of Z_s for fuses, circuit breakers and RCBOs. Protection against direct shock is by insulation, barriers or enclosures.

Section 412 – Double or reinforced insulation

As described in Chapter 3 of this book, the protective measure of double or reinforced insulation is commonly used for protection against both indirect and direct contact electric shock. The particular requirements relating to electrical equipment, enclosures, installations and wiring system are covered in this section.

Although wiring systems that have protection against mechanical damage are considered to have the same level of protection as equipment with double or reinforced insulation, Note 2 to Regulation 412.2.4.1 advises that they do not require to have the ⧈ symbol applied.

Section 413 – Electrical separation

The requirements for electrical separation, in which a circuit is electrically separate from earth and other circuits, are in Section 413. The requirements relate to single items of equipment supplied from an unearthed source such as a transformer. If more than one item of equipment is being supplied, the provisions of Regulation 418.3 come into play; these include ensuring that exposed conductive parts of the separated circuits, such as metal enclosures, are connected together by insulated bonding conductors. However, the bonding conductors must not be connected to earth, any circuit protective conductors or exposed conductive parts of any non-separated circuits or to any extraneous conductive parts such as water or gas pipes.

Section 414 – Extra-low voltage provided by SELV or PELV

The intention of SELV is to minimise the shock hazard by voltage limitation. This section sets out the parameters, which include a safe source of supply such as a safety transformer to BS EN 61558–2–6 or BS EN 61558–2–8 and other precautions to avoid the SELV circuit becoming live at a higher voltage. If the SELV does not exceed 25 Va.c. or 60 V ripple-free d.c., i.e. not more than 10% ripple, and if the equipment is not immersed, the direct contact shock risk is regarded as negligible and exposed live parts are allowed except in locations of enhanced shock risk, such as most of those in Part 7 where the conductors have to be protected against direct contact by a barrier, enclosure or insulation. The SELV circuit is not earthed, and the cables are not metal-sheathed. As conductive parts of the installation are not deliberately or fortuitously earthed, it will usually be more practicable to use insulated rather than metallic conduit and ducts.

Where it is not possible to meet all the requirements for SELV, but the voltage does not exceed extra-low voltage, the protection is called protective extra-low voltage (PELV). The difference between an SELV and PELV system is that the latter is connected to earth but in other respects meets the SELV requirements. Other extra-low voltage systems are classed as FELV. The socket-outlets in SELV, PELV and FELV systems have to be different from those used for higher voltages so that the plugs used to connect the apparatus cannot be inserted into a higher voltage socket-outlet.

Section 416 – Provisions for basic protection

This section includes requirements for the prevention of direct contact with live parts by the use of basic insulation or barriers and enclosures. The intention is to prevent any part of the body coming into contact with live parts at dangerous potentials.

Regulation 416.2.4 refers to situations where barriers have to be removed or enclosures opened, perhaps for maintenance work, and live parts may be exposed to touch. The requirement is that access may be gained only by using a key or tool, or that supplies may be reenergised only after the barrier or enclosure has been replaced or reclosed, implying the use of an interlocking arrangement.

Section 417 – Obstacles and placing out of reach

This protective measure, which does not apply to DNO apparatus such as overhead lines, applies to installations that are controlled or supervised by skilled persons. A skilled person is defined as *a person who possesses adequate education, training and practical skills in relation to the electrical work being undertaken, and who is able to perceive risks and avoid hazards which electricity can create.* In those terms, it is the same as a competent person as envisaged by Regulation 16 of the Electricity at Work Regulations. However, it is not obvious how a skilled person is able to exercise adequate control or supervision in all circumstances all of the time to prevent unskilled persons touching bare live conductors. The important preventive measure is simply to ensure by physical measures, such as guards and adequate distances, that any such conductors cannot be reached.

The term 'placing out of reach' means not within arm's reach, which is meant to be explained in a less-than-clear illustration in Figure 417. However, it also needs to include a dimension that takes account of any conductive objects that may be carried or handled in the area, such as ladders and poles.

Section 418 – Protective measures for application only where the installation is controlled or under the supervision of skilled or instructed persons

This section describes three protective measures that are not for general application and which should only be implemented under the direction of a skilled person, as previously defined. The measures are

(i) non-conducting location such as an electrical test area with insulated floors and walls and where a person cannot simultaneously touch two exposed conductive parts or an exposed conductive part and an extraneous conductive part;

(ii) locations with earth-free equipotential bonding; and
(iii) systems using electrical separation but where more than one item of equipment is being supplied.

Chapter 42: Protection against thermal effects

This chapter deals with the thermal effects associated with fire, burns and overheating. Its scope includes the effects of thermal radiation; the ignition, combustion or degradation of materials; flames and smoke propagating from an electrical fire; and the potential for safety services such as fire and burglar alarms to be cut off by the failure of electrical equipment. The treatment is quite brief because if the regulations in the other chapters are observed, these risks should be minimal.

Regulation 421.1.2 requires fixed equipment which focuses or concentrates heat to be sufficiently far away from any other fixed object to avoid damage to it, or to be mounted, supported or screened in such a way as to prevent thermal damage. This would apply to recessed downlighters, radiant electric fires and fan heaters, for example. Regulation 421.1.3 requires enclosure materials of equipment to be able to withstand the heat generated by the equipment.

The new Regulation 421.1.201 introduced in Amendment 3 requires, with effect from 1 January 2016, consumer units in dwellings to be made from non-combustible material or to be mounted inside a non-combustible cabinet or enclosure. This forces a move away from the plastic-encased consumer units that have dominated the sector for the past 20 years or so and towards the use of metal-clad consumer units. This is not without risk because metal-clad units come with their own problems, including the fire risk from cables damaged by unprotected sharp-edged cut-out holes and the shock risk from units that have not been properly earthed. It can only be hoped that the standards makers have got their assessment of the balance of risks between the two types of materials correct.

Regulation 421.1.5 is applicable to oil-filled switchgear, controlgear and transformers where precautions such as a retention pit are required to contain an oil leakage or fire.

Section 422 contains additional requirements for locations with specific fire risks. These include conditions for emergency evacuation; and risks due to the nature of stored or processed materials, combustible construction materials, fire propagating structures, and locations of national, commercial, industrial or public significance where improved fire protection would be justified.

Regulation 422.2 deals with wiring systems in escape routes and the need to ensure that they do not create a hazard in the event of a fire. There have been serious incidents in which fire-damaged cables have dropped down from roof spaces and blocked escape routes, with firefighters becoming entangled in the dropped cables and trunking – the installation standards aim to prevent this happening.

Regulation 422.3.2 sets an upper limit of 90°C on equipment surface temperature under normal operating conditions to prevent ignition of accumulated dusts or fibres. Regulation 422.3.9 requires RCD protection not exceeding 300 mA of wiring systems in high fire risk premises with TN and TT supplies, other than where mineral-insulated cables or busbar trunking or powertrack systems are used. The RCD's rated residual operating current should be reduced to 30 mA if a resistive fault that may cause a fire is foreseeable.

Chapter 43: Protection against overcurrent

The intention is to use automatic interruption of the supply to protect live conductors of the installation against overheating from overload and fault currents and against mechanical damage from electromagnetic stress. This entails the provision of fuses or circuit breakers fitted with overload trips and adequate supports for the conductors. For TN and TT systems, these need to be placed in the phase, or line, conductors but will need to be installed in neutral conductors only if they have a smaller cross-sectional area, and hence current-carrying capacity, than the line conductors, which is not normally the case. In IT systems, any neutral conductor should have overcurrent protection installed.

The protection is not intended for connected equipment and its wiring, which should generally have its own protection, e.g. the fuse in a BS 1363 plug. The devices must operate safely and may be for overload or short circuit protection or both. Devices used for both overload and fault current protection must be coordinated. Circuit breakers used for overload and short circuit protection or short circuit protection only must be capable of making on to a short circuit as well as of interrupting it.

Section 433 – Overload protection

Section 433 is for overload protection only. It requires the characteristic of the protective device to match the current rating of the conductors so that it will operate on overcurrent before the safe temperature limit of the conductors or their insulation is exceeded. This is characterised by the fundamental relationship:

$$I_b \leq I_n \leq I_z$$

where

I_b is the design current of the circuit, i.e. the expected normal load current.
I_n is the nominal rated current of the protective device (fuse or circuit breaker).
I_z is the current-carrying capacity of the conductor in the installation's conditions, determined by its material, cross-sectional area, various rating factors covering features such as grouping and ambient temperature. This is the current a cable can carry before becoming hot enough to damage its insulation, commonly either 70°C or 90°C depending on the insulation.

In addition, the current (I_2) that causes effective operation of the protective device should not exceed 1.45 times the lowest of the current-carrying capacities (I_z) of the conductors in the circuit; this changes to 0.725 times I_z where semi-enclosed fuses to BS 3036 are used. This means that the conductors may carry small overloads, but it is anticipated that these will rarely cause any problems. There should be a sufficient number of socket-outlets to discourage the use of adapters which could result in overloading the socket-outlet receptacle contacts, and ring main socket-outlet spur circuits should be fused.

An exception to this general rule is the British design of ring main circuits using 2.5 mm^2 2-core and earth PVC-insulated cables protected by 30 A or 32 A fuses or circuit breakers; this configuration is in widespread use in the UK. These cables may have a current-carrying capacity of less than 30 A or 32 A, but the ring configuration coupled with the low probability of the cables' current-carrying capacity being exceeded for extended periods of time mean that this design configuration is acceptable.

Regulation 433.2.1 requires a device to be placed at any point where a reduction occurs in the current-carrying capacity of the conductors. This often occurs where the conductor size is reduced, but it can be due to a variety of other causes which affect the conductor current rating, e.g. wiring on boilers or other locations of high ambient temperatures. It is permissible to position the device along the conductor run instead of at the current rating reduction point provided there are no branch circuits or outlets for the connection of current-using equipment between them and the installation is not in an abnormal fire or explosion risk location.

Regulation 433.3 details the cases where the omission of overload protection is allowed and, in some circumstances, should not be provided because of the possible danger caused by its operation, such as generator exciter circuits and secondary circuits of current transformers.

Section 434 – Fault current protection

Section 434 covers protection against fault current, which can be current arising from a negligible impedance fault between live conductors or between live conductors and earth. Rapid disconnection of circuits subject to fault current is essential, although there are similar provisions in Regulation 434.3 for the omission of fault current protection as described previously for overload protection. In circumstances where separate devices are used for overload protection and fault current protection, their characteristics must be coordinated, as described in Section 435.

The prospective short circuit currents have to be determined at all relevant locations in the installation by calculation or measurement and protective devices selected to protect all conductors against thermal and mechanical effects. For new installations, the designer will first have to ascertain the loop impedances and characteristics of the excess current protection at the intake from the DNO, except where private generation is proposed. In large installations with their own substation, no problem should arise, but where the supply is to be taken from the DNO's low voltage network, it may be possible to obtain only anticipated maximum and minimum loop impedance figures.

To find the loop impedance at any point, the designer has to calculate the value from the intake to that point and then add the external loop impedance; this can then be used to calculate the prospective short circuit current. For small wiring, up to 35 mm^2, the inductance may be ignored, so the loop impedance is the external impedance plus the resistance of the internal wiring, but for larger sized conductors, the cable impedances, obtainable from the makers, should be used.

A check of the characteristics of the protective devices that are in series in the circuit will indicate whether or not they are suitable for clearing the fault. For example, a motor starter may be capable of breaking the overload current of the motor but not of clearing a short circuit in the wiring between it and the motor. In this case, protection by suitable HBC (high breaking capacity) fuses may be provided so that the energy let-through of the fuses and starter will be limited to protect the starter from serious damage; see Regulation 435.2 and British Standard BS EN 60947–4–1:2010+A1:2012 *Low-voltage switchgear and controlgear. Contactors and motor-starters. Electromechanical contactors and motor-starters*, which is the standard for LV a.c. motor starters. The standard requires the back-up protective device to operate only on a short circuit and to be coordinated with the starter, so that on the occurrence of a short circuit fault on the load side of the starter it is cleared either without, or with limited, damage to the starter. This is the process of coordinating devices that are separately provided for overload protection and fault current protection.

Regulation 434.4, supported by Appendix 10, deals with fault current protection of conductors connected in parallel. It allows for a single fuse or circuit breaker to be used to protect the cables so long as it will provide efficient protection for a single fault at the most onerous position in one of the parallel conductors, taking account of the way in which fault current will be shared and distributed. Where this cannot be guaranteed, the available options are to reduce the risk of insulation failures by mechanical protection and other means; to provide fault current protection in each of the conductors, at the supply end when two conductors are paralleled and at both the supply and load ends when three or more conductors are paralleled.

The rated short circuit breaking capacity of the fuse or circuit breaker providing fault current protection must be equal to or greater than the maximum prospective fault current at the position at which it is installed, and any fault occurring on the circuit must be cleared before the fault current can thermally damage any conductors or cables. Thermal effects are related to the conductor's impedance and the magnitude of the current. The calculated impedance of the wiring is usually based on the full load operating temperature of the conductor, i.e. the specified ambient temperature plus the temperature rise due to the I^2R loss. The conductor resistance is proportional to its temperature, so a fault current exceeding the full load current will increase the resistance and thus the impedance and diminish the fault current.

BS 7454:1991+A1:2008 *Method for calculation of thermally permissible short circuit current taking into account non-adiabatic heating effects* indicates how to calculate the fault current. A less accurate method is shown in Regulation 434.5.2. It is not normally necessary to do this calculation where a fault current causes rapid operation of the protection in, say, not more than 0.1 second, but where there is a significant time delay the calculation should be done.

To be satisfactory, the energy withstand of the cable k^2S^2 must not be less than the energy let-through of the protective device I^2t, where k is a material-dependent factor and S is the cross-sectional area of the conductor in mm^2. If it is not, the conductor size should be increased, a heat-resisting cable used or the proposed protective device changed.

Regulation 434.3 allows for fault current protection to be omitted in certain circumstances.

Chapter 44: Protection against voltage disturbances and electromagnetic disturbances

This chapter deals with four specific issues associated with variations in voltage:

Section 442: Protection of low voltage installations against faults in the high voltage system.
Section 443: Protection against overvoltages of atmospheric origin or due to switching.
Section 444: Measures against electromagnetic disturbances.
Section 445: Protection against undervoltage.

For most designers and installers where a low voltage supply is taken from a DNO or IDNO, the provisions of Section 442 can be ignored. The requirements concern consumers who take supplies at high voltage and install a substation on their premises.

Section 443 is mainly concerned with the effects of lightning. The main provisions have already been described, but essentially, they require that where thunderstorm activity is expected to exceed 25 days per year, additional protection must be provided against high voltage lightning transients. In most places in the UK, this level of thunderstorm activity does not occur, so the surge protection is not needed – it can therefore be omitted, but only if the equipment connected in the installation has impulse withstand characteristics set out in Table 44.3. Most equipment designed and built to harmonised standards will achieve the specified levels of withstand voltage.

Section 444 provides design guidelines on methods of reducing electromagnetic emissions from low voltage installations and their associated equipment such as lighting, motors, transformers, switchgear and power electronic devices such as inverters and variable speed drives. Following the guidance will provide a presumption of conformity with the fixed installation requirements of the Electromagnetic Compatibility Regulations 2006, as described in Chapter 8 of this book.

Section 445 requires protection from any danger which may arise from a reduction in the supply voltage, either partial or total, such as when there is a supply failure or dip in the supply voltage. This usually means that motors must not restart automatically when the voltage is restored if this would entail danger. Motor starters should, therefore, have automatic no-volt protection which operates on low or no volts by disconnecting the supply from the motor. If damage, rather than danger, could arise from voltage reduction, the designer has the choice of tolerating it or avoiding it by, for example, the provision of standby no-break facilities.

Part 5: Selection and erection of equipment

Part 5 deals with the selection and erection of equipment. The intention is to ensure fitness for purpose. All apparatus has to comply with the relevant British Standard or comparable foreign standard that provides an equivalent level of safety. If there are no standards, the specifier should satisfy himself that the equipment meets the other requirements of the Sections, including the compatibility requirements of Regulation 512.1.5.

The erection requirement includes the location of equipment in positions where it is readily accessible for maintenance and safe from mechanical damage or adequately protected against it. Dissimilar metals should not be placed in contact with each other where this might cause electrolytic corrosion and a failure of earth continuity. The owners of a new installation would be well advised to insist that the contractor provides the labelling, wiring diagrams, etc, specified in Section 514, before taking over the installation, as this data is needed for safe usage and to facilitate subsequent maintenance.

Chapter 52: Selection and erection of wiring systems

Chapter 52 deals with cables, conductors and wiring materials including conduit and trunking. Regulation 521.5 forbids the use of single-core steel wire or tape armoured cables in a.c. circuits and the ferrous enclosure of a single a.c. conductor because the eddy currents produced from the electromagnetic field surrounding the conductor can cause overheating.

Regulation 521.9 allows flexible cables to be used for fixed wiring so long as it satisfies other parts of the standard. Flexible cable can be used to supply stationary equipment, and must be used to supply movable equipment. It is also convenient to use short flexible cable tails and a plug and socket to connect fixed electrically powered apparatus which may have to be replaced from time to time by a non-electrically qualified person. The small, motor-driven beer pumps in hotel cellars, which are usually replaced on failure by a serviceman who is not an electrician, are an example. The serviceman is competent to disconnect and connect by withdrawing or inserting a plug but not to open terminal boxes and interfere with the wiring therein. Where temporary fixed wiring is installed and then recovered for use elsewhere, as on construction sites and outside broadcast lighting installations, flexible cables are preferable, as they are better able to withstand the flexing from handling during installation and dismantling.

Where non-sheathed cable is used, such as single-core flexible cable, it must be provided with mechanical protection by being installed in conduit, ducting or trunking.

Regulation 521.11.201 is a British national provision that requires cables and cable management systems on escape routes to be supported such that they will not collapse prematurely in the event of a fire and block the escape route or create an entanglement hazard. Plastic cable ties, clips, conduit and trunking, for example, should not be used as the sole means of supporting cables in those locations because they will melt in a fire and potentially allow the cable to fall into the escape route.

Section 522 – External influences

Section 522 refers to the environmental conditions classified in Appendix 5 which have to be considered when selecting a wiring system. For installations in new premises, the designer should ascertain the likely environmental conditions and locate the wiring, if possible, where it will not be adversely affected, or otherwise match the type of wiring to the adverse conditions expected. For example, it is better to avoid the high ambient temperatures in the vicinity of radiators rather than having to increase the conductor size and/or use a heat-resisting cable insulation. Hot spots in appliances and luminaires may require the addition of heat-resisting insulation over the cable cores to ensure safety should the original insulation fail. Very low ambient temperatures adversely affect some insulating materials. The PVC insulation and sheaths of flexible cables and cords used to supply portable apparatus out of doors in cold weather harden and are liable to crack if flexed so should be replaced by insulating and sheathing materials able to tolerate low temperatures.

Protection is also required against wet and/or corrosive conditions, dirt and pollutants. Regulation 522.5 warns against electrolytic action of dissimilar metals in contact where dampness accelerates the corrosion process leads to hazardous degradation. An example is the degrading of PVC cable insulation when in contact with certain types of plastic loft insulation materials. Wiring accessories such as connection boxes may need sealing to exclude fine dust, and where wiring is subject to accumulations of dirt and fluff, the thermal rating of the wiring may be impaired. It will certainly be affected where it is in contact with the thermal insulating materials used in cavity walls and roof spaces. See Regulation 523.9 and Table 52.2 for the derating factors.

It is not always appreciated that some cable sheathing materials are adversely affected by the ultraviolet radiation in sunlight and, to counter this effect, a black compound is sometimes added. Unfortunately, this increases the absorption of infrared radiation, and this solar heating can increase the cable temperature considerably, so precautions need to be taken in accordance with Regulation 522.11. Outdoor cables should either be screened from sunlight or derated to allow for an increase in ambient temperature of about 20°C.

Regulations 522.6 to 522.8 refer to mechanical protection against impact, vibration, abrasion, sagging and strain. Regulations 522.6.201 to 522.6.204 contain British national requirements relating to the routing of cables under floors, above ceilings and in walls, especially in walls where they are at a depth of less than 50 mm.

Cables concealed in walls at a depth of fewer than 50 mm must be routed inside safe zones, as illustrated in Figure 12.1.

Cables routed within the safe zones which do not have some form of earthed metallic protection against the penetration of nails or screws, must be protected by an RCD with a rated residual operating current of 30 mA. Permissible forms of protection are listed in Regulation 522.6.204 and include earthed metallic conduit, ducting, capping and trunking, or cables that incorporate earthed metallic screening, or cables that form part of an SELV or PELV circuit. Where a cable is not inside a safe zone it should be provided with earthed metallic protection as per Regulation 522.6.204.

Cables concealed in walls constructed with metallic parts, such as metal partition walls, must have both 30 mA RCD protection and earthed metallic protection regardless of the depth to which they are buried in the wall. Cables in such walls at a depth of fewer than 50 mm should also be routed within the safe zones. This is in response to a fatal accident that occurred when a cable was buried inside a metal partition that became live under a fault condition.

For cables buried in the ground (which are frequently struck by careless excavators) Regulation 522.8.10 requires those that are not run in a duct or conduit to have earthed metal armouring or sheathing or both. The cables must be covered with a cable marking tape or be placed underneath cable tiles – a requirement that is frequently overlooked. The cables must also be buried at a sufficient depth, usually taken to be at least 400 mm for low voltage cables and 600 mm for high voltage cables.

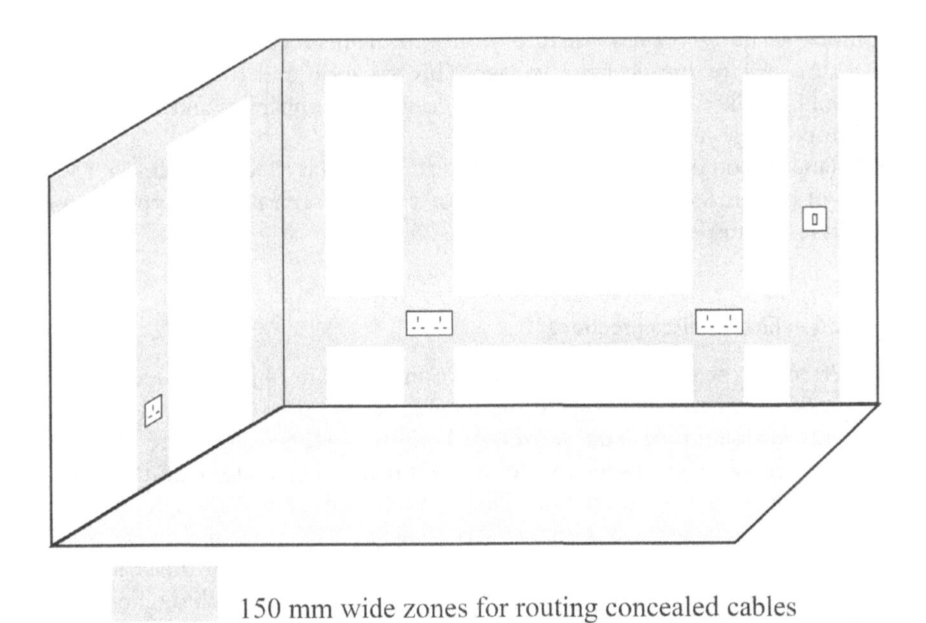

Figure 12.1 Safe zones for cables in walls

Section 523–525 – Current-carrying capacity of cables

Section 523 deals with the factors that affect the temperature rise of current-carrying conductors. Table 52.1 identifies the maximum operating temperature limits for cables with different forms of insulation, an example being 70°C for thermoplastic insulation. The cables must not be loaded to the extent that these temperatures are exceeded for sustained periods of time.

This overriding requirement can be met by using the tables in Appendix 4 to select cable sizes according to the load current and taking into account the ambient temperature, group rating factors, and the extent to which the cable is embedded in, or covered by, thermal insulation. Table 52.2 gives the derating factors for cable lengths of less than 0.5 m when totally surrounded by thermal insulation, and Table 52.3 gives the minimum conductor sizes for various applications. The calculation method is dealt with in the Appendix.

A designer will select a cable type to suit the environmental conditions. The load current and type and rating of the protective device should then be determined, then the current rating of the protective device is divided by the applicable correction factors. The factors are for ambient temperature (Table 4B1 or 4B2), grouping (Table 4C1 to 4C6), thermal insulation and, if semi-enclosed fuses are to be used, a factor of 0.725 is applied.

The next step is to ascertain the conductor size for this current from Tables 4D1A to 4J4A under the appropriate installation method column – the various installation methods are illustrated in Table 4A2. Next, the volt drop should be determined from Tables 4D1B to 4J4B to ensure the conductor size is adequate for a volt drop from the supply terminals to the equipment not exceeding 4% of the declared supply voltage or the lower limit specified in the British Standard for the apparatus. Greater volt drops are permissible during motor starting. Where a more accurate volt drop calculation is necessary to take account of operating temperature and power factor, the formulae are given in Section 6 of Appendix 4. The DNO has a statutory duty to maintain the voltage at the intake to within +10%/-6% of the declared voltage. This variation and the volt drop from the intake should be taken into account for voltage-sensitive apparatus and the provision of an automatic voltage regulator considered.

Finally (and if thought necessary) the cable size should be checked to ensure its insulation will not be damaged on the occurrence of a short circuit fault by applying the formula given in Section 434 for fault protection.

Section 526 – Electrical connections

Section 526 covers the suitability of electrical connections. Its object is to require all joints and terminations to be properly made and secured so that they can safely carry load and fault currents, withstand mechanical stress or be in fire-resistant enclosures. They should preferably be accessible for inspection, testing and maintenance, although there are some obvious exceptions, such as compound-filled and soldered joints. Although not a requirement, it is good practice to avoid unnecessary joints in cable runs, and where conductor sizes have been increased to cater for high ambient temperatures or thermal insulation, consideration should be given to using the larger size conductor throughout to avoid joints on either side of the hot spot.

Regulation 526.9 addresses the jointing of cables with multi-wire, fine wire or very fine wire conductors. It prohibits the tinning of the ends of such wires if they are to be used in screw terminals because of the problem of solder flow and the consequential potential for the terminal becoming loose.

Section 527 – Minimising the fire risk

The requirements in Section 527 are compatible with the building regulations throughout Great Britain and specify that the wiring system and its method of installation will not materially add to the fire risk, so the materials used should not readily ignite or propagate a flame, and joints and terminations should be in non-ignitable enclosures as already mentioned in 'Electrical connections' cited previously. Wherever wiring passes through the structure (floors, ceilings, walls, roofs and barriers), non-ignitable sealing is required to prevent the spread of fire. Internal sealing of ducts and conduits is also necessary. The sections do not extend to flammable and explosive locations, where additional fire safety measures are necessary as described in Chapter 15 of this book on flammable atmospheres.

Section 528 – Proximity to other services

Section 528 is largely concerned with the potential for interference between adjacent electrical services and between electrical services and telecommunication cables.

For electrical services, the general principle is that Band I and Band II circuits must be segregated from each other and from high voltage circuits. Band I effectively means extra-low voltage (ELV), including telecommunications, signalling, bell, control and alarm circuits. Band II covers all voltages used in low voltage electrical installations not included in Band I. The requirement for segregation does not apply if every cable is insulated for the highest voltage present, or each conductor in a multicore cable is insulated for the highest voltage present, unless conductors of the two bands are separated by an earthed metal screen, they are installed in separate compartments of a trunking or ducting system, they are installed on a tray with a partition providing separation, or a separate conduit or ducting system is provided for each band.

Note, however, that even though the cables may be installed next to each other, assuming they meet these criteria, there is still a need to ensure electromagnetic compatibility. Also, special requirements for the separation and segregation in relation to safety services are given in BS 5266–1:2011 *Emergency lighting. Code of practice for the emergency escape lighting of premises* and BS 5839–1:2013 *Fire detection and fire alarm systems for buildings. Code of practice for design, installation, commissioning and maintenance of systems in non-domestic premises.*

Regulation 528.2 requires that underground power and telecommunications cables are separated by at least 100 mm. Subsection 528.3 is concerned with non-electrical services which produce adverse environmental conditions. Cables that are not part of a lift installation are not allowed in lift shafts.

Section 529 – Maintainability and cleaning

The last section, Section 529, of Chapter 52 requires the wiring system to be selected and erected to facilitate cleaning and maintenance, so easy and safe access is required. This is particularly relevant to overhead systems for lighting, cranes and busbars. It must also be possible to dismantle and reassemble wiring accessories and the covers over wiring terminations without detriment to the original protection, so damaged cover gaskets and missing screws should be replaced, and protective conductors and neutral links disconnected for testing must be reconnected. The bridging conductors sometimes used to short out RCDs for earth loop impedance testing must be removed before restoring the circuit to service.

Chapter 53: Switchgear

Chapter 53 covers switchgear for protection, isolation, switching, control and monitoring. Many of the provisions repeat the content of Part 4, so the following text highlights content worthy of note.

Residual current devices

It is perhaps instructive at this point to identify those circuits in which the standard recommends that RCD protection is installed, as follows:

- all socket-outlet circuits rated at up to 20 A and those rated up to 32 A supplying mobile equipment for use outdoors;
- all circuits within a bathroom;
- concealed cables that are fewer than 50 mm deep or are not run in safe zones, or are not protected by earthed metal such as conduit and trunking;
- cables concealed in metal partition walls;
- circuits in which the required disconnection times cannot be achieved using overcurrent devices such as fuses and circuit breakers; and
- special locations such as swimming pools, saunas and caravan parks.

Section 531.2 describes in general terms the use of RCDs with respect to fault protection and automatic disconnection.

Fuses

A common misconception is that rewirable fuses are not permitted in low voltage installations. Regulation 533.1.1 confirms that, whereas cartridge fuses are preferred, rewirable fuses to BS 3036 are acceptable. Table 53.1 lists the nominal diameter of the fuse wire for a range of rated currents.

Protection against overvoltage – surge protective devices

Surge protective devices (SPDs) are not generally required in the UK because, as noted in the discussion under Section 443, the level of lightning activity in many places does not justify it. Nonetheless, Section 534 offers designers who choose to install them advice on how the devices should be selected and connected.

Isolation and switching

Section 537 covers protection by non-automatic isolation and switching. There are requirements relating to the following switching functions: (i) isolating the whole or parts of the installation or equipment when not in use, usually to enable work to be carried out on it safely; (ii) switching off for mechanical maintenance on electrically powered apparatus; (iii) emergency switching off to remove a hazard; (iv) functional switching to control parts of an installation; (v) motor control; and (vi) firefighter switching operations. The types of devices that can be used to satisfy these requirements, and their associated standards and functionality are described in Table 53.4.

For the first requirement, all live conductors have to be isolated. This does not include the neutral conductor in TN-S and TN-C-S systems, which is regarded as being connected to earth.

The standard defines isolation as "a function intended to cut off for reasons of safety the supply from all, or a discrete section, of the installation by separating the installation or section from every source of electrical energy". This contrasts with Regulation 12 of the Electricity at Work Regulations 1989, which defines isolation as "the disconnection and separation of the electrical equipment from every source of electrical energy in such a way that this disconnection and separation is secure". From a legal perspective, the reference to the isolation needing to be 'secure' should not be ignored; it is of fundamental importance and it is a pity that the BS 7671 definition does not include it in its definition.

Regulation 537.2.1.5 stipulates that, where the means of isolation is remote from the equipment being isolated, means must be provided to secure the isolation. This is to prevent inadvertent reenergisation and is most commonly achieved using a padlock or other locking device to secure the disconnector or switch in the open position and attaching a caution notice, as described in Chapters 3 and 6 of this book.

The isolation device has to be capable of carrying the circuit load and fault currents but is not required to interrupt them when operated. Therefore, isolation may be effected, off load, by the removal of links, fuse links and plugs from socket-outlets as well as by opening disconnectors, switches and circuit breakers.

Devices used for isolation must be clearly marked or labelled to indicate the circuits or equipment they isolate – this is also a requirement of Regulation 12 of the Electricity at Work Regulations 1989.

For the second requirement covering switching off for mechanical maintenance, uninstructed persons are likely to effect the switching and may perhaps operate the switch when the apparatus is on load, so switches and circuit breakers provided for mechanical maintenance purposes have to be of the load breaking type. Regulation 537.3.2.2 requires an externally visible contact gap or a reliably indicated off or open position; as the switching may be effected by non-electrical personnel, the alternative is preferable. Regulation 537.3.2.3 requires the switching device to be designed or installed so as to allow prevention of inadvertent or unintentional switching on; this means that the device should be lockable in the open position to prevent, for example, the type of accident which can occur on process plant which is being worked on by maintenance staff when a production worker, unaware of their activities, goes to start up a plant machine.

For the third requirement for emergency switching off, covered in Regulation 537.4.2, the switch, contactor or circuit breaker must be capable of interrupting any load or overload current including, for example, the stalled current of an induction motor. Circuit breakers may be manually operated or opened by means of stop push button(s) or switch(es). Emergency switching devices should be clearly identified and preferably be coloured red. Not only should emergency stop push buttons be provided within reach of operatives at potentially hazardous positions, but also consideration should be given to the provision of others remote from these positions and available to other people to operate, should they see an operator in difficulties or other cause for emergency stopping. Resetting an emergency switching device should not by itself lead to reenergisation of the part of the installation that has been switched off.

Regulation 537.4.2.8 prohibits the use of a plug and socket-outlet for emergency switching. This is because the emergency might be an electrical fault entailing an excess current which could not be safely interrupted by plug withdrawal. Uninstructed persons, however, are unlikely to be aware of this, and it may be prudent to install switched socket-outlets with interlocked plugs in appropriate locations.

The sixth requirement covering firefighters' switches is covered in Regulation 537.6. Such switches are required where there are low voltage circuits supplying external high voltage installations and interior high voltage discharge lighting. Exterior installations include covered markets, arcades and shopping malls. Regulations 537.6.3 and 537.6.4 provide detailed requirements for the location and identification and marking of the switches.

Monitoring

Section 538 describes methods by which the insulation properties and leakage current in installations can be monitored automatically and continuously. The principle is that the monitoring devices set off alarms when insulation or leakage current values reach preset alarm levels.

Chapter 54: Earthing arrangements and protective conductors

Chapter 54, covering earthing arrangements and protective conductors, should be considered together with Chapter 41 with respect to prevention of indirect electric shock. Sections 541 and 542 are concerned with methods of earthing the consumer's protective conductors and are dependent on the type of supply system earthing that is illustrated and described in Chapter 3 of this book.

Section 542 – Effective earth connections

Regulation 542.1.3.1 means that the connections to earth must be effective and the earth loop impedance low enough for the protective devices to operate within the prescribed time in the event of an earth fault.

Section 542.2 deals with earth electrodes typically used in TT installations, for which earth rods, tapes and plates; underground structural metalwork; concrete reinforcing rods; and lead sheaths of cables may be used. Water and gas pipes should not be used as earth electrodes. Precautions must be taken against deterioration caused by excessive corrosion of the electrode. In particular, it is important to avoid corrosion of the electrode terminal; the joint should not employ metals in contact which are widely separated in the electrochemical series, and moisture and damp air should be excluded from the joint by taping or painting.

Corrosion can be caused by electrolysis such as that which occurs in d.c. supply systems where earth leakage can cause corrosion from electrolysis at the positions where the current enters or leaves metallic parts. This problem occurs in d.c. traction systems and certain types of electrochemical plants and is often associated with cathodic protection systems which impress d.c. currents in the ground.

Section 543 – Protective conductors

Section 543 deals with the types, sizes and preservation of electrical continuity of protective conductors. Regulation 543.1.1 requires a mechanically protected protective conductor that is not part of a cable, or is not formed by conduit, ducting or trunking to have a cross-sectional area of not fewer than 2.5 mm^2. If the conductor is not mechanically protected, its cross-sectional area must be at least 4 mm^2. A buried earthing conductor, which is the conductor that connects an installation's main earth terminal to the means of earthing (such as an earth electrode or a neutral/earth block), must have the minimum cross-sectional areas set out in Table 54.1.

Regulation 543.1.3 allows for the cross-sectional area of the protective conductor to be calculated using the formula $S = (I^2t)^{½} /k$. 'S' is the cross-sectional area in mm^2. 'I' is the fault current. 't' is the operating time in seconds of the disconnecting device for a current of I amps. 'k' is a factor that takes account of resistivity, temperature coefficient and heat capacity of the conductor material, and the appropriate initial and final temperatures; values can be taken from Tables 54.2 to 54.6. Alternatively, the size can readily be selected from Table 54.7.

So far as the earthing conductor is concerned, the prospective earth fault is assumed to occur on a phase conductor between the distribution company's cut-out and the consumer's main switchgear, so the earth fault loop impedance consists of the external impedance Z_E, obtainable from the distributor, plus the impedance of the earthing conductor, which is generally small and can be disregarded. The prospective earth fault current to be included in the formula for S is, therefore, the supply voltage divided by Z_E.

The time 't' in the formula is found from the time/current curves for the distributor's cut-out. For small installations, this is likely to be a BS 1361 fuse with the characteristics shown in Appendix 3. 'k' is obtainable from Table 54.2 assuming an initial temperature of 30°C. These values are then used in the formula to determine the required cross-sectional area of the earthing conductor. The formula may also be used to calculate the minimum size of the circuit protective conductors, but the earth loop impedance will now consist of the external plus the internal earth loop impedances. For cable sizes not exceeding 35 mm^2, their inductance is small and can be ignored, and only their resistances are used in the calculation. The impedance of protective conductors consisting of conduit, trunking or the metalwork of apparatus is normally ignored because it is negligible, provided the joints are properly made.

The difficulty about the calculation method is in determining the prospective fault current when the supply is derived from a distributor's LV network. Changes to the network and to the installation can affect the earth loop impedance and the prospective fault current. On the whole it is probably better to use Table 54.7 than do the calculation and employ a smaller section conductor at what may be a small cost saving, only to find later that it was a false economy.

The minimum size of main equipotential bonding conductor for PME systems is given in Table 54.8 and is related to the size of the supply neutral. For new PME installations, the designer should ascertain the intended size of the supply conductors from the distributor and then select the size of the bonding conductor from Table 54.8.

Earth continuity

As a discontinuity in a protective conductor is not self-revealing in service, except where circulating current earth monitoring is used, Section 543.3 lists the precautions needed to ensure the initial and continuing preservation of electrical continuity. Joints, in particular, need careful preparation and assembly. Components of switchboards are often pre-painted before construction, necessitating the removal of the paint at joints to ensure bare metal-to-metal contact. Hinged and drawout sections should have flexible bonding conductors. A precaution often overlooked is the provision of insulating sleeving at terminations on the emergent bare protective conductors of small insulated and sheathed cables; see Regulation 543.3.201.

Installations with equipment having high protective conductor currents

Section 543.7 is mainly concerned with electronic apparatus, such as information technology equipment fitted with suppressors with an earth connection, but it is equally applicable to equipment with high leakage currents through the insulation, such as electric furnaces with resistance wire heaters in ceramic insulation. The section deals with the hazards that arise from large currents flowing in the protective conductors of such installations and is mainly concerned with the maintenance of the integrity of the protective conductors.

If the protective conductor current is below 3.5 mA, no special precautions are required. If it is between 3.5 mA and 10 mA, equipment must be permanently connected or connected using industrial-style connector complying with BS EN 60309–2. Where the current exceeds 10 mA, the equipment must be permanently connected, or connected using BS EN 60309–2 connectors subject to restrictions on the protective conductor's cross-sectional area, or there must be an earth monitoring system that automatically disconnects the supply in the event of an open circuit protective conductor being detected. There are also specific requirements for maintaining the integrity of the protective conductor where the current exceeds 10 mA; see Regulation 543.7.1.203 and Regulation 543.7.2.201 in relation to socket-outlet final circuits.

Section 544 – Protective bonding conductors

Section 544 covers main and supplementary bonding conductors, which are single-core cables with green/yellow insulation. Their purpose is to maintain touchable metalwork in the equipotential zone at the same potential so as to avoid the possibility of electric shock to anyone touching different metalwork items at the same time. It is necessary to check that joints in metal pipes are metal-to-metal to ensure low resistance; otherwise bonding across is needed. Look out for plumbing in which both plastic and metal pipes are used and check the earth continuity of the metal pipes.

In general, the main bonding conductor must have a cross-sectional area not fewer than 6 mm^2, and should be not less than half the cross-sectional area of the installation's earthing conductor. The size of the main equipotential bonding conductors in TN-C-S installations is given in Table 54.8; in most dwellings it will need to be at least 10 mm^2, increasing to at least 50 mm^2 where the supply neutral conductor is larger than 150 mm^2.

Unless the circuits in bathroom and shower rooms have 30 mA RCD protection, supplementary bonding must be provided even if there are satisfactory metal-to-metal joints because of the enhanced shock risk. Consideration should be given to the provision of supplementary equipotential bonding in kitchens, sculleries and laundry rooms with conducting floors, such as quarry tiles, particularly if they are likely to be wet, again because of the enhanced shock risk. Similar locations in commercial and industrial premises should receive the same consideration.

Chapter 55: Other equipment

Section 551 – Generating sets

Section 551 applies to low voltage and extra-low voltage installations that incorporate generating sets running either continuously or in standby mode for permanent and temporary installations. This would include, for example, generators supplying temporary accommodation on construction sites, diesel generator sets operating in standby mode in hospitals and photovoltaic cells in factories and dwellings.

Regulation 551.2.2 requires the short circuit and prospective earth fault rating of the generator to be determined, as well as the ratings of other supply sources. This will allow suitably rated protective devices to be selected. Regulation 551.2.3 covers the rated capacity of the generator and the need to have automatic load-shedding facilities to cater for circumstances where the load exceeds the supply capacity.

Regulation 551.4.3 addresses fault protection for static inverters, typically used for uninterruptable power supplies in installations where continuity of supply is crucial. Where the disconnection times of Section 413 cannot be achieved, supplementary bonding must be used to minimise the risk of a shock between exposed metalwork. A warning is provided in Regulation 551.4.3.3.2 about the possible deleterious effects on the operation of protective devices, such as circuit breakers, of direct current generated by the static inverter or filters.

Regulation 551.6.1 covers the interlocking arrangements that must be put in place to prevent unintentional paralleling of generators with the public supply. These can include electrical, mechanical or electromechanical interlocks on the changeover switches; locks with single transferrable keys; a 3-position break-before-make changeover switch; and an automatic changeover switch with a suitable interlock. Subsection 551.7 gives requirements relating to generators running in parallel with the public supply.

Section 552 – Rotating machines

Section 552, referring to motors, requires the circuit equipment including cables to be suitable for the starting, accelerating and load currents of a motor, so the designer needs to know the characteristics of the motor, its controlgear and duty cycle to enable him to determine the cable sizes and characteristics of the protection needed. Where, for example, an induction motor is to be used and started direct to line, the starting current may be some seven times the full load current so the circuit fuses or excess current trips of the circuit breaker have to be suitably selected to cater for this. Regulation 552.1.2 requires motors rated at greater than 0.37 kW (0.5 HP) to be provided with overload protection, unless the motor forms part of a machine that complies with a British or harmonised standard.

Section 553 – Plugs and socket-outlets

Section 553.1 sets out some general requirement for plugs and sockets, including specifying that the accessories in low voltage installations must comply with BS 1363, BS 546 or BS EN 60309–2. However, accessories not complying with these standards may be used for electric clocks, shavers with shaver supplies to BS EN 61558–2–5 or BS 4573, or for circuits with special characteristics that justify the use of special-to-type accessories.

Regulation 553.1.6 requires that socket-outlets are mounted at a height above the floor or working surface sufficient to prevent mechanical damage. However, it should be kept in mind that the approved documents and technical standards supporting the building regulations in Great Britain require socket-outlets to be at a height of between 450 and 1200 mm above floor level (see Chapter 9).

Regulation 553.1.7 is a plea to avoid the tripping hazard of long, flexible cables and cords trailed across the floor. As portable apparatus may be used anywhere, socket-outlets need to be fairly closely spaced, and where there is much portable apparatus to be used at one location, multi-way socket-outlets are preferable to the use of adapters.

There is no mention in the standard of the positioning of socket-outlets near kitchen sinks; sockets are quite frequently installed directly above sinks with an obvious risk of water/steam contamination of the socket or the risk of apparatus powered from the socket falling into the sink. The NICEIC offers guidance to its members that sockets should be located at least 300 mm horizontally offset from the edges of the sinks, and this seems to be good advice.

Section 554 – Electrode water heaters and boilers

Electrode water heaters and boilers are covered by Section 554.1; these are heaters in which the water is heated by electric current flowing between the electrodes immersed in it. The boiler must be supplied through a circuit breaker that operates on all the phases and which has overcurrent detection in each phase. As some earth leakage is inevitable, earthing is important, so the boiler shell has to be connected to the protective conductor and to the metallic armour and sheath, if any, of the incoming supply cable. If the boiler has a 3-phase low voltage supply, the neutral must also be connected to the boiler shell. The neutral does not have to be connected to the boiler shell for boilers not piped to the water supply; see Section 554.1.6.

Earthing the neutral will affect the operation of an RCD, as the earth leakage current will be shared by the neutral and protective conductors. Regulation 554.1.4 refers to an electrode boiler supplied directly from an HV supply. In this case there is no neutral connection, but an RCD is a requirement with some specific requirements relating to the residual operating current.

Section 554 – Instantaneous water heaters

Instantaneous water heaters, i.e. where the element is uninsulated and in direct contact with the water, again need additional earthing by bonding the metal water supply pipe to the main earthing terminal via a separate protective conductor; see Section 554.3. The equipment must be supplied through a 2-pole switch positioned incorporated in or in close proximity to the boiler. Supplementary equipotential bonding will also be needed if the heater is in a bathroom or shower cubicle and there is no RCD protection on the supply.

Section 557 – Auxiliary circuits

Section 557 on auxiliary circuits was introduced in the 3rd amendment in January 2016. These are defined as circuits for the transmission of signals intended for the detection, supervision or control of the functional status of a main circuit, such as circuits for control, signalling and measurement. Auxiliary circuits for fire and intruder alarms, traffic lights, and so forth (where specific standards exist) are excluded. The following matters are covered:

- power supplies for auxiliary circuits dependent on the main circuit;
- auxiliary circuits supplied by an independent source;
- auxiliary circuits with or without connections to earth;
- a.c. and d.c. supplies;
- protective measures, including protection against overcurrent;
- types and sizes of cables for auxiliary circuits;
- special requirements for auxiliary circuits that are used for measurement;
- functional considerations, including loss of functionality; and
- connection to the main circuit.

Section 559 – Luminaires and lighting installations

Section 559 covers interior lighting installations that are part of the fixed installation but excludes high voltage discharge lighting such as neon tubes and temporary festoon lighting. There are quite a few very detailed requirements relating to the different types of lampholders, fixing and connecting techniques, and lamp control. There is a general requirement in Regulation 559.4.1 for due account to be taken of thermal effects, particularly with regard to the fire-resistance of adjacent materials and the separation distance to combustible materials.

Chapter 56: Safety services

Chapter 56 is concerned with the requirements for electricity supplies for safety services, which are defined as electrical systems provided to protect or warn persons in the event of a hazard or essential to their evacuation from a location. These are supplies to fire and gas detection and alarm systems, firefighting and evacuation systems, emergency lighting installations, smoke ventilation systems, essential medical systems and industrial safety systems. Installations in hazardous areas are excluded from the scope as they are covered in other standards. Some of these installations are subject to statutory requirements which the designer must observe by having regard to applicable standards and codes of practice.

The systems are classified into two main types: those that are initiated by an operator and those that start automatically independent of any operator input. The automatic systems are further sub-classified according to the maximum changeover time, ranging from no-break to long break where the automatic supply to the safety service becomes available in more than 15 seconds.

The paramount requirement is reliability. There have to be alternative power sources, therefore, which automatically provide a supply on mains failure; these can be storage batteries, primary cells, generators, or a separate feeder from the distribution network that the DNO confirms is independent from the normal feeder such that they are unlikely to have common cause failures. These sources can be used to supply other services so long as this does not impair the reliability of the safety service. Regulation 560.6.2 requires that these sources are in locations accessible only to skilled or instructed persons. Further requirements for the sources are given in Section 560.6.

Section 560.7 contains requirements for the circuits of safety services. For example,

- Safety circuits must be independent of other circuits, such that a fault in one system does not affect the other.
- Safety circuits passing through locations subject to a fire risk must be fire resistant. Regulation 560.8.1 describes the types of mineral-insulated and fire-resistant cables that may be used. The circuits must not pass through zones that have an explosion risk.
- Overload protection (fuse or circuit breaker) may be omitted if the loss of the supply could create a greater hazard. This is a balance-of-risk assessment, but if the protection is omitted, the presence of an overload has to be monitored and a warning provided.
- Safety circuit cables that are not metallic screened or fire-resistant must be separated from other cables either by distance or barriers.
- Safety circuits must not be run in lift shafts or other flue-like structures unless they are provided specifically for fire and rescue and/or for lifts with special requirements.
- There must be detailed design schematics, an equipment inventory showing the electrical specifications for current-using equipment, and operating instructions for the safety circuits.

Components of assured quality are to be preferred, where available, and installation work should be carried out by persons who are competent in the field. Subsequent periodic maintenance is essential for ensuring continuing reliability.

Many safety supplies are afforded at SELV, which has the advantage of avoiding the possibility of harmful shock in the event of an earth fault. In cases where the supply is provided from an isolating transformer, the connection to the incoming LV mains is often arranged on the supply side of the consumer's switchboard so as to ensure that the safety supply remains on when the rest of the installation is dead. It will usually be found convenient to use HBC fuses at this position, to provide the required short circuit protection.

Part 6: Inspection and testing

Introduction

There are detailed requirements in this part for inspection and testing of an installation. This is to be undertaken at the time of initial installation before the system is put into service, called Initial Verification; after any additions or alterations to an installation, called Verification; and periodically to ensure the installation's continuing safety and functionality, called Periodic Inspection and Testing. Detailed guidance is published in IET Guidance Note No. 3 *Inspection and Testing* and the types of inspections and tests are also explained more fully later in this book (see Chapters 16 and 18).

Initial verification

Initial verification consists of a formal inspection of the installation prior to it being energised, followed by the conduct of prescribed tests, all of which must be completed satisfactorily before the installation is handed over to the client.

The content of the wide-ranging inspection is set out in Regulation 611.3 and the types of tests to be conducted by competent persons are described in Section 612. The tests include

- continuity of ring final circuit conductors;
- insulation resistance, including in SELV and PELV systems and those incorporating electrical separation;
- polarity;
- earth electrode resistance, where appropriate;
- voltage;
- earth fault loop impedance and prospective short circuit current; and
- functional testing of switchgear, controlgear, drives, controls and interlocks.

In the past, the installation had to be completed prior to inspection and test, but Section 610 recognises that it is often more convenient and efficient to adopt a more flexible approach. In fact, an installation in which cables are going to be concealed in walls and partitions must be inspected during the construction activity before the walls are completed to ensure, for example, that the cables are run in safe zones and suitably protected, especially if they are less than 50 mm deep.

It is routine on larger projects for some testing to be carried out during the progress of the work; this is appropriate for those parts of the installation that will be concealed in the structure and may also have some quality control advantages. Such projects are often completed in sections with the requirement to energise particular circuits for other contractors, such as lift installers, before final completion. The successive energisation of circuits during the construction phase needs to be carefully managed to ensure the safety of the workers, and it would be good practice to cover this in a risk assessment and associated method statement, with all the workers being instructed on the risk control measures. All energised circuits should be clearly labelled and access to live parts suitably safeguarded to prevent danger. Safe isolation procedures should be implemented with rigour, with incomplete circuits that are connected into energised distribution boards being securely isolated.

The objective is to check compliance with the regulations and the safety of the installation. The tests are carried out in a particular order to reveal faults before the installation is energised from the supply. Before connecting the supply, the distributor may require a certificate from the contractor stating that the first stage tests, which indicate that the installation is safe to connect, have been done. The distributor may inspect the installation to satisfy itself that it is safe to connect, but it is not obliged to do so as the contractor is entirely responsible for the safety of the installation.

On completion of the installation work the contractor must provide the client with an Electrical Installation Certificate which describes the extent of the installation and its main characteristics and is signed by the persons responsible for the design, construction and inspection and testing. The certificate, which confirms that the installation complies with BS 7671 and notes any departures from the standard, is normally submitted to local authority building control officers as evidence for compliance with the Building Regulations.

The certificate should have attached to it the schedules of inspections and test results and should include a recommendation on the interval to the next inspection and test. Guidance on this interval is available in Guidance Note 3 for a variety of types of premises (see Chapter 18). Appendix 6 of BS 7671 contains model forms for the Electrical Installation Certificate and the schedules of inspections and tests.

Additions and alterations

Where a new circuit is added to an existing installation, an Electrical Installation Certificate should be raised as described for initial verification. However, where modification work does not involve a new circuit being added but is restricted to work such as adding a new socket-outlet or lighting point to an existing circuit or repositioning a light switch, a Minor Electrical Installation Works Certificate should be raised and handed to the client. Some inspection and testing of the modified system is required to ensure that it can take the additional load and is in a serviceable and safe condition.

Periodic inspections and tests

BS 7671 recommends that electrical installations are periodically inspected and tested in order to determine, so far as is reasonably practicable, whether the installation is in a satisfactory condition for continued service (i.e. Is it safe and does it work?). An inspection and test programme implemented in accordance with the standard's recommendations is likely to secure compliance with Regulation 4(2) of the Electricity at Work Regulations. However, duty holders do not have to follow the recommendations to secure legal compliance and are free to adopt other maintenance strategies that achieve the same goal. Indeed, Regulation 622.2 advises on one other possible strategy, continuous monitoring and maintenance, albeit without providing any description of what this means in practice.

The frequency at which an installation is inspected and tested is recommended by the person who carried out the previous inspection and test, or initial verification, and should be based on factors such as the type of installation, its use and operation, its age and condition, and the external influences to which it is subjected. IET Guidance Note No. 3 gives the maximum recommended initial intervals between inspections and tests; some examples are every 5 years for offices, every 3 years for industrial premises and annually for public premises such as cinemas and theatres. These intervals may need to be reduced, or could even be increased, to reflect the usage, environmental conditions and potential hazards. A risk assessment should be done in collaboration with the user, taking these factors into consideration so as to arrive at an agreed period, which should be subsequently reviewed and altered where necessary in the light of experience.

The inspection and test consists of a detailed examination of the installation, mostly without dismantling equipment but with some partial dismantling of accessories if required. These inspections are supplemented by tests of the same form as those undertaken in the initial verification but focussed on ensuring that the disconnection times for protective devices such as fuses, circuit breakers and RCDs continue to be complied with. In larger installations it is acceptable for sample inspections and tests to be carried out, but the extent of the sample and any limitations must be recorded.

The person carrying out a periodic inspection and test will normally be an electrician; a formal qualification in inspection and testing, such as City & Guilds 2394/5, provides good evidence of competence.

The results of the inspection and test should be recorded; the standard recommends the use of an Electrical Installation Condition Report (EICR), a model form of which is included in Appendix 6. The EICR should have attached to it schedules of inspections and tests and should identify if the installation's condition is either satisfactory or unsatisfactory. It should also list any remedial works that are required, using the prioritisation system described in Chapter 18 of this book, and recommend the interval to the next inspection and test.

Although BS 7671 does not recommend it, duty holders should conduct periodic visual inspections of the installation at intervals between the formal inspections and tests. This is to ensure that any hazardous defects such as damaged enclosures, wear and tear, overheating, burning and missing covers are spotted and corrected. These inspections can be carried out by a trained member of staff and do not require an electrician.

Part 7: Special locations

Part 7 covers so-called special installations or locations where there is an enhanced risk of electrical injury and where special, additional or modified measures are required to ensure safety. There is an ever-increasing number of these 'special locations' in the standard, the main provisions for which are briefly described in this section.

Section 701 – Locations containing a bath or shower

Rooms containing baths or showers justify special consideration because of the increased risk of electrical injury to people using them. The increased risk stems from the combination of electrical equipment such as heaters, showers, and lighting, and people who will be wet and naked. The only exception noted is emergency showering facilities used in laboratories and industry.

The standard's approach is to divide rooms containing baths or showers into zones and then to specify what type of equipment can be installed in those zones. This is not restricted to bathrooms and shower rooms; any room, such a bedroom that contains a bath or shower, is within scope.

There are three zones, zones 0, 1 and 2, which are defined in regulations 701.32.2–4. Unlike BS 7671:2001, the 16th edition, there is no zone 3 defined. Essentially, zone 0 encompasses the volume in a bath or shower tray; zone 1 is the volume in which an individual is bathing or showering (apart from the zone 0 volume); and zone 2 is the volume defined by the horizontal distance from the zone 1 boundary extending to a distance of 0.6 m. Figure 12.2 illustrates one example of the zonal classification – in this case the straightforward example of a bath in a bathroom. More examples are contained in the standard.

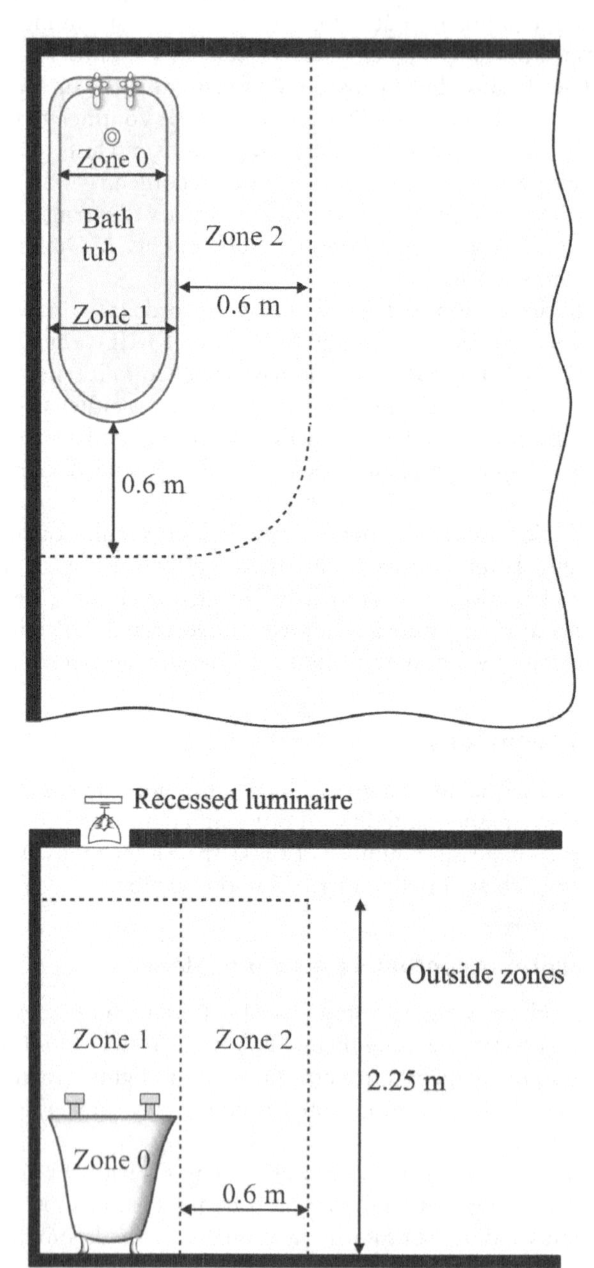

Figure 12.2 Example of zone dimensions for bath located in a bathroom

A radical change in the 17th edition is the requirement in Regulation 701.411.3.3 for 30 mA RCD protection on low voltage circuits supplying the location and, where this RCD protection is provided, the permission in Regulation 701.425.2 to omit the supplementary equipotential bonding that traditionally has been required in bathrooms and shower rooms. However, for those many locations in buildings that pre-date the 17th edition in which RCD protection is not provided, supplementary equipotential bonding will still be necessary. This does not include plastic pipes or metal components, including short lengths of 'cosmetic' copper pipework supplied though plastic pipes.

Regulation 701.415.2 specifies the requirements for supplementary equipotential bonding. This requires the terminals of protective conductors associated with Class I and Class II equipment in a room containing a bath or shower to be connected together and for them to be connected to accessible extraneous conductive parts. The latter parts are listed as including metallic items such as service pipes (gas, water etc), waste pipes, central heating pipes, air conditioning duct work and accessible structural parts of buildings.

The most important requirements relating to the zones are detailed in Table 12.1.

Table 12.1 Summary of requirements for electrical equipment in zones 0–2

Zone 0 Requirements
▪ Electrical equipment in the zone must have degree of protection of at least IPX7. Only equipment which can reasonably only be located in the zone is permitted, it must be suitable for the conditions of use in the zone, and it must be fixed and permanently connected. Moreover, no switchgear or accessories are permitted unless they are incorporated in fixed current-using equipment suitable for use in the zone. ▪ The only permitted protection against electric shock is SELV at a nominal voltage not exceeding 12 V rms a.c. or 30 V ripple-free d.c. so long as the safety source is outside zones 0, 1 and 2. ▪ Only wiring systems feeding fixed electrical equipment in the zone are permitted.
Zone 1 Requirements
▪ Electrical equipment in the zone must have degree of protection of at least IPX4. However, where water jets are likely to be used for cleaning purposes in communal baths or showers, a degree of protection of at least IPX5 must be used. ▪ Only wiring supplying fixed electrical equipment located in zones 0 and 1 must be installed. ▪ Only switches and socket-outlets of SELV systems at a nominal voltage not exceeding 12 V rms a.c. or 30 V ripple-free d.c. are permitted so long as the safety source is outside zones 0, 1 and 2. ▪ The following equipment, including incorporated switches and controls, may be installed so long as it is suitable for the zone: - Whirlpool units. - Water heaters. - Electric showers and shower pumps. - Ventilation equipment. - Towel rails. - Luminaires. - SELV equipment.

(*Continued*)

Table 12.1 (Continued)

Zone 2 Requirements
▪ Electrical equipment in the zone must have degree of protection of at least IPX4. However, where water jets are likely to be used for cleaning purposes in communal baths or showers, a degree of protection of at least IPX5 must be used. ▪ Only wiring supplying fixed electrical equipment located in zones 0, 1 and 2 must be installed. ▪ Switchgear, accessories incorporating switches or socket outlets must not be installed with exception of: - switches and socket outlets that are part of SELV circuits in which the safety source is not inside zones 0, 1 and 2; and - shaver supply units complying with BS EN 618-2-5.
Outside Zone 2
▪ Socket-outlets are prohibited within a horizontal distance of 3 m from the boundary of zone 1, apart from SELV socket-outlets and shaver supply units.

This requirement for supplementary bonding is frequently misunderstood. Although the exposed metalwork should already be earthed by the main equipotential bonding, additional safety precautions are needed to make sure that all touchable metalwork is at the same potential because of the increased risk of electric shock in the wet environment of the bathroom. Connecting together the metalwork is called 'bonding', and this is done using a 2.5 mm^2 sheathed conductor with green/yellow insulation (4 mm^2 if mechanical protection is not provided). It is called 'supplementary' bonding because it supplements the main equipotential bonding. A typical supplementary bonding configuration is illustrated in Figure 12.3.

Section 702 – Swimming pools and other basins

The enhanced shock risk in swimming pools is similar to bath and shower locations, so similar precautions are specified. Section 702 covers requirements for swimming pools and the basins of fountains and paddling pools; natural pools, lakes in gravel pits, and the like are not within scope.

There are three zones (0, 1 and 2) in descending order of risk. Zone 0 is in the pool, fountains, basins for foot cleaning, and in water jets or waterfalls and the space below them. Zone 1 is from the poolside up to 2.5 m vertically and 2 m horizontally from the rim of the pool, and zone 2 is the 1.5 m zone surrounding zone 1.

In zone 0, only SELV equipment at a nominal voltage of 12 V a.c. rms or 30 V ripple-free d.c. and IPX8-rated is permitted, with the source being positioned outside the zoned areas; these are usually luminaires behind a glass panel in the pool wall with access for maintenance in a passage outside the pool wall where the transformers are located. A similar requirement exists for zone 1, except the SELV voltage can be increased to 25 V a.c. rms or 60 V ripple-free d.c. and the equipment can be IPX4-rated, or IPX5 in the case of equipment subject to cleaning by water jets. If fixed low voltage equipment other than equipment supplied at SELV has to be installed in a zone 1 area, such as filtration systems and jet stream pumps, the supply must either be 30 mA RCD-protected or electrically separated with the source located outside the zones areas and supplying only one item of equipment.

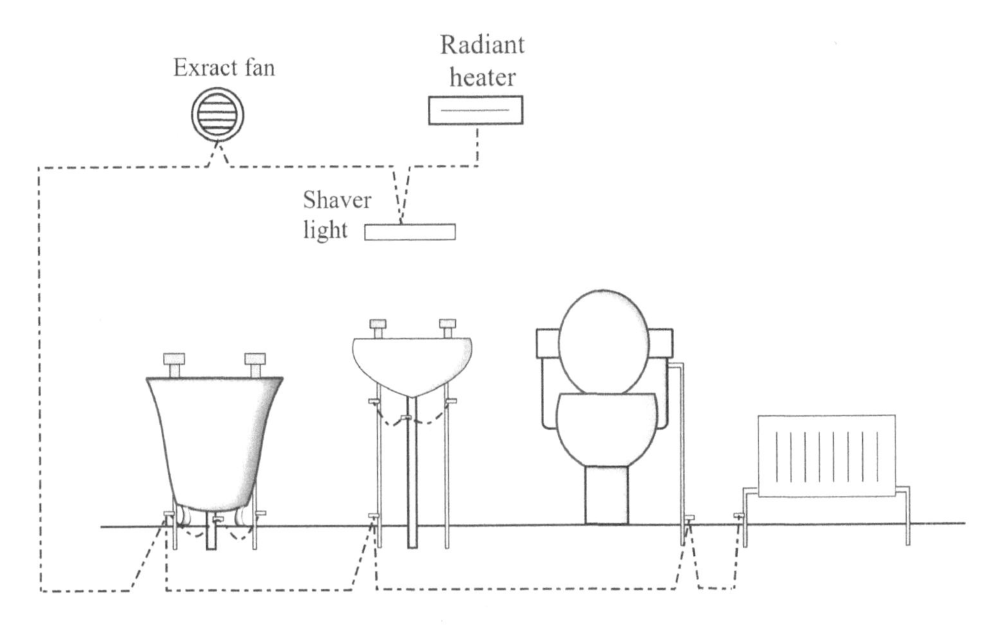

Figure 12.3 Supplementary bonding of touchable metalwork in a bathroom

Equipment for use inside the zone 0 and 1 areas of basins when nobody is using the basins have different requirements, covered in Regulation 702.410.3.4.1.

There must be no switchgear, controlgear or socket-outlets located in zones 0 and 1. However, a socket-outlet may be installed in zone 1 if it is not possible to locate one outside the zone, but it must be positioned at least 1.25 m horizontally from the border of zone 0 and at least 0.3 m above floor level, and must be protected by SELV, or a 30 mA RCD, or by electrical separation; such socket-outlets are normally provided to supply pool cleaning equipment and are to BSEN 60309–2 IPX4 or IPX5 if water jets are used.

Any equipment located in the zone 0 and 1 areas of fountains must have an SELV or electrically separated source situated outside the two zones, or be supplied through a 30 mA RCD. There is no zone 2 relating to fountains.

Equipment located in zone 2 must be IPX2-rated for indoor locations, IPX4 for outdoors, and IPX5 if cleaned using water jets. It may be supplied at SELV, and the source may be located in zone 2 as long as the supply to it is RCD-protected; otherwise it must be located outside the zoned areas. Alternatively, the equipment may be supplied by a 30 mA RCD-protected low voltage circuit or an electrically separated source with a source in zone 2 that has an RCD-protected supply.

Supplementary equipotential bonding is specified for all three zones.

Section 703 – Rooms and cabins containing sauna heaters

The enhanced shock risk in hot air saunas is due to reduced body resistance because the skin is wet from perspiration. It would, therefore, be desirable to avoid contact with mains voltage apparatus as far as possible.

Three zones are described: zones 1, 2 and 3. Zone 1 embraces the sauna heater and extends to a distance of 0.5 m; no other equipment is to be installed in this zone and the heater must comply with BS EN 60335–2–53, the product standard for sauna heaters which also specifies requirements for switchgear and controlgear. Zone 2 is outside zone 1 and extends from the sauna floor to a height of 1 m. Zone 3 is outside zone 1 and extends from the upper boundary of zone 2 to the ceiling; this is the hottest part of the occupied section and any equipment (and presumably the users) must be capable of withstanding a minimum temperature of 125°C, or 170°C in the case of cable insulation and sheaths.

Circuits supplying the sauna must be 30 mA RCD-protected, although the heater need not be unless recommended by the manufacturer. Equipment must be protected against ingress to IPX4, or IPX5 if likely to be cleaned using water jets. Apart from the heater, other equipment should be either excluded or mounted at a high level out of reach. Although it is preferable for wiring to be installed outside the sauna, any installed inside the sauna must be heat resistant with no metallic sheaths or conduit ordinarily accessible, meaning either sheathed cables or unsheathed cables in plastic conduit.

Section 704 – Construction and demolition site installations

The special requirements of Section 704 are for the temporary installations used on site for any type of building and civil engineering work, including demolition but excluding installations in site offices and other accommodation where the general Sections apply.

The electrical distribution assemblies need to conform to BS EN 61439–4:2013 *Low-voltage switchgear and controlgear assemblies. Particular requirements for assemblies for construction sites*, and although not mentioned in this section, contractors should carry out the installation work in accordance with BS 7375:2010 *Distribution of electricity on construction and demolition sites. Code of practice* as described in Chapter 13 of this book where more information is provided.

If the supply system is PME, the DNO will generally not provide an earthing terminal for outdoor sites because of the difficulty of bonding and earthing all extraneous conductive parts, so the contractor has to provide his own earth. He may well then find that the earth fault loop impedance is too high for operation of the excess current protection on the occurrence of an earth fault within the specified time, and he will have to resort to using an RCD to achieve the required disconnection times. Indeed, Regulation 704.410.3.10 requires that any socket-outlet circuit rated up to and including 32 A protected using automatic disconnection must be protected by an RCD with a rated residual operating current of 30 mA. Moreover, Regulation 704.411.3.2.1 requires RCD protection on any socket-outlet circuits rated above 32 A, including those at reduced low voltage, in order to ensure a 5 second disconnection time for TN systems and 1 second in the case of TT systems; the rated residual operating current must not exceed 500 mA.

The same regulation allows such socket-outlet circuits to be protected using reduced low voltage, which means in the majority of circumstances 110 V CTE single phase supplies. This is by far the most common and preferred technique for protecting supplies to equipment and lighting. However, SELV is preferred for handlamps used in confined spaces and damp locations.

Section 705 – Agricultural and horticultural premises

Section 705 prescribes safety precautions to protect people and especially livestock from fire and electric shock in and around farm and horticultural buildings within an equipotential zone. The special requirements apply largely because livestock such as cattle, sheep and pigs tends to be more susceptible to the effects of electric shock than humans, mostly because they have four unprotected feet in simultaneous contact with the ground.

A general requirement is that, for protection against electric shock, all final circuits supplying socket-outlets rated up to and including 32 A must be protected by one or more 30 mA RCDs. Circuits supplying socket-outlets rated higher than 32 A must be protected by one or more RCDs, with rated residual operating currents not exceeding 100 mA, and all other circuits must have fire protection by RCDs rated at no greater than 300 mA.

Another important measure, set out in Regulation 705.415.2.1, concerns supplementary equipotential bonding. All exposed and extraneous conductive parts accessible to livestock have to be bonded together by supplementary equipotential bonding conductors. Where there are metal grids in conductive floors, they must be included in the bonding arrangements.

In addition to RCD protection, fire precautions require heaters to be separated from livestock and combustible material. Such heaters should also comply with BS EN 60335–2–71:2003 *Household and similar electrical appliances. Safety. Particular requirements for electrical heating appliances for breeding and rearing animals.*

Enclosures have to be IP44 or better. Farms tend to have very corrosive atmospheres, so equipment, switchgear, enclosures and cable containment systems should be resistant to the effects of corrosion. Socket-outlets should be of the industrial variety to BS EN 60309, or BS 1363 or BS 546 where the rated current does not exceed 20 A.

Regulation 705.560.6 provides particular requirements for automatic life support safety services in premises undertaking high intensity livestock rearing. These are mainly concerned with ensuring reliable supplies to ventilation, food, water and lighting services.

Section 706 – Conducting locations with restricted movement

The more typical description for Section 706 is confined conductive locations. These are spaces where freedom of movement is restricted and the body is likely to be in contact with exposed and extraneous conductive parts. This section covers work inside boilers, metal ventilation ducts and tanks, for example, where extensive contact with the metalwork increases the electric shock hazard; for there to be a shock hazard, there has to be some fixed or mobile electrical equipment. The risk is enhanced if these interiors are wet or so hot that the operator's clothes are soaked with perspiration, thereby lowering the total body impedance. Incidentally, although not mentioned in the sections, the precautions listed are advisable in unconfined wet and hot locations.

The best precaution is not to use electrical apparatus in these types of locations, substituting electrical apparatus for pneumatic or hydraulic equipment wherever possible, or using battery-powered apparatus such as drills. However, there are occasions when there is no option but to use electrical equipment.

For protection against both direct and indirect electric shock when using hand-held tools, mobile equipment and handlamps, the specified supply system is SELV, although electrical separation may be used for tools and mobile equipment so long as each source such as a secondary winding on a transformer supplies only one item of equipment. Where a functional earth is needed, for certain instruments for example, it may be utilised provided all exposed and extraneous conductive parts are bonded together and to the protective conductor.

For fixed equipment, Regulation 706.410.3.10 allows mains voltage supplies with automatic disconnection and with supplementary equipotential bonding with exposed conductive parts connected to the extraneous conductive parts in the location. Alternatively, Class II equipment may be used provided its enclosure is suitable for the location and it has 30 mA RCD protection. Electrical separation, SELV and PELV supplies may also be used but the PELV option requires earthed equipotential bonding.

Unless it is part of the fixed installation, the power source for an SELV or PELV supply has to be situated outside the restricted conductive location.

Section 708 – Electrical installations in caravans / camping parks and similar locations

Section 708 deals with electrical distribution systems in caravan and camping parks and similar sites. The electrical installations in caravans are covered in Section 721.

In the UK, the supplies to caravan and camping pitches will invariably be at 230 V single phase 50 Hz a.c. and caravans with mains voltage installations will usually have some Class I equipment, so the protection system will be earthed equipotential bonding and automatic disconnection of the supply.

The distribution system should be treated as a distribution network and comply with the relevant parts of the Electricity Safety Quality and Continuity Regulations 2002 (as amended), as well as with the specific requirements of this section.

The conductors of overhead distribution circuits must be insulated and at a minimum height of 3.5 m, apart from where they cross areas where there will be movement of vehicles where they should be at a height no lower than 6 m. Underground cables must be buried to a depth of at least 0.6 m and kept away from the pitches or be mechanically protected against penetration by tent pegs or ground anchors.

DNOs will normally provide a TN-S supply for a caravan park but where the supply earthing arrangement is TN-C-S (PME) or TT an earth electrode has to be installed by the occupier for those circuits supplying the pitches. The PME earthing must not be connected through to the individual pitch supplies and must be broken at a convenient part of the circuit.

The principal requirements for the supplies to individual pitches are as follows:

- The connection point must be no further than 20 m from the pitch and should contain no more than four socket-outlets with at least one socket-outlet per pitch.
- The socket-outlets must be of the round-pin industrial variety to BS EN 60309–2 rated at 16 A or higher and at least IP44 to protect against water ingress. They must be at a height of between 0.5 and 1.5 m above ground level, or higher if the environmental conditions such as the possibility of flooding or deep snow justify it.
- The supply to each pitch must be individually protected by an overcurrent device and a 30 mA RCD. Note that the 16th edition allowed RCD protection to be grouped, so the requirement for individual protection is new to the 17th edition.

Section 709 – Marinas and similar locations

This section covers single and three-phase shore supplies to small leisure craft and houseboats moored in marinas and similar locations. The requirements are very similar to those in Section 708.

In the same way as described in Section 708 for caravan pitches, the individual shore supplies must not be earthed from the distributor's PME earth. Moreover, the shore-based supply equipment must be suitable for the wet and corrosive conditions that may prevail as well as for potential impact damage. Plastic enclosures to at least IP44 (up to IPX6 if likely to be subjected to waves) will be the norm.

Socket-outlets must be of the industrial variety to BS EN 60309 with no more than four per group and be individually protected by 30 mA RCDs and suitable overcurrent protection. They must be at a height of no less than 1 m above the highest water level, although this can be reduced to 300 mm on floating pontoons and walkways as long as they are protected against splashing.

The section contains an example of a set of instructions for connecting to shore supplies that marina operators could provide to boat operators on berthing.

Section 710 – Medical locations

This new section applies to electrical installations in medical locations (hospitals, clinics, dental practices and the like). It is aimed at the safety of both staff and patients, many of whom in the latter case will be especially vulnerable to electrical risks, particularly when undergoing invasive medical procedures.

The section describes the classification of locations and the allocation of group numbers according to the type of procedure taking place in the particular location; there are three types of groups, numbered as Groups 0, 1 and 2 according to the nature of the treatment being provided and which have increasing risks. Group 0 locations are where no low voltage medical electrical equipment is being used and failure of the supply will not create harm; Group 1 is for locations where such equipment is intended to be used but not for procedures which involve invasive use of conductors near to the heart (so-called intracardiac procedures) or where loss of supply could cause harm; and Group 2 locations are where medical electrical equipment is intended for use in intracardiac procedures and loss of supply could cause harm.

Locations may be, for example, an operating theatre, a delivery room, and intensive care room, or a hydrotherapy room. This allows the required protective measures to be specified, which may involve altered disconnection times, reduced fault voltages on exposed metalwork and limitations on the types of RCDs that may be used. The classification and grouping processes are not particularly clear, and designers and specifiers will need to work with their clients to get a clear understanding of the requirements; the National Health Service, for example, has a comprehensive set of Health Technical Memoranda that provide greater clarity.

One of the more specialised features of the section is the specification of an IT earthing system with insulation monitoring for final circuits supplying medical equipment intended for life support and surgical applications in Group 2 locations such as operating theatres. The aim is to prevent the loss of supplies in the event of a first fault and the detailed requirements are covered in Regulation 710.411.6.

Other noteworthy requirements are the need for supplementary equipotential bonding in Group 1 and 2 locations as well as SELV voltages not exceeding 25 Va.c. or 60 V ripple-free d.c. in the same locations.

The presence of flammable gases needs to be taken into account by the designer, and Regulation 710.512.2.1 specifies a separation distance of at least 0.2 m between socket-outlets and switches and gas outlets.

Medical locations have a clear need for continuity of supply, and the detailed requirements for safety services such as power supplies and emergency lighting are covered in Section 710.56.

Section 711 – Exhibitions, shows and stands

This section covers temporary electrical installations in exhibitions and shows, and the installations in stands and equivalent (e.g. vehicle or wagon or caravan or container) erected at the event. It does not cover the fixed installations at venues, nor does it cover the type of temporary installations addressed in BS 7909:2011 *Code of practice for temporary electrical systems for entertainment and related purposes*, which covers musical concerts, sporting events, theatres and TV outside broadcasts.

The main protection against electric shock is afforded by all supply cables being protected by RCDs with a rated residual operating current not exceeding 300 mA. These RCDs should have a time delay to allow them to discriminate with the 30 mA RCDs required on all final circuits supplying stands and socket-outlet circuits not exceeding 32 A. Additionally, structural metal parts associated with the stand must be bonded back to the main earthing terminal.

Cables, which must have conductors with a minimum cross-sectional area of 1.5 mm^2, must be armoured or protected against mechanical damage, usually using cable protectors that cover the cables. They must not create a tripping hazard. If there is no fire alarm system at the venue, the cables must be either flame retardant and low smoke or located in IPX4 conduit or trunking; see regulation 711.521.

Each stand, or other type of exhibit, must have its own means of isolation, as must any motors associated with the equipment.

Section 712 – Solar photovoltaic (PV) power supply systems

This new section sets out the safety requirements for the installation of PV cells, or array of cells, in a property, the output of which is connected into the property's a.c. electrical system to run in parallel with the DNO's supply. It needs to be read in conjunction with Section 551 on the connection of low voltage and extra-low voltage generating sets.

The output of a PV array is a d.c. voltage which must be transformed to a.c., most commonly through a power electronic device known as an inverter or converter. The system therefore has a d.c. side on the supply to the inverter, and a.c. side between the inverter and the point of connection into the property's a.c. system.

One of the features of PV arrays is that they will generate electricity when illuminated – they cannot be switched off so they must be considered always to be energised, with special care being needed with the means of isolation when work is to be carried out on them.

As described in Part VI of Chapter 7, the appropriate DNO will need to be consulted (or simply informed if the output is fewer than 16 A per phase) before a PV system is connected to the distribution network. Most small-scale PV domestic arrays will be handled under the terms of ENA Engineering Recommendation G83 *Recommendations for the Connection of Type Tested Small-scale Embedded Generators (Up to 16 A per Phase) in Parallel with Low-Voltage Distribution Systems*, but the larger installations will be covered by the more complicated ENA Engineering Recommendations G59 *Recommendations for the*

Connection of Generating Plant to the Distribution Systems of Licensed Distribution Network Operators.

The main technical requirements in Section 712 are summarised as follows:

- The a.c. side of the PV output must be connected to the supply side of an overcurrent protective device (fuse or circuit breaker) for automatic disconnection of connected equipment.
- An overcurrent device is not required on the d.c. side so long as the cable is rated continuously to carry at least 1.25 times the short circuit current.
- A Type B RCD to BS EN 62423 with a rated residual operating current of 30 mA must be connected between the inverter and the overcurrent device unless the inverter cannot by design feed d.c. fault current into the a.c. installation.
- Isolation devices must be provided on both the d.c. and the a.c. side of the inverter to allow maintenance work to be carried out safely.
- All junction boxes must carry a label warning that the internal conductors may still be live after isolation from the inverter.

In addition, and in accordance with Regulation 551.7.4, automatic switching must be incorporated so as to disconnect the array from the DNO's system in the event of the loss of the DNO's supply.

One of the most important factors in these systems is that they are installed and connected by competent people. There is a competence scheme in the UK known as the Microgeneration Certification Scheme (MCS) which allows customers to select MCS certified installation companies to carry out the installation work. These companies have to have demonstrated suitable competence and other criteria to an accredited certification body before they can be certified.

It is worth commenting that this section does not address energy storage systems, such as lead acid or lithium ion batteries, associated with PV and other renewable sources. These will become increasingly common over the next few years as battery energy densities increase and costs reduce, and as worries about continuity of supply increase. The fire, burn and shock risks associated with this emerging technology will need to be included in the standard in due course.

Section 714 – Outdoor lighting installations

This section covers outdoor lighting installations and applies both to the distribution wiring and connected apparatus. It covers highway power supplies and street furniture and off-highway areas including the outdoor parts of public parks, car parks, sports fields, bus shelters, advertising boards and the like, where movement lighting, illuminated signs and other electrical apparatus are used.

The principal protective measure is automatic disconnection, with a 5 second maximum disconnection time. A distributor's PME earth can be used as the earthing conductor so long as the conductor has a minimum cross-sectional area of 6 mm^2. 30 mA RCD protection is required on supplies to the likes of telephone kiosks, bus shelters, advertising panels and illuminated signs such as town plans.

The doors in lamp posts and some street pillars are usually easily openable by vandals and mischievous children, so Regulation 714.411.2.201 requires those that are accessible to be secured such that they can only be opened using a key or special tool and there must be a fixed internal barrier to screen bare live parts to achieve a level of protection of at least IP2X.

Subsection 714.537 requires that every circuit must be capable of being isolated. However, there is recognition that many installations do not have isolating switches and are fed from the distribution company's combined cut-out, neutral link and sealing box on the end of the service cable, so in such cases electrical maintenance is restricted to qualified and instructed persons who may achieve isolation by removing a fuse carrier or switching off a circuit breaker.

Supplies are sometimes taken from street furniture to feed temporary installations such as market stalls, Christmas decorative street lighting and small roadworks. There must no impairment of the safety of the permanent installation so it must not be overloaded and the connection must be safe.

Section 715 – Extra-low voltage lighting installations

The particular requirements in this section apply to installations that are supplied from SELV and PELV sources with a maximum rated voltage of 50 V a.c. rms or 120 V d.c. It covers protection against electric shock (SELV); protection against the risk of fire due to short circuit; requirements for isolation, switching and control; the acceptable types of wiring systems, including special requirements with bare conductors supplied at 25 V a.c. or 60 V d.c.; the types of transformers and converters; and suspended systems.

Section 717 – Mobile or transportable units

The term 'unit' in the title of this section refers to vehicles and/or mobile or transportable structures in which all or part of an electrical installation is contained. Examples given are technical and facilities vehicles for the entertainment industry, medical or health screening services, welfare units, promotion and demonstration, firefighting, workshops, offices, and transportable catering units. In other words, a very broad range of specialist vehicles, caravans, trailers, and skids incorporating electrical power systems (but not the vehicle's own systems). Some flexibility will be required in interpreting whether or not a particular vehicle or structure is covered by this section, although in most instances it should be obvious. In that regard, the main exclusions are generating sets, marinas and pleasure craft, caravans, and machinery when covered by BS EN 60204–1.

There are six diagrams at the end of the section which attempt to shed some light on the detailed technical requirements, using the different forms of supply that may be available such as a generating set mounted on or in the vehicle or an external supply from a low voltage network or generating set.

Where the vehicle has an external supply with internal distribution that is not electrically separated through, for example, a transformer, automatic disconnection must be achieved by an RCD with a rated residual operating current not exceeding 30 mA. However, some users for which continuity of supply is important, such as in outside broadcast or satellite communication vehicles, may wish to exclude RCDs so as to prevent nuisance tripping, in which case they must ensure that the installation otherwise satisfies the full safety requirements of the standard.

PME earthing may be used only when the installation is under the supervision of a competent person and the earthing has been tested; in practice, it is rarely used and should be avoided. Accessible conductive parts should be bonded to the main earthing terminal, using finely stranded conductors.

An installation having an IT earthing system may be used, with the supply provided by an isolating transformer or generating set with insulation monitoring designed to disconnect the supply automatically in the event of a second fault. Alternatively, a transformer providing electrical separation may be used, but there must be an RCD and earth electrode providing automatic disconnection in the event of a transformer fault and each socket-outlet intended to supply external equipment must be protected by a 30 mA RCD.

Apart from socket-outlets in SELV and PELV circuits and electrically separated circuits incorporating insulation monitoring, all socket-outlets intended to supply external equipment must be protected by 30 mA RCDs. In general, external socket-outlets should be industrial assemblies to BS EN 60309–2 and have ingress protection of P44 when in use and IP55 when not in use.

Section 721 – Electrical installations in caravans and motor caravans

Regulation 721.1 makes it clear that the section applies to touring and motor caravans and not to fixed structures such as mobile homes and residential park homes. The supply may be 230/400V a.c. or up to 48V d.c.

Alternating current power is brought into the caravan via a BS EN 60309–2 IP44 appliance inlet and connected to a double pole isolator from which final circuits are routed through a fuse or circuit breaker providing overcurrent protection. Automatic disconnection is to include a 30 mA RCD protecting all live conductors, and there must be protective conductors connected to the earthing contacts of the inlet connector and socket-outlets, and to exposed conductive parts.

Wiring must not be in metallic conduits, and conductors must have a minimum cross-sectional area of 1.5 mm^2. Flexible conductors and shake-proof terminals are preferable for withstanding the vibration in transit, and cables should be protected against damage by sharp edges or abrasive parts. Provision needs to be made for securing pendant luminaires if damage is to be prevented.

The special requirements include the provisions of Section 701 if there is a bath or shower.

The appendix to Section 721 provides guidance on the installation of extra-low voltage d.c. systems.

Section 722 – Electric vehicle charging installations

This new section applies to charging circuits for electric vehicles. It focuses on situations where cars are charged through cables, and specifically excludes inductive charging where energy is transferred to the car via magnetic induction rather than through cables connected using plugs and sockets. A future amendment will include the safety requirements for circuits feeding electricity from an electric car into a distribution network.

Where an installation is to be used for charging the batteries in an electric vehicle, there must be a separate final circuit provided for that purpose. More than one charging point can be supplied from the circuit. Because of the extended time over which charging takes place, there is no diversity allowance in these circuits for the calculation of cable sizes.

In TN systems, which will apply in the majority of cases, PME earthing must not be used on the supply to external charging points or to charging points that might be expected to connect to vehicles located outside a dwelling or garage. This is because these will be located outside the building's equipotential zone and may therefore export dangerous fault voltage from the building to the exposed metalwork of the charging point, which may put at risk anyone standing on the ground and touching the charging point or any car connected to it. However, this restriction would not apply if measures are taken to prevent voltages in excess of 70 V a.c. appearing between the charging point and earth for longer than 5 seconds; three techniques that can be used to achieve this are described in Regulation 722.411.4.1.

As an alternative to automatic disconnection, an unearthed electrically separated supply from a fixed isolating transformer may be used to charge one vehicle at a time. Charging points must be protected by a 30 mA RCD, with due consideration being given to any possible d.c. component in the residual current.

Regulation 722.55.201.1 covers the types of socket-outlets that may be used in charging points, including specialist vehicle connectors described in BS EN 62196–2:2012+A12:2014 *Plugs, socket-outlets, vehicle connectors and vehicle inlets. Conductive charging of electric vehicles. Dimensional compatibility and interchangeability requirements for a.c. pin and contact-tube accessories.*

When electric vehicle charging modes 3 and 4 are used, as defined in the definitions, means must be incorporated to prevent plugs being withdrawn unless the supply has been switched off.

Section 729 – Operating and maintenance gangways

This section applies to basic protection, avoiding direct contact with live parts, in the operation of switchgear and controlgear in gangways where access is restricted to competent persons. It is useful guidance for securing compliance with some aspects of Regulation 15 of the Electricity at Work Regulations. It is targeted mainly at areas where there are live overhead busbars, such as internal substations.

The precautionary measures are the use of barriers, obstacles and enclosures. Gangways must be at least 700 mm wide, and the distance to live parts placed out of reach must be at least 2.5 m. The dimensions are illustrated in the standard.

Section 740 – Temporary electrical installations for structures, amusement devices and booths at fairgrounds, amusement parks and circuses

This section applies to the electrical installations of temporarily erected mobile and transportable machines and structures such as fairground rides and booths and machines used in circuses and amusement parks. They are characterised by being regularly dismantled, transported to a new location and re-erected, most often by people who are not electrically qualified but who are skilled at their jobs, such as showmen and show-women.

Given the nature of the equipment, its usage, and the fact that it is used by members of the public who may not be especially concerned about electrical safety, it is surprising that the number of electrical injuries at these locations is very low. To some extent this is a credit to the owners and operators but also to the effectiveness of the Amusement Device Inspection Procedures Scheme (ADIPS) which ensures that amusement devices are regularly inspected by registered inspection bodies and certified as safe for use. A fairground ride, for example, may be put into use by its controller only if it has a current Declaration of Operational Compliance issued by a registered ADIPS inspector indicating that the ride is safe; for electrical safety purposes, that means it complies with the provisions of BS 7671 and BS EN 60204-1.

PME earthing systems must not be used, so most supplies are TN-S. They must be protected at their origin, most usually a generator which may or may not be referenced to earth using an earth electrode (depending upon location and the ability to obtain a suitable connection to the general mass of earth), by an RCD with a rated residual operating current not exceeding 300 mA with a time delay where necessary for discrimination with down-stream RCDs. Final circuits for lighting, socket-outlets rated up to 32 A and mobile equipment connected by cables rated up to 32 A must be protected by 30 mA RCDs; the only exceptions are circuits protected by SELV, PELV or electrical separation, and lighting circuits that are out of reach (>2.5 m) and not supplied through domestic or BS EN 60309–1 socket-outlets.

In locations such as circuses where there may be livestock present, the supplementary equipotential bonding requirements of Section 705 will be needed. A particular problem at these locations is cables snaking along the ground and potentially being damaged by vehicular and human traffic. Regulation 740.521.1 covers this, requiring cables to be protected against mechanical damage using conduit, trunking and ducting systems or armoured cable.

Electric dodgem rides, with their large expanses of exposed live metalwork, must be operated at voltages not exceeding 50 V a.c. or 120 V d.c., with the source electrically separated from the mains by an isolating transformer to BS EN 61558–2–4 if a generating set is not used. One common problem with dodgem rides is the power pick-up feather at the top of the car pole shorting the live roof panels to the earthed structure of ride, showering the car's occupant with hot sparks – this should be prevented by the use of barriers or separation.

Section 753 – Floor and ceiling heating systems

This new section covers heating systems buried under floors or above ceilings, most commonly in dwellings and similar places, such as care homes and hospices. The main hazards are physical damage by penetrating objects and overheating.

The main precautions are automatic disconnection by 30 mA RCDs and the prevention of overheating to limit the temperature to a maximum of 80°C by design and/or the use of protective devices.

Appendices

The standard contains fourteen appendices, some of which are of considerable importance to designers, installers and maintainers.

Appendix 3, for example, contains the time/current characteristics of the most commonly used fuses, circuit breakers and RCDs. Appendix 4 contains comprehensive information on the current-carrying capacity and voltage drop of a variety of types of cables, including guidance on grouping and rating factors and the method of calculating cable sizes, as well as the effect of different installation methods and harmonics on current-carrying capacity. Appendix 5 classifies the external influences to which installations may be subjected.

Appendix 6 provides the model forms for the certification and reporting of installations, which are in widespread use by electrical contracting companies.

Appendix 7 describes the harmonised cable core colours that must be used: brown, black and grey for line conductors in a.c. systems, blue for neutral and green/yellow for protective conductors. Tables compare these new (since 2005) colours to the previous colours that remain in widespread use. Table 7E sets out the required core colours for 2-wire and 3-wire d.c. circuits.

Appendix 10 provides guidance on the overcurrent protection of conductors connected in parallel.

Appendix 15 describes the ring and radial final circuits commonly used in installations, with excellent diagrammatic summaries of the main requirements.

The 18th edition

The 18th edition of the Wiring Regulations is scheduled for publication in mid-2018, so it will be published as BS 7671:2018. The various BSI panels responsible for the standard have been working on the amendments, and the IET has published some information on the likely changes. It would appear as though the major changes are likely to be in the following areas:

- Clause 443 concerning overvoltage protection against the effects of lightning is likely to be amended and significantly revised to clarify when such protection should be provided.
- New information is likely to be provided on the use of arc fault detection devices (AFDDs), also known as arc fault circuit interrupters, as a means of fire prevention in final circuits. These devices analyse voltage and current waveforms to detect hazardous arcing activity and switch the affected circuit off. Their value is that they can detect arc faults that will not be detected by fuses, circuit breakers or RCDs. An impediment to their widespread adoption will be their relatively high price, but this can be expected to reduce as their take-up increases.
- Section 753 on floor and ceiling heating systems is likely to be amended to apply the requirements to embedded heating systems for surface heating and other types of electric heating systems.
- There is likely to be a new section on energy efficiency to bring the standard into line with international efforts to maximise energy efficiency.

Some of these proposed changes stem from the UK's membership of the EU and the need for BS 7671 to match the content of European harmonised standards. The British standards makers will have the opportunity to adopt alternative strategies once the UK leaves the EU, and it will be interesting to see to what extent they take advantage of this opportunity.

13

Construction sites

Introduction

The construction industry has an unenviable accident record and in the not-too-distant past had more electrical fatalities per year than any other industry. The accident rate has dropped substantially in recent years, largely as a result of the emphasis placed on the use of reduced low voltage supplies to hand tools, equipment and lighting; the use of safer distribution systems; and improvements in the safety competences of electricians working in the electrical contracting industry. The Construction Design and Management (CDM) Regulations have also played a part because they have driven considerable improvements in the safety management of construction activities. However, serious accidents and fatalities still occur, with electric shock accidents during final fit-out, contact with overhead power lines, and underground cable strikes being all-too-frequent occurrences.

The accidents tend to dominate on building and demolition sites where there is a temporary electrical distribution system rather than at works of engineering construction such as road building where little or no electrical apparatus is used. There is, however, a significant number of burn accidents in street works when underground electricity cables are damaged during excavation activity. This is, perhaps, not surprising given that the number of cable strike incidents is in the order of 50,000 per year.

From a historical perspective, the problem with electrical safety on construction sites started largely in the 1950s when portable electric tools, electrically powered plant and electric lighting came into general site use. Up until the 1960s, there was no purpose-designed electrical distribution system available, so the main contractor would usually ask the electrical subcontractor to provide a minimum installation at minimal cost.

The usual practice was to erect a few switchfuses, distribution boards, socket-outlets and lighting points on an ad-hoc basis; connect them with unprotected PVC-insulated conductors draped over the structure; and operate the installation at mains voltage. Non-weatherproof apparatus, of the type used for indoor fixed wiring installations, was employed, usually with indifferent quality to save cost and often with insufficient robustness for site use.

As the work progressed, the temporary installation had to be altered to suit, involving repositioning apparatus and wiring. This was done with scant regard for safety. The apparatus and wiring were seldom effectively secured, were vulnerable, and sustained a great deal of damage. At the end of the contract very little of the temporary installation was serviceable and recoverable for use elsewhere, so it was not surprising that most site installations were illegal, cheap and nasty and a recipe for electrical accidents.

To counter the rising accident toll, in the early 1960s the Building Research Establishment designed and built a safe site distribution system which set the standard against which several manufacturers began to produce equipment specifically for use on construction sites.

The system was accepted by the National Joint Council of the Building Industry in 1965, who recommended its progressive adoption over a 5-year period, with a final implementation date of 1 January 1970. This system is now accepted as the benchmark for safe electrical systems on construction sites and is reflected in existing standards and guidance material.

Safety concept

The safety philosophy is based on dividing the site installation into two distribution systems: a 400/230 V system where there is a comparatively high shock risk because the voltage to earth is 230 V, and a safer reduced low voltage 110 V system where the voltage to earth does not exceed 63.5 V. The 400/230 V system is intended to be installed and maintained only by authorised and competent persons, i.e. those who are electrically qualified. The construction workers' role is confined to operating the distribution equipment, which is metal-clad, weatherproof and earthed. The portable apparatus which they constantly handle is all connected to the 110 V system. The overall system is illustrated in Figure 13.1.

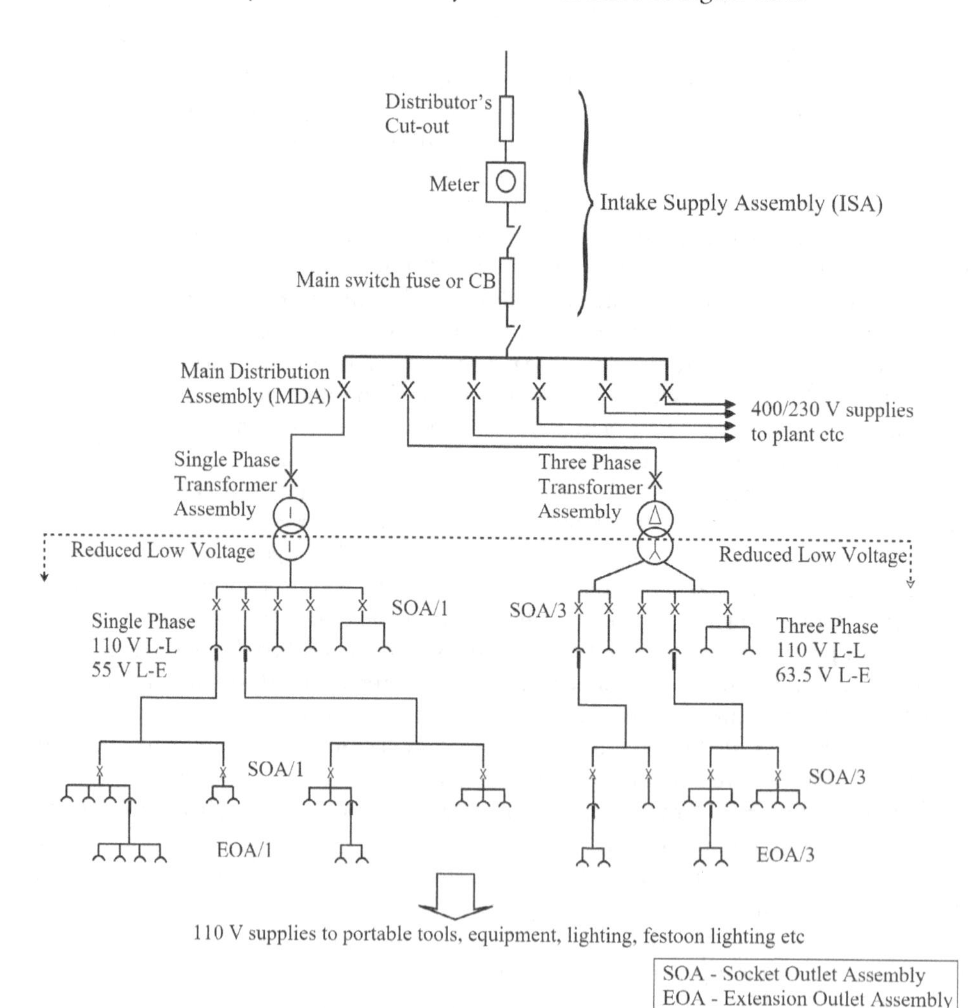

Figure 13.1 Schematic diagram of construction site distribution system

The standards

BS 4363 *Distribution units for electricity supplies for construction and building sites* was first issued in 1968. A new edition was published in 1998 and was amended in 2013; it complements and is compatible with the recently published BS EN 61439–4:2013 *Low-voltage switchgear and controlgear assemblies. Particular requirements for assemblies for construction sites (ACS)*. These standards specify the power distribution equipment for use on construction sites.

In 1969 BSCP 1017 *Distribution of Electricity on Construction and Building Sites* appeared. This was the code of practice which described how the BS 4363 apparatus was to be installed and used. This code was subsequently amended and renumbered as BS 7375 *Distribution of electricity on construction and demolition sites*; the latest version of this standard was issued in 2010. The HSE also published a guidance booklet, HS(G)141 *Electrical Safety on Construction Sites*, but this was withdrawn as part of the HSE's unfortunate withdrawal from providing helpful technical guidance.

As noted in Chapter 12, Section 704 of BS 7671 contains specific requirements for low voltage distribution on construction and demolition sites.

The distribution equipment comprises a number of metal-clad, fireproof and weatherproof units with air break switches for isolation and circuit breakers for overcurrent protection. Although not a requirement in the standard, the transformers used in practice are generally of air-insulated or resin-encapsulated construction. The apparatus is robust to withstand rough handling, free-standing and often mounted on ground clearance supports with facilities for anchoring. To facilitate handling, the heavier units are provided with lifting lugs and the lighter units with handles. The apparatus is totally enclosed, and unauthorised access is prevented by bolted-on or lockable covers, although it should be noted that the recently introduced BS EN 61439–4 requires that switchgear is accessible at all times.

Intake supply assembly

Where a construction site is to be supplied by a DNO, rather than by the construction company's own generator, an Intake Supply Assembly (ISA) is installed to cater for the DNO's requirements. Generally, it has two compartments with lockable access, one for the incoming cable termination, service fuses, neutral link, current transformers and meters, and another for the consumer's main switchfuse or circuit breaker fitted with excess current protection and, where appropriate, earth fault protection. The latter may be needed to achieve the required disconnection times on the low voltage system, especially if the earthing arrangement is TT, which it is in many cases because the DNOs will not provide a PME (TN-C-S) supply to construction sites. An example of an ISA is shown in Figure 13.2, courtesy of Blakley Electrics Ltd. This model contains combined current transformer (CT) chambers and RCDs and is rated from 200A to 2000A. The CT chamber is designed to house current transformers used to meter LV supplies and the RCD provides fault protection for the site.

Main distribution assembly

The main switchboard is called a Main Distribution Assembly (MDA) and consists of an incoming switch or circuit breaker suitable for isolation which feeds a set of busbars to which are connected a number of moulded case circuit breakers (MCCBs). For smaller sites and/or where it may be convenient, the ISA and MDA can be joined to form a combined Intake Supply and Distribution Assembly (ISDA). In this case, the 'on load' isolator is omitted. An example of an MDA is shown in Figure 13.3, courtesy of Blakley Electrics Ltd.

Figure 13.2 Intake Supply Assembly with CT chambers and RCDs (Courtesy of Blakley Electrics)

Figure 13.3 Main Distribution Assembly rated at 1600 A (Courtesy of Blakley Electrics Ltd)

There is an increasing move to construct MDAs as Form 4 enclosures with internal separation of the busbars from all switching, isolation and control items and outgoing terminations and separation of all items and outgoing terminations from each other. This allows for access to any single item, such as an MCCB and its outgoing terminations, to enable work to be carried out whilst the assembly remains live and operational. This improves the safety of MDAs, particularly on large and complex sites where cables may need to be connected or disconnected during the project.

Another recent development is the fitting of test sockets that allow live tests such as voltage, earth fault loop impedance and RCD trip time to be carried out without opening the assembly and exposing live parts.

Transformer assembly

Transformer assemblies (TAs) are used to generate 110 V supplies for portable electric tools, small plant items and the temporary lighting installation. The transformer is controlled by means of an incoming circuit breaker with excess current protection. The

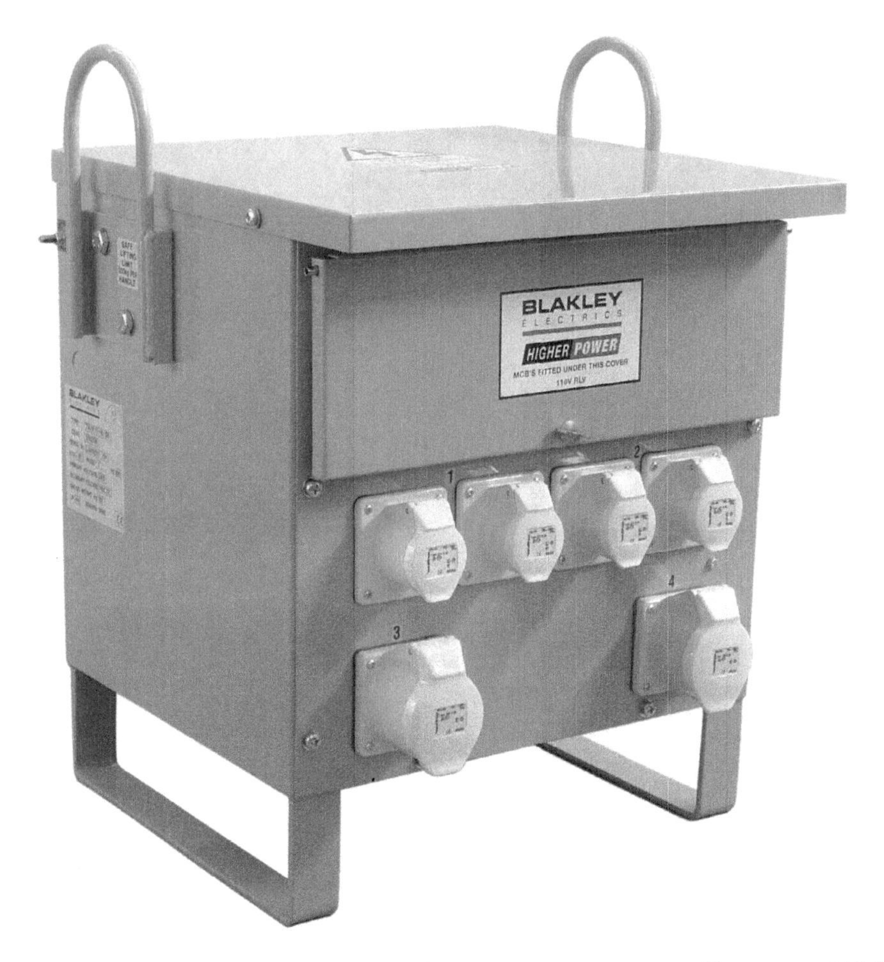

Figure 13.4 230/110 V transformer assembly for construction sites (Courtesy of Blakley Electrics Ltd)

secondary output circuit feeds a number of BS EN 60309–2 socket-outlets controlled by circuit breakers with excess current protection. Single- and three-phase TAs are available. The secondary winding has its centre point, for single phase units, or its star point, for three-phase units, earthed to limit the electric shock voltage to earth to approximately 55 V and 63.5 V respectively. Single phase 110 V supplies may be obtained from three-phase units by utilising two phases.

For jobbing work, small, single phase, portable TAs are available up to around 2 kVA capacity and with one or two BS EN 60309–2 socket-outlets. A length of flexible cable is provided, usually terminating in a BS 1363 fused plug for insertion into 230 V domestic socket-outlets. The output circuit is provided with overcurrent protection, usually in the form of a resettable thermal cut-out.

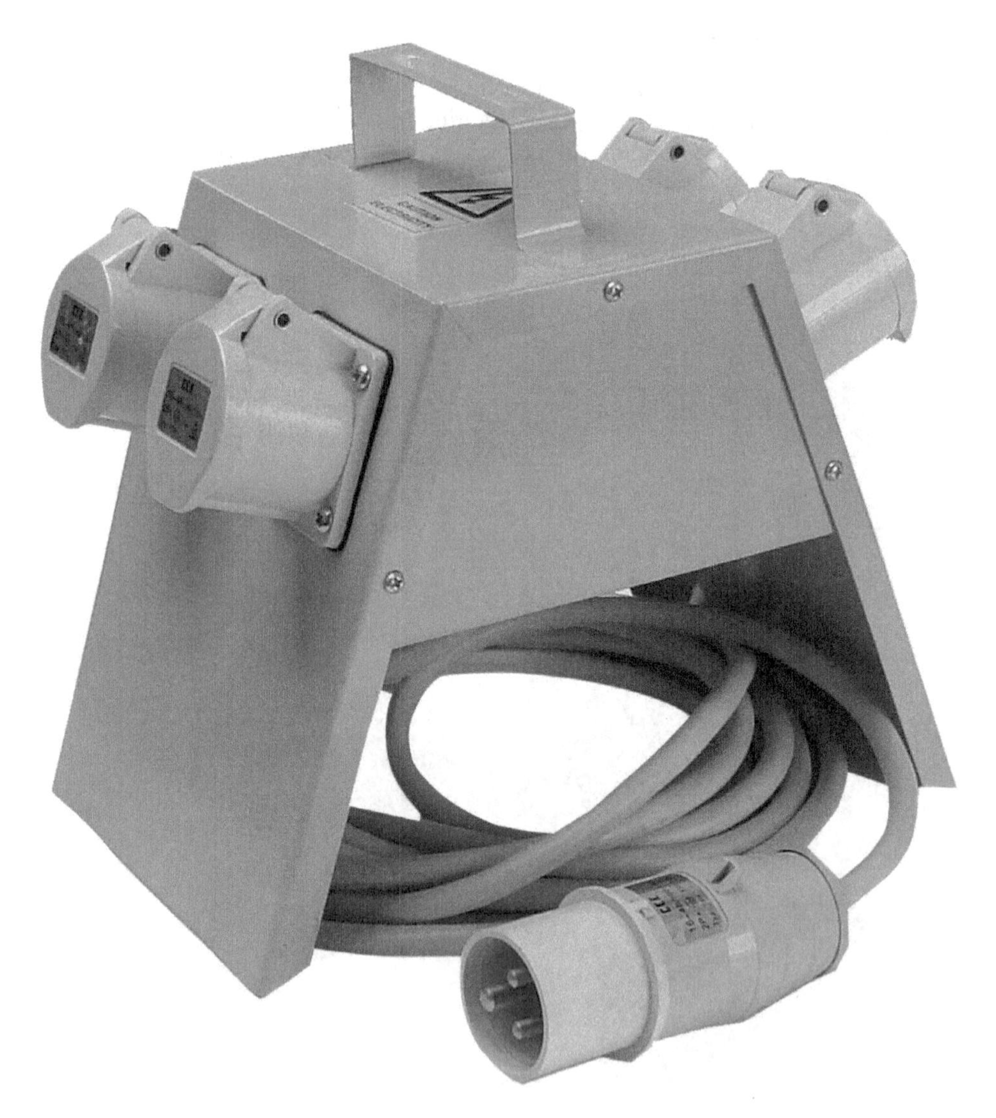

Figure 13.5 Example of an EOA (Courtesy of Blakley Electric Ltd)

Figure 13.4 illustrates a single phase transformer assembly in which each of the 110 V outputs is protected by a circuit breaker.

Socket-outlet and extension outlet assemblies

To improve the versatility of the TAs and facilitate the rapid provision of 110 V supplies anywhere on site, multiple Socket-Outlet Assemblies (SOAs) and Extension Outlet Assemblies (EOAs) are specified. An SOA has an incoming length of flexible cable terminating in a BS EN 60309–2 plug for insertion into a 110 V socket-outlet on a TA or on another outlet assembly. The cable is connected to busbars feeding a number of BS EN 60309–2 socket-outlets controlled by circuit breakers with excess current protection. The EOAs are similar but without the circuit breakers. The EOAs can be fed from the socket-outlets on either the TAs or SOAs. The SOAs and EOAs are light enough to be carried by one man. Figure 13.5 shows an example of an EOA.

Installation practice

Initial planning

The CDM Regulations require the installation designer to cooperate with other designers, such as the architect, other contractors, and the client to ensure that the installation will be safe both initially and subsequently as it is altered to suit the various phases of the construction work. This entails a consideration of the risks and a determination of the means to be used to minimise them. When the site work begins, the electrical contractor responsible for the temporary installation has to plan the work so that it can be done safely.

At the initial planning stage and as soon as the maximum demand has been assessed, negotiations with the electricity distribution company should start so as to ensure the supply will be available when required. In some cases the electricity company may have to obtain wayleaves for an overhead line to supply the site, and this can take months. For larger sites with a substantial demand, the company may have to supply at high voltage and will then require a site location for a temporary substation. When it is known that a permanent substation will be required to supply the development, it will sometimes be economic to build and equip the substation first and use it to supply the site.

The position of the supply intake where the ISA is situated should be agreed with the distribution company and should be located where it will not be disturbed for the duration of the project and where it is not liable to be damaged by vehicles and mobile civil engineering plant. The location will usually be on the site periphery and should be selected to minimise the length of the distribution company's service cable or overhead line and its consequential costs.

If work has to commence before a DNO supply is available, a temporary supply may be obtained from a mobile engine-driven generating set which could be hired.

The electricity distribution companies generally use single-core PVC-insulated and sheathed conductors without further protection for any connections between their cut-outs, metering equipment and the consumer's main switchfuse or circuit breaker if all the equipment is not contained within a single ISA, and very often this wiring is not secured. It is not, therefore, as safe as the site installation wiring, so access to it should be restricted to authorised and competent persons. Any building or hut in which the equipment is installed should not be used for other purposes, such as an electrician's store, and the door should bear an electrical danger notice.

The MDA is akin to a main switchboard and should be located at the load centre. On scattered and large sites it may be convenient and more economical in sub-circuit cabling to employ several MDAs at local load centres. The locations chosen should, preferably, be in positions where the equipment can remain for the duration of the contract as relocating could prove inconvenient, time consuming and expensive. To facilitate connecting the MDAs on a ring main, or several on a single cable, double incoming terminals should be provided for looping.

The location of the TAs is not as critical as the MDAs because they are more readily disconnected and can conveniently be moved from time to time, to suit the needs of the users of the 110V portable tools and apparatus.

Distribution cables

BS 7375 recommends plastics-insulated armoured cables for site distribution, except where the risk of mechanical damage is slight. There are not likely to be many locations free from the risk of mechanical damage, so it is probably wise to standardise on PVC-insulated, single-wire-armoured and PVC-sheathed cable and to equip the distribution units with glands and armour clamps suitable for this cable. Screwed spout entry, bolted-on glands are better than securing the gland with lock nuts on each side of a hole in the gland plate. The former method provides a joint that is less likely to corrode and/or loosen. Glands with earthing terminals are available and are recommended as they enable a direct bond to the earthing terminal or bar to be made.

There are differing views on installing the cables. On balance, it is probably better to run the cable on the surface where it is visible rather than bury it. Buried cable routes are supposed to be surface marked, but on construction sites it is difficult to preserve the markers and the cables then become vulnerable to damage from excavators. They are more difficult to recover on project completion for use elsewhere, whereas surface cables are easily recovered. Although surface cables are more readily damaged, they are tough and resistant to mild abuse. Where they cross site roads, however, damage should be prevented by installing them in buried ducts or in steel pipes within surface ramps.

Cables subject to frequent movement, such as those used to supply mobile cranes, should be flexible to BS 6708. This standard calls for a protective conductor other than the armour and for circulating current earth monitoring to prove its integrity (see Chapter 3). Cables in the 110V system are usually unarmoured plastics-insulated and -sheathed. The safety feature is the comparatively low voltage to earth, so damage does not cause a serious shock hazard.

Overhead lines

Overhead lines for site distribution at mains voltage are discouraged because of the shock hazard from contact with the jibs of cranes and excavators, bodies of tipping lorries, ladders, scaffold poles and so forth. It is not easy to avoid such contacts, which are responsible for a number of electrocutions and electric burn fatalities annually. The Electricity at Work Regulations 1989, Regulation 14(c), stipulate that when work is being carried out adjacent to live overhead lines suitable precautions must be taken against injury.

An overhead line distribution system is unlikely to be economic as any saving in initial cost is likely to be more than lost by the cost of fencing off the route to comply with the law. Furthermore, an overhead system and its fence is not readily rerouted from time to time to suit the progress of the work. Overhead line systems are, therefore, best avoided.

See Chapter 14 for information on the avoidance of danger from overhead lines.

Electrical protection

Short circuit rating

Fault protection against overcurrent and short circuit current is required and is mainly provided by circuit breakers fitted with overload trips. At the initial planning stage the distribution company should be asked for the prospective fault level at the supply intake so that adequately rated equipment may be selected. At this position, if a switchfuse is used to control the installation and BS 88 HBC fuses are employed, there should be no problem, as the breaking capacity is not fewer than 80 kA at 400V. If a moulded case circuit breaker is used, however, it is necessary to select one with adequate fault breaking capacity as they are made with a range of ratings.

If the distributor provides a supply from its LV network, the quoted fault level is unlikely to exceed 16 kA, but a higher fault level could occur on large sites where a substation is needed – about 27 kA, for example, at the output terminals of a 1000 kVA transformer with 5% impedance. The prospective fault level becomes progressively lower as the distance from the intake increases, mainly due to the impedance of the distribution cables, so a lower fault rating for the MDAs and MCCBs is permissible.

At the initial planning stage the prospective fault level should be calculated for several locations in the 400V distribution system so that apparatus of adequate fault rating may be used. Inadequately rated apparatus, for instance an automatic motor starter, can often be protected by back-up HBC fuses, providing they have a suitable time/current characteristic so that they will rupture before the contactor is damaged by arcing in the event of a short circuit.

Fault protection – disconnection times

Fault protection using overcurrent protective devices will be satisfactory if the earth fault loop impedance is sufficiently low for enough fault current to flow to operate the protective device within the recommended time. BS 7671 used to require that construction sites had lower disconnection times than other lower risk locations, but the required disconnection times are now the standard times set out in Chapter 41 of BS 7671:2008 (see Chapter 12 of this book).

To ascertain the earth fault loop impedance, Zs, the designer needs to know Z_E, which is the external earth loop impedance measured at the supply intake. The earth loop impedance between the intake and the relevant item of equipment can then be added, I_F calculated, and the suitability of the protective device ascertained from its time/current curves.

If the distributor provides a separate neutral and earth TN-S service from its LV network, it should permit the use of its own earth terminal for the site earth and ought to be able to estimate Z_E at the intake to enable the designer to proceed with the calculations. If the service is combined neutral and earth, TN-C-S, the distribution company will not permit the use of its terminal for earthing, and the contractor will have to provide a site earth electrode and terminal, in which case the value of Z_E cannot be determined without site measurement.

A site earth electrode will not provide as low a value of Z_E as a distributor's earth terminal. In many cases the contractor will find it uneconomic to construct an earthing facility able to provide a low enough value of Z_E to achieve the required disconnection times using overcurrent protective devices. In these circumstances the remedy is to use in the ISA an MCCB fitted with earth fault protection or a separate RCD. As there is bound to be some earth leakage in the site installation from capacitive currents, electric heating elements, and radio interference filters, the residual operating current should not be set at too low a value; otherwise nuisance tripping of the whole installation may occur. The 300 mA to 500 mA range should be satisfactory. As required by BS 7671 Regulation 411.3.2.4 for TT systems, the MCCB or RCD should be capable of disconnecting the installation in not more than 1 second when the selected residual operating current is exceeded. For the majority of sites this arrangement will suffice, but for very large sites it may be considered prudent to sectionalise the system and provide local earth fault protection capable of discriminating against the intake protection so as to shut down the relevant section, instead of the whole site, in the event of an earth fault. To attain this, time-delayed protection will be needed at the intake.

Where the distributor provides a site substation there should be no problem about using its earth terminal. Z_E will be very low and the overcurrent protection should be adequate for fault protection except for the larger installations where very long cable runs may introduce impedance that adversely affects the disconnection times and necessitates the use of an RCD, although care should be taken to ensure the volt drop does not create problems.

All portable apparatus on construction and demolition sites should preferably operate on the comparatively safe 110 V system where the tripping time, for shock protection, is not so important because the primary safeguard is the low voltage to earth. However, BS 7671 Regulation 411.8.3 requires an automatic disconnection time not exceeding 5 seconds, and in many cases this will require the use of an RCD as well as an overcurrent device in each line conductor, particularly where there are long cable runs introducing impedance into the earth fault loop.

If 230 V portable apparatus is used, fault protection should be provided by means of overcurrent protective devices with an additional RCD with a rated operating current of not more than 30 mA. Such devices also afford protection against direct electric shock, although RCDs must not be treated as a main and sole means of achieving basic protection against direct contact injuries. This facility can be provided by fitting an MCCB in the MDA with sensitive earth fault protection or by interposing a sensitive RCD in the supply from the MDA to the portable apparatus. Although the use of mains powered apparatus should be discouraged, where there is no alternative to its use, in addition to using RCD protection it must be maintained in good condition to minimise the risk of injury.

Confined conductive locations

In confined and conductive locations the potential electric shock hazard is increased and special precautions are needed, as covered in BS 7671 Section 706 (see Chapter 12). Examples of confined, conductive spaces are inside boilers and other metal vessels or inside metal pipes, flues and ducts where the area of body contact with earthed metalwork is likely to be substantial. Even if the interior is dry, the shock risk is enhanced, but if it is damp it is worse. In these circumstances the 110 V system is not considered safe and pneumatic, hydraulic or battery-powered tools are advocated.

For lighting, battery-powered cap and handlamps could be used or the luminaires could be supplied from an SELV or electrically separated source. The source transformer should not, of course, be taken into the confined space.

Fixed equipment may be supplied at low voltage with automatic disconnection and with supplementary equipotential bonding with exposed conductive parts connected to the extraneous conductive parts in the location. Alternatively, Class II equipment may be used provided its enclosure is suitable for the location and it has 30 mA RCD protection. Electrical separation, SELV and PELV supplies may also be used but the PELV option requires earthed equipotential bonding. Unless it is part of the fixed installation, the power source for an SELV or PELV supply has to be situated outside the restricted conductive location.

Lighting

Site lighting may be divided between the fixed lighting provided for security and/or safe working and movement and local lighting used by individuals or small groups. In the first category there are floodlighting systems where the luminaires are mounted at high level on poles, cranes and other structures and usually operated at mains voltage. Robust, totally enclosed, weatherproof (IP55) luminaires are preferable. Open types, where the lamp is exposed, should only be used where they are not vulnerable to damage from a carelessly handled scaffold pole or length of conduit, for example. During non-working hours, it is better to switch off the installation at the intake, but if it is necessary to leave any circuits energised, they must be reasonably safe and protected against mischievous children, vandals and thieves. Security lighting luminaires, for example, should be pole-mounted or fixed in relatively inaccessible positions.

Luminaires that are within reach and/or used and handled by site personnel should be operated on the 110V system. Festoon lighting, unless out of reach, should also operate on the 110V system. If operated at mains voltage, only the type with moulded-on lampholders, designed to seal against the glass bulb of the lamp, should be used. The detachable types, which have spikes to penetrate the cable, are not suitable because water may enter the joint and provide a conducting path between a live conductor and anyone handling the wet cable. Moreover, when the lampholder is removed, the holes left by the spikes again permit the ingress of water.

There are obvious difficulties in designing a lighting installation to cater for the changing needs on a construction site but, even so, good practice, such as described in lighting guides published by the Chartered Institute of Building Services Engineers (CIBSE), should be observed to ensure that the lighting contributes to, rather than detracts from, site safety.

Installation and maintenance

Safety is dependent on the quality of the installation work and the subsequent maintenance. If the initial installation work is competently carried out to the requirements of

BS 7671 and a planned maintenance system inaugurated on its completion, the installation should be safe and will have a better chance of remaining safe for the duration of the project.

On completion of the initial installation work, the initial verification tests specified in BS 7671 should be carried out, including checks on the actual prospective fault levels (there is an instrument available for this purpose) to ensure that the circuit breakers originally selected, at the planning stage, are adequately rated.

A planned preventive maintenance system for the fixed installation should be drawn up and should include frequent visual inspections and combined inspections and tests; IET Guidance Note No. 3 *Inspection and Testing* recommends a maximum of 3 months between the inspections and between combined inspections and tests. The aim is to detect damage and defects which may create danger and which should be repaired immediately. If repairs have to be deferred, the defective apparatus should not continue in use but should be disconnected from the system until it has been made good.

In addition to the fixed installation, all other electrical equipment and apparatus used on the site should be included in the preventive maintenance programme. This will include portable electrical tools, handlamps, lighting equipment and so on. By far the most important element of this maintenance is a routine visual examination of the equipment, which will detect most faults that can lead to danger. These examinations should be carried out as pre-use user checks and then periodically as part of a formal visual inspection, typically at a frequency of once every week for 230 V hand-held equipment and extension leads, and monthly for 110 V equipment and fixed 230/400 V equipment. Guidance on the frequency of these examinations, and the frequency of tests aimed at detecting defects, is published in the HSE's guidance note HSG 107 *Maintaining portable electrical equipment* and the IET's *Code of practice for in-service inspection and testing of electrical equipment.*

The persons carrying out the maintenance need to have the competences to carry out the work safely and to detect failures or deterioration that may lead to injury. Whereas visual examinations do not need to be carried out by electrically qualified persons, testing work carried out on the fixed installation should be carried out by electricians or similarly qualified tradesmen.

Systems of work on the permanent installation

Work on the permanent system being installed during the construction project is not electrically hazardous so long as it is not energised. Problems can arise, however, as a project nears completion and other trades ask both the principal contractor and the electrical contractor to energise the supplies to their apparatus even though the electrical installation is not complete and commissioned. The following example of a fatal accident illustrates the point.

Case study

The accident occurred at a city centre construction site in Dundee where a row of two-storey shops was being built, one of which was destined to be a sports store and gym complex. It happened at the end of the construction project, the day before handover to the client was scheduled to take place and when there was considerable pressure on the principal contractor and subcontractors to meet the project deadline. To ensure completion on time the electrical contractor, Mitie Engineering Services (Edinburgh) Ltd, called extra electricians to the site to work on the final fit-out of the low voltage installation.

A number of low voltage distribution boards with circuit breakers had been installed in the premises, some of which had already been energised both to supply power required by other contractors and to allow electrical testing to take place as part of the installation's initial verification. Where circuits fed from these distribution boards needed to be isolated, the circuit breakers were switched off and insulation tape applied across the actuator. At that time, this was common practice in the electrical contracting industry, although it did not conform to the safe isolation procedures using locking kits and padlocks that electricians were taught during their apprenticeships.

One of the electricians called to site to assist with the fit-out was an experienced and graded approved electrician. He was tasked with connecting up a single phase low voltage circuit supplying a security system. The 3-core cable he was told to use as the power source had already been connected into one of the live distribution boards and, at the field end, it was coiled above a false ceiling with a handwritten label attached to it marked 'Not in Use'. He needed to strip the insulation off the cores and connect them to another cable supplying the security system. He gained access to the cables above the false ceiling using an aluminium ladder.

Some time before the accident the supply to the cable marked 'Not in Use' had been mistakenly switched on, but this configuration was not reflected in the marked-up circuit schedules for the energised distribution board. Without having tested the cable with test equipment to prove that it was dead, the electrician cut the end off the cable and then started to strip the insulation off the live wire using his cable snips. When the snips touched the live conductor their metalwork would have become live at about 230 V with respect to earth. He suffered a fatal electric shock, with the most likely shock path being between a finger touching the metal parts of the snips and some earthed metalwork, most probably the ceiling's earthed support frame.

The accident occurred before the Corporate Manslaughter and Corporate Homicide Act was enacted. Mitie Engineering Services (Edinburgh) Ltd, the electrician's employer, was prosecuted under the Health and Safety at Work Act Section 2(1) for failing to manage the safety of its employees, found guilty and fined £300,000. Two directors and a manager of the company were also prosecuted, under Section 37 of the act, but they were acquitted.

The circumstances of this tragic accident highlight the need for careful management of electrical safety in construction projects where distribution boards are energised before the installation is complete, especially where people are working in a pressurised environment to meet project milestones. There must be a suitable risk assessment covering energisation procedures, with clear delineation of responsibilities and information provided to the electricians working at the site. Effective management and supervision of the implementation of safe and secure isolation procedures using proprietary locking devices, especially on incomplete circuits, and correct maintenance of circuit records are of paramount importance. Moreover, the competence of electricians working on the job is of crucial importance, particularly with regard to implementing safe isolation procedures and proving that conductors are dead before touching or working on them. Employers have a duty to ensure competence, as well as to provide their employees with the information, tools and test equipment (such as a voltage detector needed to prove dead) required to carry out their work safely.

Existing installations

It is more difficult to avoid electrical safety problems on alteration, extension and refurbishment work if the existing installation has to be kept energised. Adequate safety rules and work planning are not only legally required to comply with the Electricity at Work Regulations, the Management of Health and Safety at Work Regulations and the CDM Regulations but also are essential if the hazards of live working are to be contained. As well as achieving legal compliance, the effort is usually economically worthwhile.

Safety responsibility

The responsibility for site safety is often complicated where a number of different firms are present, each of which will have its own safety policy and some of which will have their own safety rules. The CDM Regulations allocate safety responsibility, with the principal contractor, designer and the client having key roles in coordinating health and safety during the planning and constructions stages of a project. In particular, they should lay down the site safety rules and enforce them.

The principal contractor normally provides the temporary electrical installation. If he has an 'in-house' electrical department, he may well install and maintain it and, in such a case, he is responsible for its safety. The electrical staff member in charge of the site installation will exercise the technical responsibility. This official should ensure that the planned maintenance is done and that defects and breakdowns are reported and rectified immediately. Some main contractors hire the distribution equipment and arrange for the hirer to install and maintain it. As the occupier, they still retain responsibility for the safe use of the electrical installation, but responsibility for its maintenance in a safe condition rests on the hirer who should appoint competent, electrically qualified persons to do the work.

Responsibility for the safety of connected apparatus generally rests on whoever has responsibility for its presence on site. The principal contractor, however, would be well advised to stipulate, in the site safety rules and perhaps in the subcontracts, that any electrical apparatus or electrically powered equipment brought onto site must be serviceable and safe. Arrangements must be made for its safe installation and maintenance and for users to be trained to use it safely. The rules should also prohibit the use of apparatus that becomes defective until the defect has been repaired.

If an electrical subcontractor energises part of the permanent installation before commissioning and handover to the client, he is responsible for its safety and for coordinating with other companies on the site, and must take the precautions already described. Although he is not responsible for the safety of the connected apparatus installed by other trades, he has a duty to notify them both before energising and before switching off the supply to their apparatus.

Underground cables and overhead lines

Introduction

Two of the most common forms of electrical accident are underground cable strikes and contact with overhead power lines. The former mostly result in burn injuries but few fatalities; the latter result in burn and electric shock injuries and account for roughly half the total number of electrical fatalities at work. Both tend to result in societal disruption caused by the inevitable loss of supplies to customers supplied from the affected networks. This chapter describes the safety precautions that can be taken to minimise the chances of these types of accidents occurring.

Underground cable strikes

Incidents

Every year there are thousands of incidents involving damage to buried cables, some of which result in burn injuries, although fatalities are rare. Most of the incidents occur during street works where utility (water, gas, electricity, telecommunications, and cable TV) companies, construction companies and local authorities carry out excavations in the highway and damage an underground electricity cable. Other common locations are construction and demolition sites.

Most damaged cables belong to electricity DNOs but some are local authority street lighting supplies, or belong to IDNOs, railway companies, generation and transmission companies, and private companies.

The burn accidents invariably occur during manual excavation work when, for example, the tool of a pneumatic drill penetrates a concealed, buried cable. A short circuit will occur when the bit comes into contact with one or more of the live conductors and the metallic armouring and/or sheath or the neutral/earth conductor where a PME distribution system is in use. The initial phase/earth or phase/neutral fault usually develops to involve the other phases, causing arcing which may emerge as a highly energetic flame and blast from the hole made by the tool, injuring the operator.

As an example, a cable strike occurred at a construction site near a primary school where a contractor was erecting a fence. One of the company's employees suffered minor burns to his face and hands when a tool he was using to dig post holes struck and penetrated a 400 V service cable. The presence of the cable was known to the contractor but the information had not been passed to the worker and the cable's position had not been marked on the ground. The company pled guilty to breaching Regulation 34(3) of the Construction (Design and Management) Regulations 2007 and was fined £7500.

The electricity companies' low voltage mains are frequently involved in these types of incidents because there are more of them and because they are not well protected against mechanical damage. Historically, they did not have to be covered by cable tiles or warning tape or placed in ducts, the requirement for such protection not being introduced until the Electricity Safety Quality and Continuity Regulations 2002 were enacted (see Chapter 7). Also, they are often not well protected against overcurrent, with the fuse protection in the distribution secondary substation tending to be slow in operation, or failing to operate at all, meaning that the arcing persists until the fault burns itself clear.

The number of burn injuries resulting from HV distribution cables being struck tends to be lower than the injuries from LV cables being struck. This is because there are fewer HV cables, and they are usually buried at a greater depth and are more likely to be physically guarded, with cable tiles for example, and because the electrical protection on HV cables operates more rapidly so that the fault does not develop so far.

There are few reported accidents where the casualty receives an electric shock. This is probably because the faults usually involve an earthed conductor such as metal sheath, screen, armour or tape surrounding the cable so the tool does not attain a significant voltage with respect to the surrounding ground and because very often the operator is insulated from the ground by footwear such as rubber-soled boots.

In the past, most of the supply authorities used paper-insulated, lead-sheathed, single-wire or steel tape armoured and served cables and very often protected them with cable tiles. For economic reasons cable tiles are seldom used now, plastic warning tape or tiles being preferred. Also, many of the high and low voltage cables being laid nowadays do not have a layer of wire armouring, having instead an earthed copper wire screen, or a corrugated aluminium sheath, or in the case of low voltage PME Waveform cables a layer of copper neutral/earth wires; the outer plastic sheath is made of PVC or cross-linked polyethylene (XLPE). Every year more new unarmoured mains cables of this type are laid so the resultant hazard to excavators is growing, and more accidents can be expected to occur unless the construction and other industries become more effective at taking precautions against striking buried cables.

A considerable effort has been made to reduce the frequency of these accidents. There has been no lack of safety advice from the HSE and the electricity companies. In 1977 the public utility companies formed a committee called the National Joint Utilities Group (NJUG) which studied the problem and issued a series of safety guidance booklets. NJUG continues to publish guidance, a good example being its publication *Guidelines on the positioning and colour coding of underground utilities' apparatus*, the latest edition of which is the 8th edition published on 29 October 2013. Some of the larger civil engineering contractors established safety training schools for their staff, and the Construction Industry Training Board (CITB) is a major provider of safety-related training. The HSE's main guidance document on the topic is HSG47 *Avoiding danger from underground services*, which provides comprehensive guidance on the systems of work that should be adopted to prevent underground services being damaged; the latest edition was published in 2014.

Safety precautions

It is always best to avoid excavating near buried services. However, where such work cannot be avoided, the precautionary safety principles are fairly simple and comprise the following:

(i) Only trained personnel should organise and carry out excavation work in the highway or anywhere else where there could be buried cables.
(ii) No work should start until location drawings of the buried cables and other services in the locality have been obtained.
(iii) The position of the buried cables shown on the location drawings should be checked and confirmed using a cable locator, commonly called a cable avoidance tool or CAT scanner.
(iv) The route and depth of the cables should be marked on the surface of the ground.
(v) Further markings should be provided 0.5 m on either side of the cable route to indicate the danger zone.
(vi) Any excavation necessary within the danger zone must be carried out with great care to avoid damaging the cable.
(vii) The exposed cable must be supported where necessary and protected from damage.
(viii) Reinstatement of the cable should be in accordance with the owner's instructions.

Training

The training referred to in item (i) can be in-house or external but should be undertaken by all personnel before they undertake any excavation work where there may be buried cables. The training should embrace the legal requirements such as Regulation 14(c) of the Electricity at Work Regulations 1989, the risk assessment requirements of the Management of Health and Safety at Work Regulations 1999, and the duties on clients, designers, principal contractors, and contractors, under the CDM Regulations 2015.

It should also cover how the hazard arises; the type of burn injuries that can occur; accident statistics; accident and dangerous occurrence reporting requirements; cable laying practices including types and colours of cable, depth of laying, physical protection and locations; cable location plans and how to obtain them; permit to work practices and procedures; and the use of cable locators. The practical work should demonstrate the use of the cable location plans and cable locators, marking the cable route and the danger zone, the use of power tools in and near the danger zone, digging trial holes to locate cables, the use of hand tools to expose the cable, protection of the exposed cable, and backfilling precautions.

Training in the use of the cable location devices is especially important as many accidents have occurred as a result of the operator of the device not knowing its limitations or how to use it properly. Most manufacturers of cable locators offer training in the use of their products.

Drawings

In regard to point (ii), it is not sufficient to contact just the local electricity company's district or regional office to obtain cable plans as there may be buried cables belonging to other organisations, such as cables supplying street lighting, other 'street furniture', railway systems and so forth. Under the New Roads and Street Works Act 1991 the local authority has to keep a register of information concerning the apparatus in the streets and coordinate the activities of excavators digging them up. The excavator should therefore be able to obtain all the details of the buried services from the local authority without the need to approach all the undertakings before excavating. In some regions there are coordinated single-point-of-contact services covering information on buried utilities, such as Linesearchbeforeudig and the Susiephone (Scottish Utilities Services Information for Excavators) service in Scotland.

It should be borne in mind that the plan may not be entirely correct, as it can only reflect the knowledge available to the cable owner and in many cases will show the cable location as originally laid. Since then excavation work may have resulted in an unauthorised displacement of the cable by others who may not have notified the cable owner, or the ground may have been regraded thus altering the cable depth, or reference points showing the cable location dimensions, such as kerb lines, may have been changed. References on the drawing to cable tiles or marker tapes should be treated with caution, as subsequent excavation work may have displaced such protective devices. If the type of cable is not specified on the drawing, the owner should be asked for the details.

Cable locators

There is a variety of cable locators readily available, varying in sophistication and specification with the more capable, and expensive, devices being able to discriminate between different underground structures and give an indication of the depths of cables. They are known generically as CAT (Cable Avoidance Tool) scanners. There are three types of functions that can be incorporated into these devices: (i) the detection of the power frequency (50 Hz or 60 Hz) magnetic field of a live cable, (ii) the detection of re-radiated low frequency radio signals, and (iii) and the detection of metal. The latter type is of limited value in urban area highways because there is a great deal of buried metal of other services present, making it almost impossible to identify the cable being sought. Most cable locators on the market incorporate the first two passive detection techniques in one device, with them being selectable modes such as 'power mode' and 'radio mode'; the principle is illustrated in Figure 14.1.

Two other types of detector have become increasingly available in recent years. Firstly, ground mapping radar can be used to map out the distribution of underground services including electricity cables, although there are limitations in its capability. Secondly, passive Radio Frequency Identification (RFID) tags can be buried alongside cable routes to allow them later to be detected and identified using compatible RFID detectors.

The types of locator that detect a cable's magnetic field are most effective when the cable carries a substantial current because the field is proportional to the current – the higher the current, the greater the magnetic field strength and the likelihood of the cable being detected. Cables that are live but carry little or no current generate insufficient magnetic field to be readily detectable. This fact is frequently ignored or not understood, and cables that are live but off load, such as pot-ended cables or 3-phase cables with current well-balanced across the 3 phases, are often not detected until they are struck and somebody injured.

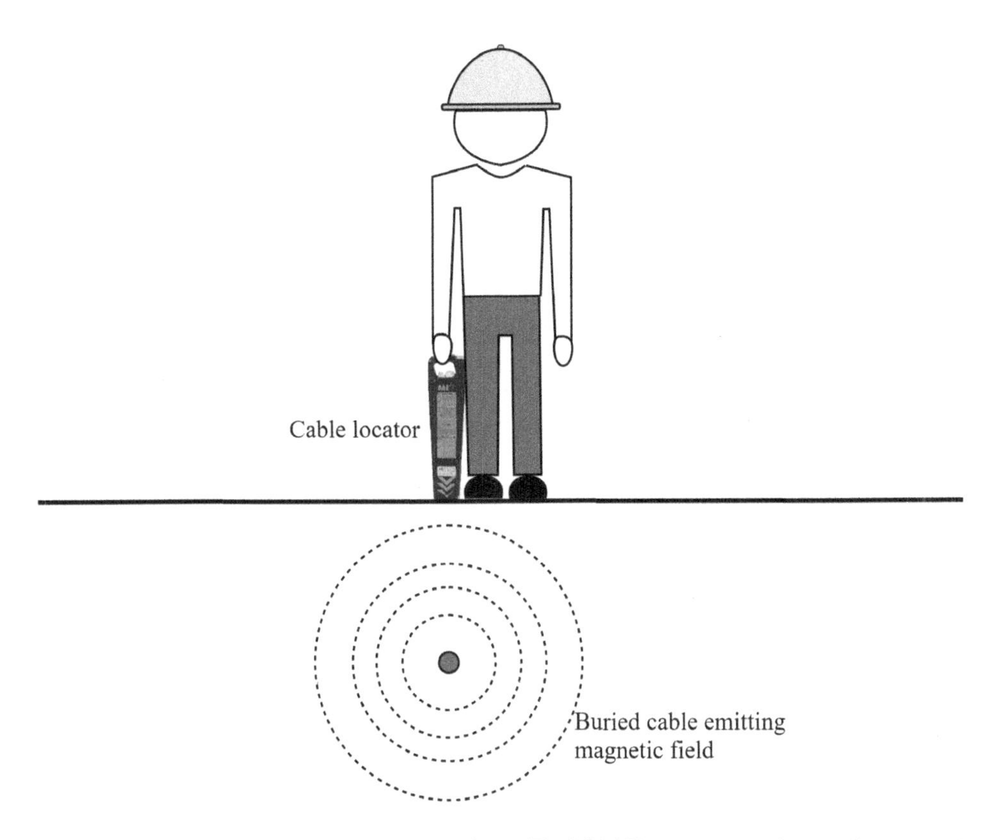

Figure 14.1 **Principle of using a Cable Avoidance Tool (CAT) scanner to detect the magnetic field emitted by buried cable**

This problem can be overcome by injecting a signal into the buried cable at a position where its location is known and then searching for it at locations where excavation works are planned. A purpose-built small portable signal generator supplied with most models of cable locator is used to inject signals in the band 30–300 kHz either by induction or by direct connection using a cable clamp or plugs. The cable locator's receiver is switched to a mode to detect the signals and the ground scanned until the signals are detected and the cable located. This active technique, commonly known as 'CAT and genny', has significant advantages in allowing cables to be detected in congested areas and in facilitating depth measurements. However, it relies on the operator knowing that the cable is there in the first place, which is not always the case. The important benefits of using a transmitter with the locator are often overlooked, and the unwillingness or inability of cable locator operators to use them is a frequent cause of cables not being successfully detected and located.

Cable locators should be used only by personnel trained in their use. It is good practice to use them from time to time as the excavation work proceeds until the cable is exposed. Quite often, and particularly in urban areas, there may be more than one cable laid at the same or different depths. Care in the use of the cable plans and locators is needed to ensure that the location of all the cables is determined and that cables located in close proximity to other cables, including above other cables, are properly resolved. Cables are not always distinguishable by sight from other services, but the use of a locator should enable the operator to be sure.

Cable route marking

On paved surfaces the cable route and danger zone lines can be chalked or painted, but elsewhere it will usually be necessary to delineate them with pegs driven into the ground. Wooden pegs are safer than metal, as they are less likely to damage the cable.

A number of serious accidents have occurred when the marks have been covered up. For example, a construction company was laying in a road and pavements to a new housing estate. The principal contractor had located an 11 kV buried cable and marked out its route and depth using yellow paint sprayed onto the ground surface. After a few days the ground was then covered by a subcontractor with a shallow layer of hard-core as a base for the road and pavement; the hard-core was laid over the yellow markings. A labourer then came along to drive short lengths of steel reinforcing bar into the ground that would act as supports for the formers for the concrete onto which the kerb stones would be placed. He had no knowledge of the buried cables and, of course, never saw the yellow markings which were underneath the hard-core. He drove one of the lengths of bar into the buried cable; the bar was blown back from the fault and hit him in the face, causing severe injuries that resulted in the loss of his left eye, and he suffered burn injuries to his chest and face. The moral of the story is, perhaps, obvious.

Excavation

Where a substantial amount of excavation is required in the danger zone, it is advisable to dig one or more trial holes to find the cable and then work along it from the trial holes to expose it. Whereas the electricity companies lay cables to standard depths, as shown in Table 14.1, it cannot be guaranteed that the cable being excavated will be buried at these depths, in which case careful manual methods should be used to avoid damage to the cable.

Surface construction may require the use of power tools to break it up for removal, in which case it is advisable to fit the bit with a collar to prevent deep penetration and no drilling should be done immediately above the cable. Sharp-pointed tools such as picks

Table 14.1 Recommended depth of underground cables (From NJUG Guidelines, volume 1)

Type of Ground	*Voltage*	
	High Voltage	*Low Voltage*
	Minimum Depth (mm)	
Footpaths, verges, uncultivated land	450-900	450
Agricultural Land	910	910
Footway	450-1200	450
Roads	750-1200	600

should not be used immediately adjacent to the cable even if it is armoured, as the pick points are capable of penetrating between the strands of wire armour and through badly corroded steel tape.

The sharp edges of shovels can damage lead and aluminium sheaths and the outer copper or aluminium conductors of CNE cables. If a cable is damaged, work should cease in the immediate vicinity of the damage, a temporary barrier with a warning notice should be erected to keep people away and the cable owner asked to repair the cable before work is resumed at the damage site. It is imperative that nobody inspects the cable until the owner confirms that it has been made safe.

Sometimes cables are buried in concrete and need to be dug out of it, requiring the concrete to be broken up. When this situation arises, the risk of damaging the cable is such that it is always best to ask the owner if the cable can be made dead for the duration of the work. If the owner cannot or will not make the cable dead, and the excavation has to proceed, a task-specific risk assessment will be required, with the content agreed with all parties. The risk assessment is likely to conclude that the use of hand-held power tools near the cable is unsafe and that there is no alternative to using a machine such as an excavator or breaker. The risk of damaging the cable will be high, in which case it may not be sensible or acceptable to proceed with the excavation while the cable is live and an alternative solution will have to be found.

As far as the person carrying out the excavation is concerned, it is safer to do the excavation using a mechanical digger with a back-hoe or similar implement. If the cable is struck, the person in the vehicle's cab is unlikely to be injured. Any banksman or other labourer in the vicinity of the digging tool may suffer an injury, though, from the explosion products of cable strikes, and this does happen occasionally. Not unnaturally, cable owners condemn the use of excavating machinery in the vicinity of their cables because of the likelihood of damage but, on the other hand, the use of such machinery does prevent injuries. It does, however, increase the number of reportable dangerous occurrences.

Cable safeguarding and reinstatement

Exposed cable spans of more than about 1 m should not be left unsupported, and such spans should not be used as steps or to support anything. It is better to avoid moving cables because of the danger of damaging them inadvertently. Old cables may have embrittled sheaths and/or corroded armour and are vulnerable to damage if moved. Cables that have been displaced should be secured where necessary to prevent them from falling back into the excavation, and suitable guards should be provided to prevent damage to them. The cable owner's advice on reinstatement should be sought and followed. If this is unavailable the contractor should level the bottom of the trench and firm it up to prevent subsequent subsidence.

The cables should be relaid in a bed of sand or riddled earth, free from sharp stones and at their original spacing from each other and from other services. The cables should be handled with care, avoiding unnecessary bending and strain. They should be covered with sand or riddled earth and tiles or warning tape replaced above them. Where there is sufficient slack, the cables should be snaked uniformly to alleviate possible strains in the event of subsidence. If the cables have been displaced from their former positions, when they are relaid the cable owner should be notified before backfilling so that the location plans can be amended.

Overhead power line strikes

Incidents

Accidents involving overhead power lines were described in Chapter 2 and the legal requirements for the safety of overhead lines were described in Chapter 7.

The accidents mostly involve uninsulated low voltage and 11/33 kV lines carried on wooden poles. Figure 14.2 shows a wooden pole supporting a 3-wire 11 kV line, a 11/0.4 kV step-down transformer, a set of low voltage fuses, and a 4-wire low voltage line. As described in Chapter 7, the minimum statutory height of these lines is 5.2 m apart from where they cross roads when it is 5.8 m. It is important that these heights are not compromised by, for example, rubbish or soil being dumped underneath a line or structures such as portable cabins that can be climbed being positioned underneath.

Accidents involving transmission lines are very rare. However, a fatal accident involving a 132 kV transmission did occur in 2011 on the Isle of Skye. The driver of a timber handling crane died when the boom of the crane he was driving along a forest track contacted the lowermost line conductor of a 132 kV overhead line that crossed the track. The power line belonged to Scottish Hydro Electric Power Transmission Ltd and supplied electricity to the Isle of Skye and the Scottish Western Isles. The electric current that flowed to earth through the vehicle ignited the combustible materials, including the tyres. The consequential fire destroyed the vehicle, and the driver died when he jumped out of the cab onto the surrounding land that was energised by virtue of the earth fault current. He had been driving along the track with the crane's boom raised to a height of about 9 m when it struck the conductor, which was above its minimum statutory height of 6.7 m.

Precautions against contact

The benchmark guidance for avoiding contact with overhead power lines is the HSE's guidance note GS6 *Avoiding danger from overhead power lines*, the 4th edition of which was published in 2012. The core advice is that where a work activity is to be carried out near a live overhead line and there is a foreseeability of the wires being struck, a risk assessment must be generated to assess the risks and work out the control measures needed to reduce those risks so far as is reasonably practicable; it is best to consult the owner of the line when drafting this risk assessment to obtain advice on required safety clearances. A risk assessment is not required where there is no foreseeable risk of contact, such as when, for example, someone is using a hand-held strimmer to cut the grass underneath a line.

It should also be kept in mind that work near a live line is covered by Regulation 14 of the Electricity at Work Regulations 1989, so any work with an associated risk of contacting or coming dangerously close to the line should go ahead only if it is not reasonable to switch the line off; and if it is reasonable to be doing the work; and if suitable precautions can be, and are, taken against electrical injury.

The requirements for a formal risk assessment do not apply to people undertaking hazardous activities near a live overhead line that are not categorised as work and therefore do not come under the jurisdiction of the Health and Safety at Work Act, such as kite flying and river fishing. Whereas some common sense on the part of the individuals carrying out these recreational activities will help to prevent contact accidents, the owners of the

Figure 14.2 Wooden pole supporting high and low voltage overhead lines and apparatus

lines do have duties under the ESQC Regulations (see Chapter 7) to conduct risk assessments for their lines and to educate the public about the dangers where they know the activities are taking place. This might include visiting fishing clubs and schools to provide information and/or placing warning notices at the location. Ultimately, they might need to reroute the lines or replace them with insulated conductors if they cannot control the risks to the public by other means.

For construction activities, if there is a DNO's overhead line traversing the site, and it is low enough to be a hazard, the distribution company should be asked at the initial planning stage to divert it or replace it with an underground cable. Not only is this desirable for the duration of the project, but it may be necessary anyway to cater for the subsequent use of the site. If the line cannot be diverted or undergrounded, the company should be asked to make it dead for the job duration. If it is impossible to do either, the line will have to be safeguarded to achieve and maintain the safety clearances advised by the DNO.

In some circumstances, it is sufficient to erect a 'goalpost' at the entrance to the site to establish the maximum heights for any vehicles entering the site. This technique is also appropriate for locations such as the entrances to forest tracks, farmyards and agricultural fields crossed by an overhead line, although if plant such as cranes, balers and tipper trucks are used within the areas, the safest and preferable option is to move the line, or replace it with an insulated line or a buried cable.

Where plant and machinery may need to operate close to an overhead line during a construction project, and there is a risk of contact, it may be necessary to erect barriers to keep the vehicles and plant away from the line and to delineate safe passage routes underneath the line. Figure 14.3 shows one method by which this can be done, using bunting supported on wooden poles and a solidly constructed goalpost. The figure depicts a 6 m separation between the lines and the barriers; this can be varied according to the specific circumstances, but it is best to consult the DNO about this.

A rigid non-conducting goalpost is shown in the schematic, with an example of temporary portable goalposts being shown in Figure 14.4, which is a reproduction of an image used in guidance note GS6.

For wider passageways it may be necessary to use plastic or tensioned steel ropes (earthed at both ends) to span the width. In this case, the goalposts may need to be

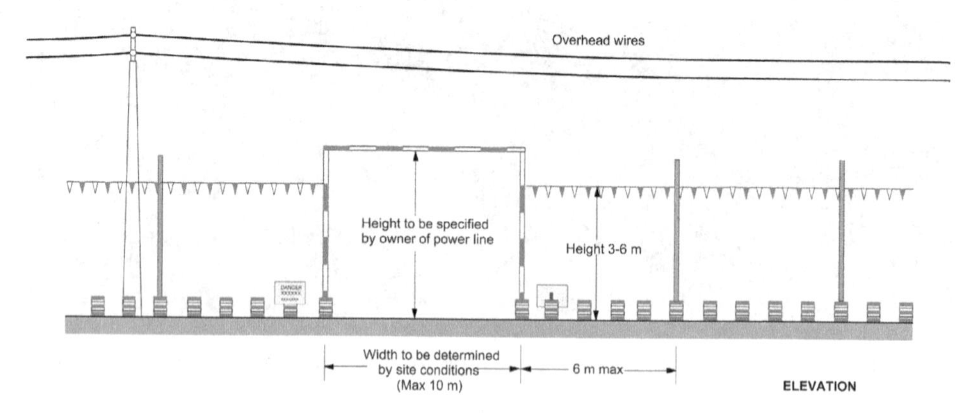

Figure 14.3 Protecting access to overhead lines

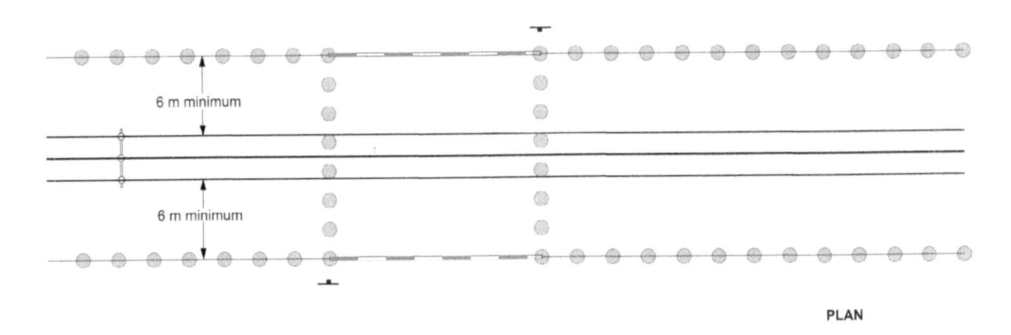

Figure 14.3 (Continued)

Figure 14.4 Temporary goalposts used to delineate passage route underneath 11 kV power line

positioned further away from the line to reduce the danger from the loss of safety clearance due to the possibility of ropes being stretched by cranes and other plant moving towards the wires.

If the line runs along the site boundary it is only necessary to provide barriers on the side where the plant is in use. If plant has to cross under the line, site roads will be required at each crossing point. The road surface should not be so uneven that the lowered plant jibs are tilted upwards, nor so rough that they bounce and hit the line conductors. Barrier metalwork which could become live either from inductive charging (e.g. a wood post and wire fence) or from contact with the overhead conductors (e.g. a wire crossbar to wide goalposts which might be caught by a jib and forced into contact with a line conductor) should be earthed.

Warning notices should be provided to instruct drivers to lower jibs and advise the height of the crossbar. The barriers should be situated far enough away from the line to ensure adequate clearance. There is no specification for the type of barriers that should be used – ditches, earth mounds, timber or concrete balks would be capable of stopping plant approaching the line, but wire or rail fences and rubble-filled drums act as warnings rather than physical barriers. If the latter type is used, bunting, at a higher level, should be displayed to attract the driver's attention. Proximity warning devices are not acceptable as a substitute for barriers but may be used as an additional warning device. For night work, the location of the barriers should be indicated by suitable lighting and the warning notices should be illuminated.

Barriers are not always appropriate for transitory work near an overhead line and when structures are being erected close to a line. There are many examples of this, some being

- cross-country pipeline laying and carriageway roadworks where the work site is constantly on the move;
- tree felling and aboricultural work near an overhead line;
- agricultural activity where farm vehicles are working fields, such as ploughing and harvesting, where barriers are simply not an option but where maps of the lines on the farm field should be generated showing maximum vehicle heights and where goalposts and warning signs could be erected at field entrances;
- refuse vehicles emptying refuse containers on streets with overhead low voltage lines;
- vehicle-mounted cranes loading and offloading loads such as building materials and portable cabins near overhead lines; and
- polytunnels, commonly used on fruit farms and plant nurseries, being erected on farmland crossed by an overhead line.

It is sometimes possible to arrange with the line owner for it to be made dead for the comparatively short period of time when the work is being carried out beneath or dangerously close to the line. If this cannot be done, a restraining device should be fitted to prevent any jibs of cranes, excavators, tippers and so forth being raised to a dangerous height. It is not usually economically practicable to erect a temporary insulating or earthed screen below the line to prevent jib contact, and the alternative of relying on a banksman to control the driver is not entirely satisfactory, as it is subject to the human error of both persons.

The Energy Networks Association (ENA) publication *Look Out Look Up! A Guide to the Safe Use of Mechanical Plant in the Vicinity of Electricity Overhead Lines* advises that, in these types of circumstances, the minimum separation from the live line should be 1 m for low voltage lines, 3 m for 11 kV and 33 kV lines, 6 m for 132 kV lines, and 7 m for 275 kV and 400 kV lines. These separation distances may need to be extended to take account of uncertainty and the possibility of unexpected movement or losing control of objects being carried, such as ladders and scaffold poles.

Tree felling and aboricultural work is commonly carried out near overhead lines by distribution and transmission companies and their contractors. This is to ensure that trees and shrubs cannot act as climbing aids that allow children to climb up to the conductors, and to prevent the trees and so forth from compromising the integrity of the lines, especially in storm conditions when they have the potential to fall onto and damage the lines, leading to widespread disruption from loss of supply. Detailed guidance on electrical safety in these circumstances is published by the ENA in its Engineering Recommendation G55 series and by the forestry industry in its publication Forestry Industry Safety Accord 804 *Electricity at Work: Forestry*. These establish safety zone requirements as well as the competence standards for people doing specialist work.

Anyone working near an overhead line must have been informed and instructed about the risks and the means of preventing contact, including the separation distances that must not be contravened. Warning notices should be posted in vehicle and tractor cabs and adjacent to operating controls on machinery such as hydraulically operated vehicle-mounted cranes.

Proximity to buildings and structures

Overhead lines can become hazards if they are dangerously close to buildings and structures such as lamp posts and mobile phone or telecommunications masts. The risk arises if, for example, a ladder can be leaned against a building or structure and climbed, perhaps by a window cleaner, technician or someone clearing the gutters, bringing the climber dangerously close to the overhead line. This includes low voltage lines that may be the service connection to a building and are attached to the building.

The required minimum clearances between overhead lines and structures are set out in an ENA publication, Technical Specification 43–8 *Overhead Line Clearances*, and should be followed by anyone erecting a structure near an overhead line, or erecting an overhead line near a structure.

The recommended clearances depend upon a range of factors including the voltage, whether or not the line is insulated, and whether or not the structure is normally accessible. For example, the required minimum clearance between an uninsulated 11 kV line and a building is 3 m, which reduces to 0.8 m if access is not required and a person cannot stand on a ladder leaning against a wall. Reference should be made to the ENA publication for detailed guidance on clearance distances.

Emergency procedures

There will be occasions when things go wrong and contact is made with an overhead line. In cases where contact is made by an object being carried, such as a ladder or scaffold pole, it is most likely that the person will be severely injured and will require urgent medical treatment to treat burns and electric shock. The emergency services will need to know the location and that overhead power lines are involved; they will contact the line's owner to warn them about the situation.

If a vehicle such as a crane or excavator comes into contact with a line, it is safest for the driver and any passengers to remain in the vehicle until the line has been made dead. However, if it is necessary to get out of the vehicle, perhaps because it is on fire, then the occupants should jump out of the vehicle as far as possible and not return to it until it has been confirmed that the line is dead.

There are two good reasons for this. Firstly, while fault current is flowing to earth from the live wires through the vehicle, the ground close to the vehicle will be energised with voltage gradients that may cause electric shocks between the feet. Secondly, even if a circuit breaker or fuse may have tripped on overcurrent and/or earth fault current to make the line dead, the company that owns the line may try to switch it back on remotely to reinstate supplies; this may cause anyone who has returned to the vehicle to be injured, an event that has happened on a number of occasions in recent years.

Many overhead lines are protected by devices called autoreclosing circuit breakers. These devices are preprogrammed to carry out one or two automatic reclosing operations in the event of a trip, although some DNOs prevent automated reclosures if the line first trips on a particular type of fault known as sensitive earth fault. This is because such faults can occur when a line breaks and falls onto dry ground, a potentially very dangerous situation. If, after the first trip, the circuit breaker continues automatically to reset and trip, it will reach a preset number of automated trips and then be prevented from carrying out any further automated reclosing operations. However, the control engineer, who will be located in a control room many miles from the scene, may take control of the circuit breaker through a telemetry link and perform another closure.

This remote switching is done because most line trips are caused by transitory events such as bird strikes or lightning activity and in the vast majority of cases the lines can be switched back on with impunity, restoring supplies to affected customers. Normally the control engineer will wait a period of time before carrying out the switching operation to see if telephone contact is made with the company or the emergency services to report that the line has been contacted; if there is no information received, the line may be reenergised, and if it trips again, arrangements will be made for an engineer to visit the scene to find out why it has tripped.

It is for these reasons that it is safest to stay away from a line that has been contacted until such time as an engineer arrives and confirms that it is dead and made safe.

15

Electrical equipment in flammable and explosive atmospheres

Introduction

The potential for electrostatic discharges and electrical equipment installed in locations where there are flammable and explosive atmospheres to act as an ignition source has long been recognised. Arcs and sparks generated in normal operation or under fault conditions, and high surface temperatures on equipment enclosures, can ignite such atmospheres, causing explosions or fires that have the potential to cause serious injury and considerable economic loss.

Over the years a considerable amount of knowledge has developed on the means by which the risks can be controlled, so much so that serious incidents arising from electrical ignition are now infrequent. However, when they do occur they can be catastrophic in nature, the Piper Alpha oil rig and BP Texas City explosions being just two well-known and tragic examples. The purpose of this chapter is to explain the principles of the main techniques for protecting against electrical and electrostatic ignition. The first section covers flammable gas/air mixtures, with the risks associated with explosions in dust-laden atmospheres considered later on.

The main legislation dealing with this topic is the Dangerous Substances and Explosive Atmospheres Regulations 2002 (DSEAR), which were briefly described in Chapter 8. This legislation deals with risks arising from a wide range of dangerous substances, including flammable solids, liquids, gases and aerosols; self-reactive substances/mixtures; and oxidising solids and liquids. In the context of electrical equipment potentially acting as an ignition source, as opposed to any other ignition mechanism, Regulation 6 of the Electricity at Work Regulations 1989 applies in addition to DSEAR. It will be recalled from Chapter 6 that electrical equipment which may reasonably foreseeably be exposed to "any flammable or explosive substance, including dusts, vapours or gases, shall be of such construction or as necessary protected as to prevent, so far as is reasonably practicable, danger arising from such exposure".

In recent years there have been significant changes to the numbering and structuring of European standards covering this topic, and the new standards are listed in this chapter. It should be anticipated that these technical standards will not be changed as a result of the UK leaving the EU, although it is possible that some of the requirements relating to the marking and conformity assessment of equipment may change.

Most applicable standards are now in the EN 60079 series with, in the UK, BS EN 60079–0:2012+A11:2013 *Explosive atmospheres. Equipment. General requirements*, providing the general requirements for preventing gas/air and dust explosions.

This chapter has the narrow focus of preventing the electrical ignition of potentially explosive gas/air and dust atmospheres. It does not cover the broader issues associated with controlling risk from flammable and explosive atmospheres, such as substituting for non-flammable materials where possible, containing the flammable materials to avoid the formation of an explosive atmosphere, and putting in place mitigation and emergency procedures. The intention is to explain what needs to be done to control the risk of ignition from electrical equipment and electrostatic discharge.

Characteristics of flammable gas/air atmospheres

A flammable atmosphere is associated with the presence in air of an appropriate concentration of flammable gases and vapours. For ignition to occur, there must be an ignition source with sufficient energy to ignite the particular gas/air mixture. The minimum amount of energy required to ignite a combustible vapour or gas is known, unsurprisingly, as the 'minimum ignition energy'. The combination of gas, oxygen and ignition source is commonly referred to as the 'fire triangle', which is illustrated as in Figure 15.1.

A flammable liquid is a liquid with a flashpoint of 60°C or below, where 'flashpoint' means the lowest temperature at which a flammable liquid gives off vapours in sufficient concentration to form an ignitable mixture with air near the surface of the liquid. This definition of flammable liquid stems from the EU Classification, Labelling and Packaging of Substances and Mixtures Regulation (no 1272/2008) (the CLP Regulation) as amended by the Classification, Labelling and Packaging of Chemicals (Amendments to Secondary Legislation) Regulations 2015.

The ignitable concentration of any flammable material in air lies between the upper and lower explosive limits (UEL and LEL). A rich (>UEL) or lean (<LEL) concentration outside these limits will not ignite. The ignition temperature is the lowest temperature at which the material will ignite due to the application of heat.

Figure 15.1 The fire triangle

Some liquids which do not emit sufficient vapour at normal ambient temperatures to be classed as flammable are ignitable when sprayed in a fine mist; Avtur, the paraffin used in jet engines, is an example.

Risk assessment and hazardous areas – gas/air atmospheres

Whenever there are dangerous substances present in a workplace, DSEAR Regulation 5 requires that a risk assessment is carried out, the results of which must be recorded if 5 or more people are employed. In the UK, the risk assessment takes the place of the Explosion Protection Document required by EU legislation. The aim is to identify the hazards and risks and what action needs to be taken to prevent fire and explosion. The risk assessment must address the following matters:

(a) the hazardous properties of the substance;
(b) information on safety provided by the supplier, including information contained in any relevant safety data sheet;
(c) the circumstances of the work, including

    (i) the work processes and substances used and their possible interactions,
    (ii) the amount of the substance involved,
    (iii) where the work will involve more than one dangerous substance, the risk presented by such substances in combination, and
    (iv) the arrangements for the safe handling, storage and transport of dangerous substances and of waste containing dangerous substances;

(d) activities, such as maintenance, where there is the potential for a high level of risk;
(e) the effect of measures which have been or will be taken pursuant to DSEAR;
(f) the likelihood that an explosive atmosphere will occur and its persistence;
(g) the likelihood that ignition sources, including electrostatic discharges, will be present and become active and effective;
(h) the scale of the anticipated effects of a fire or an explosion;
(i) any places which are or can be connected via openings to places in which explosive atmospheres may occur; and
(j) such additional safety information as the employer may need in order to complete the risk assessment.

Where measures to eliminate or substitute flammable materials have not been successful and there are locations with flammable materials that constitute an explosion risk, those locations are called hazardous areas. Regulation 7 of DSEAR requires that these areas are classified into one of three zones according to the extent of the fire and explosion risk, and the procedures for doing this in relation to gas/air atmospheres are explained in BS EN 60079–10–1:2015 *Explosive atmospheres. Classification of areas. Explosive gas atmospheres.* The zones relating to flammable gas/air atmospheres are

- *Zone 0:* A place in which an explosive atmosphere consisting of a mixture with air of dangerous substances in the form of gas, vapour or mist is present continuously or for long periods or frequently.

- *Zone 1:* A place in which an explosive atmosphere consisting of a mixture with air of dangerous substances in the form of gas, vapour or mist is likely to occur in normal operation occasionally.
- *Zone 2:* A place in which an explosive atmosphere consisting of a mixture with air of dangerous substances in the form of gas, vapour or mist is not likely to occur in normal operation but, if it does occur, will persist for a short period only.

Employers in control of hazardous areas need to determine the locations and extents of these zones in each of the hazardous areas, an exercise known as hazardous area classification. The importance of this in the context of this chapter is that only certain types of explosion-protected electrical equipment may be installed inside these zones.

Hazardous area classification

In order to determine the location and extent of the hazardous area zones it is necessary to establish the sources of release and the likely dispersion of the flammable material from the source in any direction, to the point where its concentration is below the LEL. Factors such as the rate of release, concentration, volatility, ventilation, topography and density of the gas/air mixture will need to be taken into account. A contour line joining the points so established, both vertically and horizontally, is the boundary between adjacent zones. The classification procedure needs to take account of releases during normal operation and under foreseeable fault conditions such as flange and valve failures; catastrophic failures of vessels and pipework do not form part of the assessment.

BS EN 60079–10–1:2015 provides guidance on a systematic approach that can be used to identify the volumetric characteristics of the hazardous areas, taking due account of the factors listed previously. The standard contains a number of example calculations and shows how the zones can be marked around possible sources of release. Figure 15.2 shows an example of the zones established around a tank holding a flammable liquid such as petroleum; the electrical equipment in the locality includes two lamps and associated switches, an electric motor driving a pump and its associated controlgear, and two process instruments – a level sensor and a float switch.

Establishing the boundaries between the various zones in a hazardous area is a difficult task because of the assumptions and variables involved and is best carried out by specialists who have the appropriate knowledge, such as process, production and chemical engineers and the safety officer. It will usually be found relatively easy to determine the Zone 0 and Zone 1 boundaries where the presence or release of the material is controlled or predictable. These boundaries are usually within a few metres of the release source, but Zone 2 boundaries are more difficult because the emission is not controlled as it is due to an abnormal occurrence, such as a leaking pump gland or flange, and may occur in unmanned areas where immediate discovery is unlikely.

Generally, the Zone 0 and Zone 1 areas are situated within associated Zone 2s because an unintended release of the material will almost certainly extend beyond the Zone 0 and Zone 1 boundaries. It may also be found that zones overlap, necessitating, perhaps, the regrading of a Zone 2 to a Zone 1 area. Where the flammable material is heavier than air, any pits, trenches, cable ducts or cellars without forced ventilation and in Zone 2 areas should be designated Zone 1, as the flammable material will not readily disperse from within them. For lighter-than-air gases, a similar designation is appropriate for coffered ceilings and under roofs where the gas is likely to be trapped.

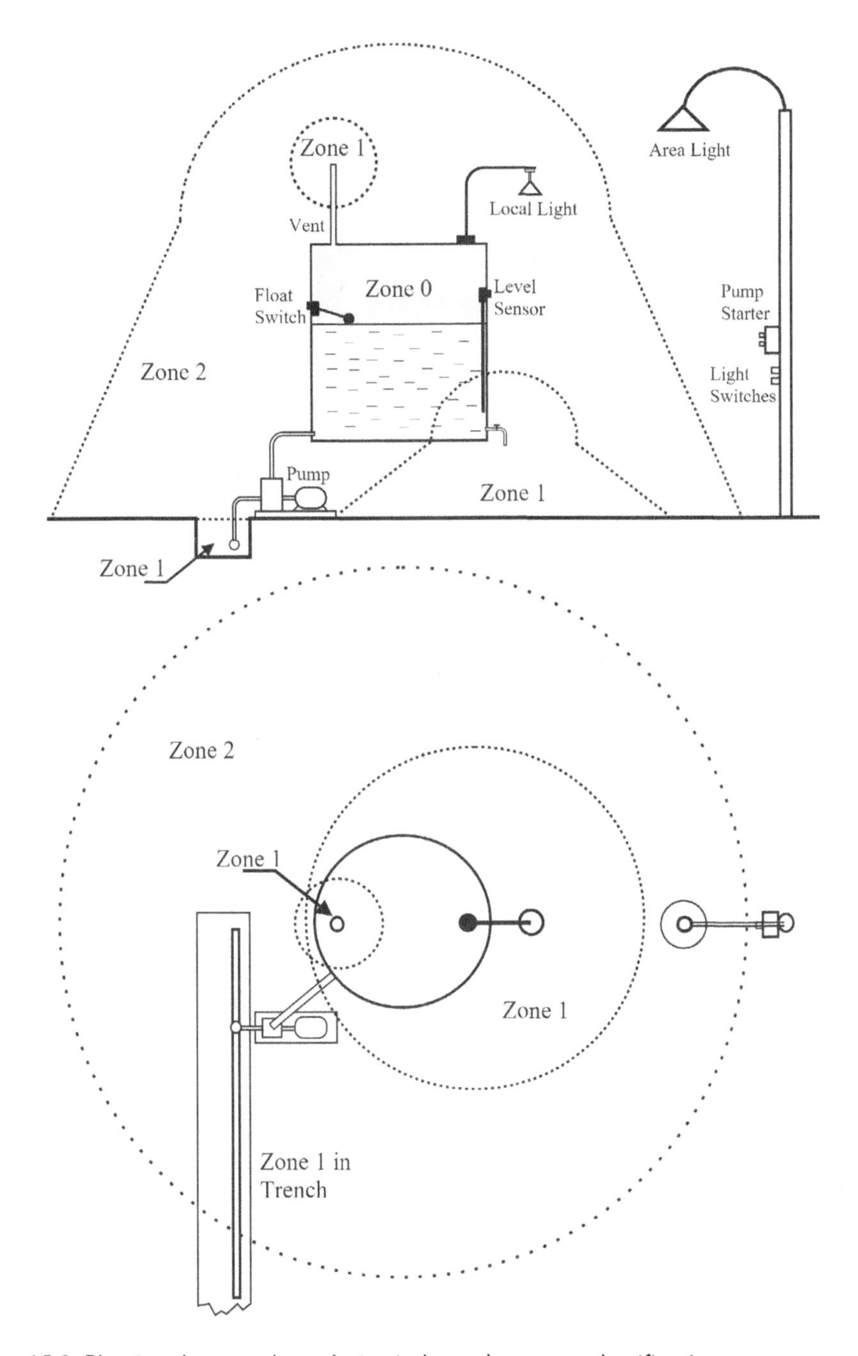

Figure 15.2 Plotting the zone boundaries in hazardous area classification

Figure 15.3 Sign to be posted at entrances to hazardous areas (black lettering on yellow background)

DSEAR Regulation 7(3) requires that a warning sign of the type shown in Figure 15.3 is posted at the entrances to any areas that have been classified as hazardous areas.

Individual companies, trade associations and professional bodies have issued guidance on area classifications for their particular sectors. For example, the Institute of Petroleum's and the Association for Petroleum and Explosives Administration's publication *Guidance for the design, construction, modification and maintenance of petrol filling stations* contains comprehensive guidance on hazardous area classification at petrol filling stations.

For petroleum and other flammable liquids in containers and bulk storage, reference should be made to HSE booklets:

- HS(G)51 – *The storage of flammable liquids in containers*, which provides guidance for those responsible for the safe storage of flammable liquids in containers at the workplace. It applies to storage of flammable liquids in containers up to 1000 litres capacity. It explains the fire and explosion hazards associated with flammable liquids and how to control the risks.
- HS(G)140 *Safe use and handling of flammable liquids*, which provides guidance for those responsible for the safe use and handling of flammable liquids in all general work activities, small-scale chemical processing and spraying processes.
- HS(G)176 *Storage of flammable liquids in tanks*, which gives guidance on the design, construction, operation and maintenance of installations used for the storage of

flammable liquids in fixed tanks operating at or near atmospheric pressure. It applies to new installations and to existing installations where reasonably practicable. It is relevant to industries such as chemical, petrochemical, paints, solvents and pharmaceutical. The guidance gives help in the assessment of the risks arising from the storage of flammable liquids, and it describes measures to control those risks such as containment (primary and secondary), separation, ventilation, substitution and control of ignition sources.

Gas groups

Engineers responsible for selecting electrical apparatus for installation in a hazardous area will need to know into which Gas Group the flammable atmosphere in each hazardous area will fall. This is because the types of gas present will affect the characteristics of some types of explosion-protected equipment that is permitted to be installed in the hazardous zones. There are two main groups: Group I, which relates to electrical apparatus for use in mines susceptible to firedamp; and Group II, which relates to electrical apparatus for use in places where an explosive gas/air atmosphere exists, other than mines susceptible to firedamp. Group II gases and vapours are further subdivided into Groups IIA, IIB and IIC, as follows:

- Group IIA representative gas: propane
- Group IIB representative gas: ethylene
- Group IIC representative gas: hydrogen

Group IIC gases are the most onerous because, having the lowest minimum ignition energies, they can be ignited more easily than the gases in the other two groups.

Temperature classification

Another consideration for electrical equipment is its temperature classification, or 'T' rating. This indicates the maximum surface temperature of those surfaces of the equipment that can be exposed to the flammable atmosphere. So the 'T' rating of electrical equipment should be below the minimum ignition temperature of the vapours or gases likely to be present in the flammable atmospheres. The 'T' ratings in terms of maximum apparatus surface temperature are as follows:

T1: 450°C
T2: 300°C
T3: 200°C
T4: 135°C
T5: 100°C
T6: 85°C

Equipment categories

As a further complication, the ATEX Directive, enacted in the UK as The Equipment and Protective Systems Intended for Use in Potentially Explosive Atmospheres Regulations

1996 (see Chapter 8), introduced the concept of Equipment Categories. In very broad terms, these categories are

- Group I Category M1 and M2 equipment, which are for mining use
- Group II Category 1, Category 2 and Category 3 equipment, which are for non-mining use

Because of their limited application, the mining equipment categories will not be considered further.

The effect of the definitions, as far as electrical equipment in flammable gas atmospheres is concerned, is that electrical equipment intended for use in Zone 0 areas would normally be classified as Category 1G; equipment intended for use in Zone 1 areas would normally be classified as Category 2G; and equipment intended for use in Zone 2 areas would normally be classified as Category 3G.

The equipment categories are similar in intent and definition to Equipment Protection Levels (EPL) introduced by the IEC and which have come into common usage in recent years. Category 1G is equivalent EPL Ga, Category 2G to EPL Gb, and Category 3G to EPL Gc.

Explosion-protected apparatus for flammable atmospheres

Concepts

Having carried out an area classification exercise, the electrical equipment to be installed in the area must then be selected for its explosion protection properties. There is a range of options to choose from, with a variety of protection techniques employed. Each of the techniques is described in detail in a European harmonised standard and is allocated a designation letter, mainly for labelling purposes. The following text summarises each of the techniques and identifies its letter designation and appropriate construction standard.

Intrinsically safe – Type 'i' – BS EN 60079–11:2010

The term 'intrinsically safe' applies to the apparatus and to the circuit in which is connected. It is normally associated with instrumentation circuits and some portable apparatus. This type of equipment may arc or spark under normal operating or fault conditions, but the design has to ensure that the energy level, even under fault conditions, is too low to produce an incendive spark that could ignite the flammable atmosphere. There are two categories, Type 'ia' and 'ib': ia is more stringent, as it allows for two faults, whereas ib allows for only one. In general, fixed intrinsically safe equipment installed inside a hazardous area is either galvanically isolated from the equipment in the safe area, commonly using optoelectronic devices, or is supplied through a device called a 'zener barrier' which serves to limit the energy delivered to the equipment in the hazardous area, the principle of which is illustrated in Figure 15.4.

Referring to Figure 15.4, the principle of operation is that the zener diodes limit the voltage on the outgoing circuit supplying a sensor in the hazardous area to the design voltage, typically in the range 15–30 V. Resistor R1 limits the current that can be supplied

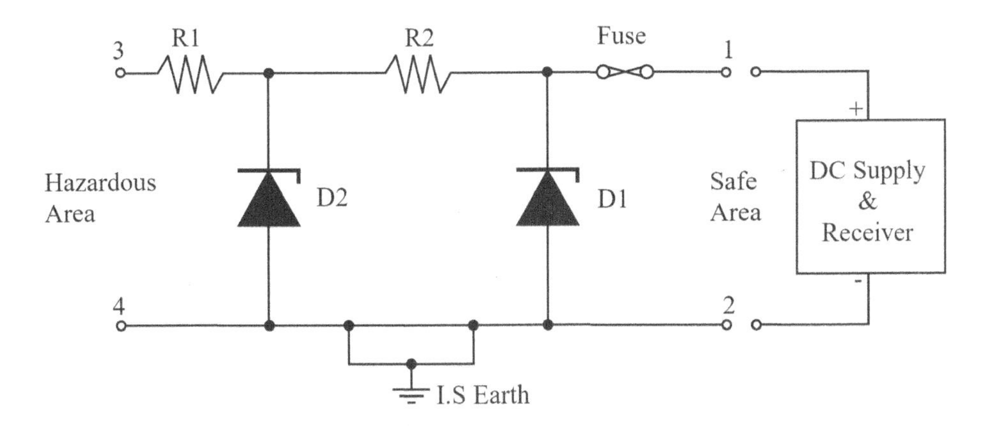

Figure 15.4 Principle of shunt zener diode barrier

to the sensor. The combination of the limited voltage and current results in a maximum value for the power that can be supplied into the hazardous area. The fuse ruptures in the event of the Zener diodes conducting when the voltage exceeds their conducting voltage, limiting the energy that will be supplied to the hazardous area to a safe level below the minimum ignition energy for the atmosphere.

There is an additional standard covering the specialist topic of intrinsically safe fieldbus, which is a type of communications network used in industrial distributed control systems, especially for instrumentation applications. The standard is BS EN 60079–27:2008 *Explosive atmospheres. Fieldbus intrinsically safe concept (FISCO).*

Pressurised or purged equipment – Type 'p' – BS EN 60079–2:2014

This technique excludes the flammable gas from the apparatus by using pressurised air or inert gas inside the apparatus that is at a slightly higher pressure than the ambient pressure outside. If the flammable gas is excluded, any internal arcs or sparks will not lead to ignition. In both cases the pressurising equipment has to be monitored and interlocked with the supply to disconnect it, should the pressurising system fail. If the system has been shut down for any length of time, arrangements are necessary to purge it of any gas that may have entered before the supply is restored.

Flameproof enclosures – Type 'd' – BS EN 60079–1:2014

Flameproof equipment is totally enclosed. The flammable gas/air mixture may enter the enclosure and be ignited by any internal arcs or sparks that may occur. However, the enclosure has to be strong enough to withstand an internal explosion without damage and without allowing the internal explosion products to leave the enclosure at a temperature that could ignite the flammable atmosphere outside. To prevent such an ignition the paths through joints have to be sufficiently long to act as flame traps. Joints are generally flanged or labyrinthine, e.g. threaded. There are strict requirements laid down in the standards about the lengths of such joints and the maximum gaps that can exist between joint surfaces; the size of the gap determines in which Gas Group the equipment can be installed.

Increased safety – Type 'e' – BS EN 60079–7:2015

Increased safety equipment is more akin to ordinary non-explosion-protected designs, but special precautions have to be taken to prevent any electrical or frictional sparking or hot spots exceeding the specified temperature limits. To prevent excessive heating the equipment is liberally rated and under fault conditions the associated controlgear has to interrupt the circuit before dangerous overheating occurs

Non-sparking apparatus – Type 'n' – BS EN 60079–15:2010

Type 'n' apparatus is designed so as to be safe in normal operation and such that a fault capable of causing ignition is unlikely to occur. There are different techniques allowable within this concept, including non-sparking (nA) and restricted breathing enclosures (nR).

Oil-immersed apparatus – Type 'o' – BS EN 60079–6:2015

In this technique, the electrical apparatus or parts of it are immersed in oil in such a way that an explosive atmosphere which may be above the oil or outside the enclosure cannot be ignited.

Encapsulation – Type 'm' – BS EN 60079–18:2015

In this type of protection, electrical parts which are capable of igniting a flammable atmosphere by either sparking or heating are enclosed or encapsulated in a compound so that the explosive atmosphere cannot come into contact with the electrical parts. There are 3 sub-categories: ma, mb, and mc, according to the level of safety integrity.

Powder filled apparatus – Type 'q' – BS EN 60079–5:2015

In this type of protection, an enclosure containing electrical parts which are capable of igniting a flammable atmosphere by either sparking or heating is filled with a material such as quartz in a finely granulated state. The design ensures that any arc produced by the electrical apparatus surrounded by the material cannot ignite the surrounding flammable atmosphere.

Special Protection – Type 's' – BS EN 60079–33:2015

Type 's' protection refers to any other protection technique, not satisfying the requirements of the other standards listed previously which a manufacturer can demonstrate to be safe.

Combinations

Some manufacturers produce equipment that combines different techniques into one package, usually in the interests of economy and ease of use. A common combination, for example, is Type 'd' and Type 'e' protection. The flameproof (Type 'd') enclosure is used to protect the sparking contents of the apparatus, and the lighter Type 'e' protection is used for the terminal box.

Selection of explosion-protected apparatus for flammable atmospheres

DSEAR Schedule 3 requires that all equipment and protective systems installed in hazardous areas must be selected on the basis of the requirements set out in the Equipment and Protective Systems Intended for Use in Potentially Explosive Atmospheres Regulations 1996 (see Chapter 8) unless the risk assessment finds otherwise. It specifies that the following categories of equipment must be used in the zones indicated, provided they are suitable for the gases, vapours, and mists as appropriate:

- in Zone 0, Group II Category 1G equipment;
- in Zone 1, Group II Category 1G or 2G equipment; and
- in Zone 2, Group II Category 1G, 2G or 3G equipment.

Not all explosion-protected electrical apparatus can be used in all zones of hazardous areas. This is because in some protection techniques the integrity of the technique is not sufficiently high to deal with the risk. The following limitations apply:

Category 1G apparatus for use in Zone 0:
- Ex ia
- Ex ma

Category 2G apparatus for use in Zone 1:
- Ex d, subject to the correct Gas Group and T rating.
- Ex p
- Ex q
- Ex o
- Ex e
- Ex ia and ib
- Ex mb
- Ex s (if suitably certified)

Category 3G apparatus for use in Zone 2:
- Ex n
- Ex mc
- All those suitable for use in Zones 0 and 1.

Equipment approval and certification

The Equipment and Protective Systems Intended for Use in Potentially Explosive Atmospheres Regulations 1996 require compliance with the Essential Health and Safety Requirements (EHSRs) and the application of the CE mark using the applicable conformity assessment procedures, which may involve the participation of a Notified Body.

This is a complicated area that will change when the 2016 regulations are enacted, but in summary,

- Category 1 electrical equipment and protective systems are subject to EC Type Examination by a Notified Body for conformity to either a harmonised standard or the EHSRs or a combination of the two, plus either production quality assurance or product verification procedures.
- Category 2 electrical equipment is subject to EC Type Examination by a Notified Body for conformity to either a harmonised standard or the EHSRs, plus either production quality assurance or conformity to type.
- Category 3 electrical equipment is subject to internal control of production by the manufacturer.
- Alternatively to the above procedures, each production item may be subject to examination by a Notified Body for conformity to either a harmonised standard or the EHSRs or a combination of the two.

The overall aim of this is to ensure both that equipment is safe and that it is not subject to artificial trade barriers within Europe.

Labelling

Manufacturers of certified explosion-protected electrical equipment fix labels to their products to indicate that it has been tested and approved for use in flammable atmospheres, and to provide details of the approval. The label identifies the manufacturer; the name or type of the apparatus; the year of construction; the construction standard against which the apparatus has been tested; the protection concept(s) used; the apparatus grouping (gas group), if applicable; the name or mark of the certifying body (Notified Body under ATEX); the 'T' rating; and any other relevant information, such as the IP rating. This was covered in Chapter 8, and for ease of reference the text is repeated here.

In addition to the CE mark, which would be needed for equipment placed on the EU market, the equipment must be marked to enable full identification of its protection category. The label must at least contain the following:

(a) The specific explosion protection mark, Ex, together with the mark indicating the equipment group and category; and, relating to equipment group II, letter 'G' (concerning explosive atmospheres caused by gases, vapours or mists) and/or 'D' (concerning explosive atmospheres caused by dust), together where necessary with all information essential for safe use, e.g. gas group, temperature class, and Ex protection concept. An example is shown below:

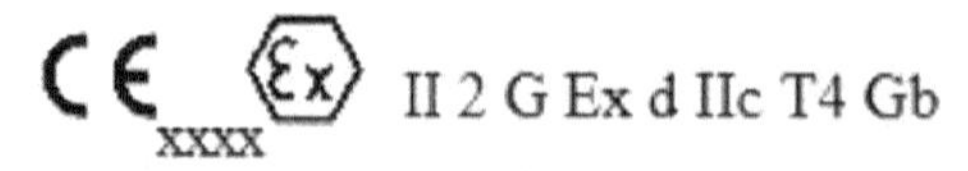

Where CE is the CE mark

xxxx is the Notified Body's registration number
Ex is the European explosion protection mark

II is the equipment group
2 is the equipment category (2 = equipment for use in Zone 1 or Zone 2 areas)
G means Gas ('D' for dust)
Ex = explosion protection (if marked EEx, this signifies compliance with an EN)
d is the protection type code ('d' = flameproof)
IIC is the gas group (IIC = acetylene or hydrogen as representative gases)
T4 is the temperature code (T4 = 135°C)
Gb is the equipment protection level

(b) The name and address of the manufacturer
(c) The designation of series or type and serial number
(d) The year of production
(e) Restricted or other safety-related conditions of use

The electrical installation

Armed with the hazardous area classification plan and knowledge of the applicable standards, the designer can design the electrical installation. In doing so he should avoid the hazard wherever possible by locating the electrical equipment and wiring outside the hazardous areas. This can be done, for example, using the following methods:

- Luminaires may be situated in a safe area outside and lighting a flammable zone interior through lay lights in the roof or walls.
- Motors can be located in a safe area, driving a machine within the flammable zone by means of a shaft extending through a sealed gland in an impermeable partition.
- Motor controlgear can often be located in a safe area remote from the motors in a hazardous zone. For many processes where there is centralised control it is often an advantage to locate the starters in motor control centres which can usually be situated in a safe area.

However, electrical apparatus and wiring that has to be in the flammable zones must be explosion protected to avoid the ignition hazard. Two examples are provided in the following sections.

Paint shop installation

Although the use of flammable and highly flammable paints has reduced in recent years, paint shops that use such paints must have explosion protection concepts incorporated in the design of the electrical installation. Much of the electrical equipment shown in the illustrative paint shop installation in Figure 15.5 is located outside the hazardous area, but some is located inside the hazardous area and would have to be explosion protected according to the hazardous area classification and zone designation.

The non-ventilated pit would be Zone 1, and the pit lights would have to be Zone 1 compatible if they weren't located behind sealed glazing; Ex d luminaires are commonly used in Zone 1 areas. Above ground level there would be a mixture of Zone 1 and Zone 2 areas. In the immediate vicinity of where the flammable paint was being sprayed, there would be a Zone 1 volume, but a metre or so above the highest spraying level it would

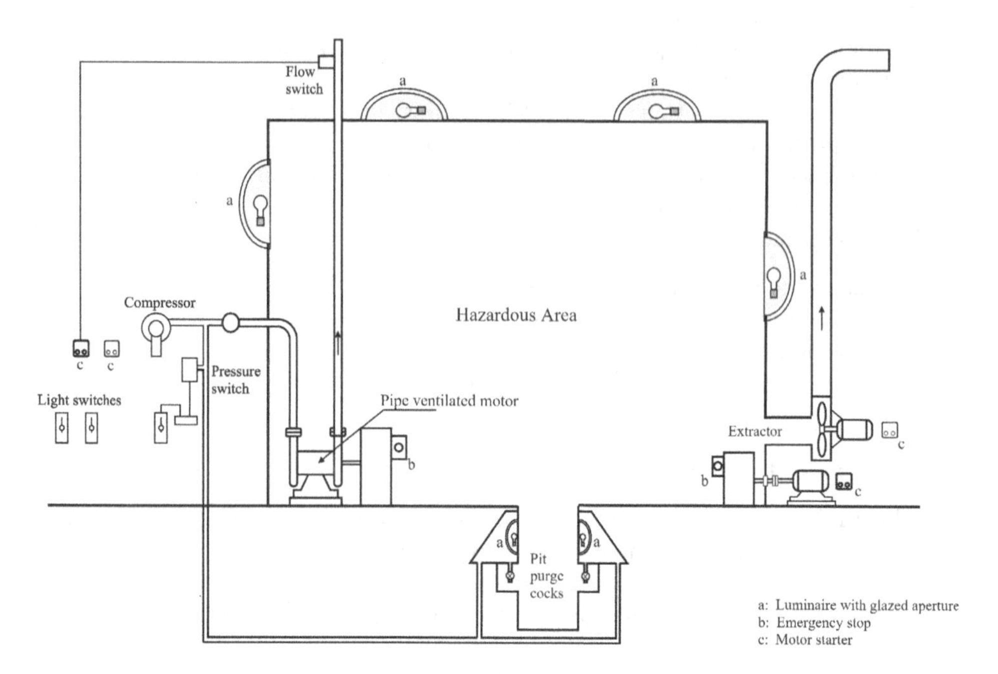

Figure 15.5 **A flammable paint spray shop for painting vehicles**

be Zone 2 assuming the paint solvent vapours are heavier than air. The motor inside the hazardous area would therefore, for example, have to be an Ex d type, or the purged Ex p type shown.

Both the motor and pit lights need safeguards against the entry of flammable gas when not in operation. Purge cocks are provided on the pit luminaires which should be opened, and then the compressor started to purge any gas from within them and the pipe-ventilated motor. The cocks are then closed which raises the pressure in the pipeline and closes the pressure switch, which in turn makes the contactor switching circuit and enables the pit lights to be switched on. The flow of air through the outlet pipe of the pipe-ventilated motor closes the flow switch and enables the motor starter. After the prescribed interval the motor starter becomes operational and the purged motor can be started. The emergency stop buttons can be explosion protected or could be non-electrical (pneumatic or hydraulic) connected to a pressure switch in the safe area which in turn is connected to the starter 'stop' circuit.

Tank of flammable liquid

The example in Fig. 15.2 shows the classified zones in the vicinity of a tank of flammable liquid which emits a heavier-than-air vapour. The space above the tank liquid is normally classed as Zone 0 because sufficient air may be drawn in through the vent when the liquid level falls or the vapour condenses or there is a fall in the ambient temperature, to form an explosive mixture. The level sensor would normally be intrinsically safe type Ex ia, and the float switch would have to be similarly protected if inside the Zone 0 volume.

Defining the Zone 1 and Zone 2 boundaries is not an exact science because there are many variables and is more a matter of judgement. The extent of the Zone 1 areas around the vent depends on the rate of emission of the flammable vapour from the orifice. A reasonable approach might be to work on the average release rate, assuming both a 100% vapour concentration and the plume being blown in one downward sloping direction by the wind, and then calculate how far it would disperse to reach the lower explosive limit. The average rate would be perhaps the mean between zero at low ambient temperatures and the maximum attained at high ambients with solar heating of the tank when it was being filled. It is usual to assume a spherical shape for the Zone 1 boundary about the vent orifice.

The emission from the tap-operated dispenser is dependent on the surface area of the exposed liquid when filling containers and the area of the container orifice. Again, it would be reasonable to assume that the vapour plume is blown in one downward sloping direction and disperses until the LEL is reached. The container orifice to this position is the radius of the Zone 1 area. The slope of the zone boundary is a function of the vapour density. The heavier the vapour, the steeper the slope.

The pump motor is located on the boundary of the Zone 1 region so it would probably be advisable to use a weatherproof Ex e rather than an Ex N motor.

The extent of the Zone 2 region is dependent on the possible mechanical failures of the pump, float switch, tank or associated pipework. The worst situation is probably an overflow through the vent tube, following float switch and level sensor failures and resulting in a fairly rapid spillage of the contents onto the ground. This would provide a large area of evaporation, and the windblown plume would disperse across the ground surface for a considerable distance before attaining the LEL. Using this distance from the tank as the radius, the extent of Zone 2 can be established. The Zone 2 boundaries from a leaking pipe or damaged pump gland are likely to be less extensive and need not, therefore, be calculated in this case.

The pipe trench, although in the Zone 2 area, is classed as Zone 1 because vapour could be trapped in it for a long time.

The local light should be suitable for Zone 2 and could be a weatherproof type Ex N. The electrical apparatus outside the Zone 2 boundary does not need to be explosion protected but would have to be weatherproof.

Wiring

BS EN 60079–17:2014 *Explosive atmospheres. Electrical installations inspection and maintenance* provides recommendations on the safety of electrical installations in hazardous areas, including recommendations for their periodic inspection and testing.

The installation should comply with the requirements of BS 7671. To avoid ignitions from the wiring installation, the necessary precautions include the following:

- The conductors should be adequately insulated for the declared voltage and to prevent leakage or short circuits between conductors or to earth.
- The conductors should be further protected against mechanical damage.
- The conductors should be protected against excess current.
- MIMS cable systems should be protected by surge arresters to prevent insulation failure if transient voltage surges are likely.

- The cables should be liberally rated to suit the load cycle without getting hot enough to damage the insulation or ignite a flammable atmosphere.
- The installation should be suitable for the environmental conditions apart from the flammable atmosphere, e.g. a wet location or a corrosive or dusty atmosphere.
- Where it is necessary to exclude the flammable atmosphere, cables, conduits and trunking and in some cases ducting should be sealed to prevent gas being transmitted through the interstices of the cable or along conduits, trunking or ducts.
- The metalwork of conduits and trunking and the armouring and screening of cables should be bonded and earthed.
- Cables should, preferably, be connected only at apparatus terminals. If intermediate joints are made, they should be of the encapsulated type or a terminal box sealed with compound should be used.
- Joints in steel conduit should be of the threaded type, tight fitting and painted to exclude moisture so as to avoid corrosion and an adverse effect on earth continuity.
- Joints between different metals, widely separated in the electrochemical series, should be avoided where the junction may attain a sufficient potential difference to cause corrosion. Where there is a risk, the joint should be made watertight by painting or other treatment.
- Terminations of explosion-protected apparatus should be such as not to invalidate the apparatus certificate.
- Flexible metallic conduit should have a non-metallic inner sheath or be suitably designed so that movement cannot cause abrasion of the cables.
- Flexible metallic conduit is not a suitable protective conductor. A flexible cable conductor should be used for this purpose to ensure earth continuity between the fixed conduit and apparatus, and the flexible conduit joints should be earth bonded at both ends.
- The use of light metal and light metal alloy components should be avoided, such as aluminium and aluminium alloy conduit, and accessories where frictional contact with oxygen-rich items, such as rusty steel, might occur and cause sparking. Alternatively, contact can be prevented using a plastic sheath or other covering.
- The use of compression glands on cables covered with insulation and sheathing which have a cold flow property where the gas has to be excluded should be avoided. Where required for cables connected into enclosures, use certified glands (Ex e, Ex d and Ex d barrier) to prevent gas seeping into cables and the enclosure's certification being compromised.
- Overhead telecommunication or power lines should terminate outside the flammable zone, be fitted with surge arresters and run underground into the flammable zone to minimise danger from voltage surges caused by lightning.
- Cable sheaths should not be used as neutral conductors.

Portable and transportable apparatus

Portable and transportable apparatus should be excluded from the hazardous areas, as far as possible, because the conditions of use make it, and particularly its flexible cable, a greater ignition and accident risk than fixed apparatus. When it is used, it should be frequently inspected, tested and maintained to avoid trouble. The flexible cable should be of the screened type. The connection to the supply should be by means of an explosion-protected plug and socket, interlocked with the supply switch, to ensure that the plug can only be inserted or withdrawn when the circuit is dead. It is also recommended that the circuit is protected by a residual current circuit breaker or circulating current earth monitoring protection or both.

A sound approach is to ensure, via suitable management procedures, that such apparatus is only used in a hazardous area after the area has been confirmed by measurement to be gas-free; work would then usually be carried out under a Permit to Work or equivalent safety document issued by the process specialist who certifies that the area is gas-free.

Gas detector protection

Another means of protecting against the explosion hazard in flammable atmospheres is to employ gas detectors. The principle is that the gas detectors detect flammable gas/air mixtures at a low percentage of the LEL, typically 10%, and cause an alarm to sound. If the concentration reaches 25% or so of the LEL, the system automatically initiates a shutdown of the process. This type of technique is commonly used in process plants such as oil and gas refineries, LPG storage tank farms, chlorine doping rooms in water treatment plants and on vehicles such as electric forklift trucks operating in Zone 1 and 2 areas.

An example of gas detector protection is found in unattended commercial and industrial freezing and refrigeration plant that uses ammonia as the refrigerant. Ammonia is both toxic and corrosive, and exposure to it is dangerous. There is practically no risk in attended plants because a few parts per million of ammonia in air is easily detectable by the pungent smell, so the attendant is aware that there is a leak and can take remedial action before the concentration becomes dangerous. The lower explosive limit is comparatively high at 16%; this concentration is likely only under abnormal conditions, such as a blown compressor head gasket, and is intolerable for the eyes and respiratory system.

Adequate protection may be obtained by the use of gas detectors, suitably located near the ceiling, at positions where flammable concentrations may be expected from gas leaks. The detectors are connected to a controller which can be set to trigger an alarm, trip the supply to the electrical equipment in the machine room and, if required, switch on an explosion-protected extractor fan to disperse the gas at a concentration of about 2%, or to actuate the alarm and fan at this low concentration and trip the supply at a higher value of say 7%. The protection will operate well below the LEL of 16%. The integrity of the protection system can be ensured by providing it with a standby supply for automatic connection in the event of its mains supply failing. The relevant standard is the BS EN 378 series covering refrigeration systems and heat pumps.

Figure 15.6 contains a representative drawing and a single line circuit diagram illustrating the principles. The gas detectors GD1 and GD2, their controllers and the alarm are powered from the supply or from an inverter which is battery energised so an alarm would be given should a gas leak occur when the plant is shut down or the supply fails. Under normal conditions, the extractor fan is controlled by the non-explosion-protected, two-way switches in the safe area and compressor room, but in the event of a leak gas detector GD1 would operate. This would shut down the compressor room installation and automatically disengage these switches and their associated non-protected wiring and connect the extractor fan to the supply via wiring of a type specified in the British Standard.

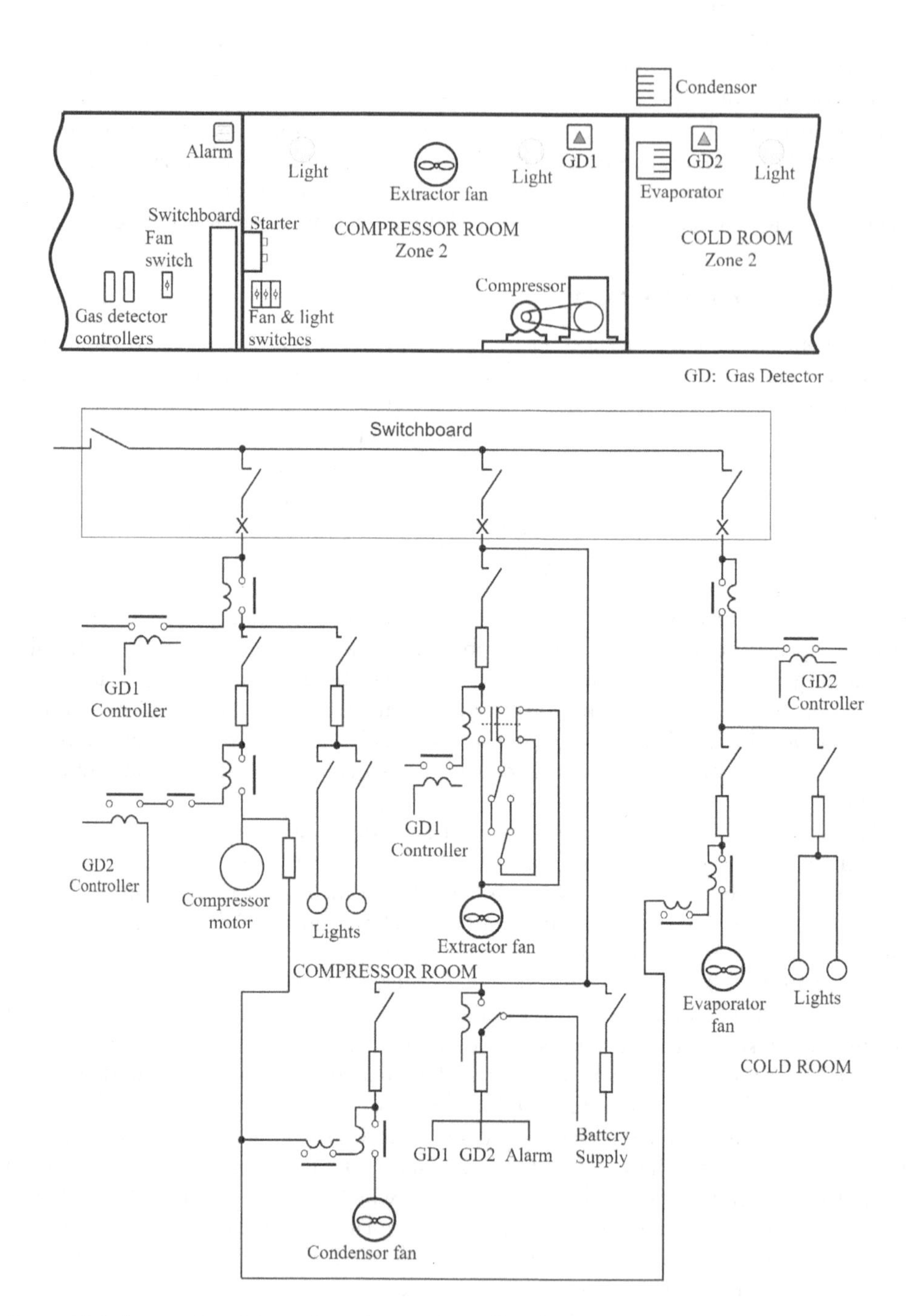

Figure 15.6 Ammonia refrigeration plant – single line diagram

A leak in the cold room would be detected by the sensor GD2, energise the alarm and shut down the cold room installation. Figure 15.6 also shows that an interruption of the compressor motor supply would shut down the condenser and evaporator fans. This, however, is an economic rather than a safety feature.

Gas detectors can also be usefully employed in other unattended Zone 2 plant areas where it is desirable to trigger an alarm in the event of a large gas leak which, if undetected and remedied, might spread beyond the zonal boundaries and reach a source of ignition in an adjoining safe area.

Avoidance of dust explosions

The hazard

There is a considerable range of materials, from powdered metals such as aluminium to flour, wood flour/dust and sugar, that will explode when mixed with air in a dust cloud and subjected to an ignition source such as a spark or arc from electrical equipment or an electrostatic discharge. Whereas gas/air mixtures will disperse with time, dust clouds tend not to disperse and will explode if the dust concentration in air is between the LEL and UEL in the presence of a credible ignition source above the dust cloud's minimum ignition energy. For many organic materials the LEL is in the range 10–50 g/m^3, which resembles a very dense fog (as indicated in the HSE publication *Safe handling of combustible dusts: Precautions against explosions*), so the presence of the cloud will be obvious to anyone in the vicinity.

In addition to the hazard from dust clouds, layers of dust can accumulate on surfaces and present a fire risk and, when disturbed, can give rise to a cloud explosion risk.

The prevention of dust explosions is a requirement of DSEAR in the same way as for gas/air explosions. As previously described for flammable gas atmospheres, employers must make an assessment of the risks of dangerous substances (including dust) exploding and causing injury and implement measures to prevent them and mitigate their effects, so far as is reasonably practicable.

Hazardous area classification

Hazardous area classification for dust atmospheres is required in the same way as for flammable gas atmospheres. Areas with a dust explosion hazard have three possible zones, as per flammable atmospheres, but their numbering and definitions are different to the flammable atmosphere zones:

- *Zone 20*: A place in which an explosive atmosphere in the form of a cloud of combustible dust in air is present continuously, or for long periods or frequently
- *Zone 21*: A place in which an explosive atmosphere in the form of a cloud of combustible dust in air is likely to occur in normal operation occasionally
- *Zone 22*: A place in which an explosive atmosphere in the form of a cloud of combustible dust in air is not likely to occur in normal operation but, if it does occur, will persist for a short period only

These designations replaced the old Zone Y and Zone Z classifications that existed under BS 6467:1988 *Electrical apparatus with protection by enclosure for use in the presence of combustible dusts*. The procedure for hazardous area classification for dust atmospheres is described in BS EN 60079–10–2:2015 *Explosive atmospheres. Classification of areas. Explosive dust atmospheres*. The zone definitions do not relate to layers or accumulations of dust, which must be avoided, or, where they cannot be avoided, the surface temperature of equipment must be kept at a safe level to reduce the fire risk so far as is reasonably practicable.

Dust groups

Materials capable of producing dust explosions are allocated to a Dust Group, in the same way that gases are allocated to Gas Groups. Dusts are classified in Group III, which has three subdivisions:

- Group IIIA – combustible flyings such as paper particles and ignitable fibres such as cotton lint
- Group IIIB – non-conductive dusts such as flour and wood
- Group IIIC – conductive metal dusts such as magnesium

Dust group IIIC is the most readily ignited and is therefore the most onerous with respect to prevention.

Temperature classification

Dust clouds have an auto-ignition temperature in the same way that flammable gas atmospheres do. For that reason, the same temperature classifications as previously described for flammable atmospheres apply to the selection of electrical equipment.

Totally enclosed equipment is often designed for a greater temperature rise than enclosed ventilated equipment. It should be selected so that the final temperature attained, i.e. ambient plus temperature rise, is less than the auto-ignition temperature of the explosive dust. Some allowance should be made for the increase in temperature rise due to the heat insulating effects of the dust coating on the apparatus which is likely to occur even with good housekeeping.

Totally enclosed fan-cooled motors are cheaper than totally enclosed machines of the same output and are, to a certain extent, self-cleaning as the airflow over the body shell from the external fan prevents the build-up of the dust.

Equipment categories

The equipment categories promulgated in the Equipment and Protective Systems Intended for Use in Potentially Explosive Atmospheres Regulations 1996 (see Chapter 8) apply to equipment for use in dust atmospheres as well as in gas/air atmospheres. Leaving aside equipment for use in mines, there are three categories of equipment for use in dust atmospheres: categories 1D, 2D and 3D. Category 1D provides a very high level of protection and is equivalent to EPL Da, Category 2D provides a high level of protection and is equivalent to EPL Db and Category 3D provides normal protection and is equivalent to EPL Dc.

Equipment selection

DSEAR Schedule 3 requires that, when electrical equipment has to be installed in a hazardous dust area, the following categories of equipment must be used, provided that the equipment is suitable for use in the particular dust environment:

- in Zone 20, Group II Category 1D equipment;
- in Zone 21, Group II Category 1D or 2D equipment; and
- in Zone 22, Group II Category 1D, 2D or 3D equipment.

Not all explosion-protected electrical apparatus can be used in all zones of hazardous dust areas. This is because in some protection techniques the integrity of the technique is not sufficiently high to deal with the risk. The following limitations apply:

Category 1D apparatus for use in Zone 20:
  Ex ia
  Ex ma
  Ex ta
Category 2D apparatus for use in Zone 21:
  Ex p
  Ex ia and ib
  Ex mb
  Ex tb
Category 3D apparatus for use in Zone 22:
  Ex tc
  Ex mc
  All those suitable for use in Zones 20 and 21

The list includes Type 't' protection which is used solely for equipment destined for dust atmospheres; it is described in BS EN 60079–31:2014 *Explosive atmospheres. Equipment dust ignition protection by enclosure "t"*. There are three levels of integrity: ta, tb and tc according to the pressure-withstand capability of the enclosure. This protection technique is based on the principle that, to prevent ignition of a dust cloud, the dust cloud can be separated from the source of ignition by making the electrical apparatus dust-tight to IP6X. To achieve this, the apparatus has to be totally enclosed, with sealed joints to limit or prevent the ingress of dust. The standard specifies maximum surface temperatures for the enclosure.

The pressurising method, Ex p, helps prevent the bearing trouble that sometimes affects totally enclosed motors with sealed joints. However, these machines are subject to pressure cycling, as they tend to breathe outwards when operating and warm and inwards when shut down and cold. Some of this breathing may be through the bearing housings and may introduce dust from the polluted atmosphere which may contaminate and damage the bearings. A bearing failure could result in overheating and/or sparking and cause an ignition. Small amounts of dust penetrating through the joints, however, are not usually hazardous because the resultant dust cloud is below the LEL. The motors should be dismantled periodically and any dust removed.

Apparatus exteriors should be designed to minimise dust accumulations and to facilitate cleaning. The maintenance instructions should include information on preserving the integrity of joints and gaskets to exclude dust and on cleaning methods to remove the dust without creating a static charge.

If ferromagnetic dust is present, precautions may be needed to avoid or minimise external electromagnetic fields generated by apparatus to prevent dust accumulating. Aluminium conductors within apparatus should be kept away from casing joints to avoid the possibility of severe arcing in the event of a fault burning through the joint and igniting the dust outside.

There should be a safety margin between the ignition temperature of the dust layer or cloud and the maximum temperature of the apparatus. For a 5 mm layer, the recommended maximum temperature of the apparatus is 75°C less than the dust ignition temperature, and for a dust cloud the maximum should not exceed two-thirds of the ignition temperature. Most dusts have relatively high ignition temperatures so there should be little difficulty in complying with the requirement.

The wiring installation should be in accordance with BS 7671:2008. Conduit should be of the solid drawn or seam welded type with screwed joints to prevent dust penetration. The joints should be protected against corrosion to preserve earth continuity.

Labelling

Apparatus labelling is the same as described earlier in the chapter for equipment destined for use in Zones 0, 1 and 2, but with the letter 'D' rather than 'G' used to designate use in dust atmospheres in Zones 20, 21 and 22. The apparatus should be labelled with the IP rating, maximum surface temperature and minimum temperature for use or storage if this is lower than -20°C. The reason for this is that low temperatures can embrittle some materials and lead to a cracking risk. The maximum surface temperature allows for a 5 mm layer of dust.

Radio frequency induction hazard

The risk of RF induction is small and likely to occur only in a few locations. It should tend to diminish as more apparatus is made to comply with EMC (electromagnetic compatibility) standards. However, installations in the vicinity of radio, TV and radar transmitters and repeaters should be checked in accordance with the guidance in PD CLC/TR 50427:2004 *Assessment of inadvertent ignition of flammable atmospheres by radio-frequency radiation. Guide.*

The lightning hazard

Again, lightning is a comparatively rare source of ignition. Outdoor installations are more vulnerable than indoor ones. The risks can be minimised by means of lightning conductors installed to BS EN 62305 *Protection against lightning.*

The electrostatic hazard

Liquids, solids, vapours, gases and dust clouds can readily become electrostatically charged to a high voltage but, with the exception of large capacitors, generally store comparatively small quantities of charge. The charge may be positive or negative and will discharge into anything of opposite polarity which it touches, or if the potential difference created is high enough, a flashover may occur when it is sufficiently close. The resultant arc or spark may well be hot enough, i.e. contain sufficient energy, to ignite an explosive atmosphere, as most gases require only fractions of a millijoule (mJ) and many dust clouds only between 10 and 100 mJ, according to the material, for this to occur.

A very good source of advice is PD CLC/TR 50404:2003 *Electrostatics. Code of practice for the avoidance of hazards due to static electricity.* This code of practice for avoiding ignition and electric shock hazards arising from static electricity covers in detail the processes that most commonly give rise to problems of static electricity and the means of dealing with them. They include the handling of solids, liquids, powders, gases, sprays and explosives.

Creating an electrostatic charge

There are a number of ways to create static charges, but most of the problems arise from two of them: (i) rapid mechanical separation of materials in contact with each other, i.e. friction; or (ii) induction, where the charge is due to the presence of an electric field. Typical examples of (i) are the finishing of paper and textiles by calendering rolls, or the separation of an insulating material web from the rollers of a coating machine, or the flow of materials in pipelines. Examples of (ii) are the charging of an operator wielding an electrostatic paint or powder spray gun.

Detection

The presence of static can be detected by instruments that measure the voltage or field strength or, more crudely, by a gas discharge lamp which will glow if held in the field, or by the use of a wide band radio receiver and listening for the crackling sounds produced by the field. In severe cases the sparking can be heard and seen.

Measuring is best done when the flammable hazard is not present, but if this is impracticable, then the tester should use certified, explosion-protected instruments and carry out the work so as to avoid creating an incendive spark. Only qualified persons with an adequate knowledge of the hazards to avoid and an expert knowledge of the difficulties that arise in measuring should do this work. Such a specialist should be able to measure with sufficient accuracy and perform the calculations to determine whether or not the static hazard is severe enough to cause an ignition and merit expenditure on remedial action. The difficulty is that the appraisal can only take into account the plant conditions at the time. Subsequent changes can alter the static hazard, so where an ignition could cause expensive damage or could be a substantial danger to the operators, it will probably be prudent to provide some precautions anyway.

Precautions

Whenever possible, the hazard should be avoided by preventing the build-up of electrostatic charges. This can be done in the case of conducting material by earthing. The resistance to earth should be not more than 1 MΩ except for locations where the capacitance is low when a higher resistant value is permissible, e.g. 100 MΩ where the capacitance of the body is not more than 100 pF (picofarads).

It is more difficult to remove the charges from insulating materials. In some cases this may be done by modifying the material to improve its conductivity. For example, flammable liquid fuels can be doped with additive and carbon black can be added to rubber during manufacture to reduce the resistivity of the material. Where this is undesirable, or uneconomic, it may be possible to carry out the process in a humid atmosphere where the water vapour acts as a leakage path to earth. This method is not suitable, however, for water-repellent materials.

To avoid the risk of electrostatic ignitions, Class II apparatus should not be used, and Class I equipment metalwork should be effectively earthed. Light metals and their alloys should be avoided where there is a risk of frictional sparking.

Limitation of energy

Limitation of energy is a method used in electrostatic spraying where a high impedance is inserted in the HV circuit to reduce the potential earth fault current and thus the spark discharge energy to a sufficiently low value to avoid ignition.

Ionisation

Another precautionary method is to ionise the air at the surface of the material to provide a conducting path. It is often used on coating machines. A row of needle-pointed electrodes is arranged near to the surface of the web. Alternate electrodes are energised at HV, and the others earthed to ionise the air and discharge the static. In another system the ionised air is produced by a radioactive sealed source and blown onto the surface of the material under an earthed electrode. These devices are generically known as 'static eliminators' and may themselves need to be built using explosion protection techniques to prevent them acting as the source of ignition.

Ventilation

In difficult cases, ventilation should be used to weaken and disperse the flammable atmosphere so that it is below the LEL, thus preventing ignition by an incendive spark.

Inert atmosphere

Ignitions can also be prevented by the use of an inert atmosphere. Nitrogen and carbon dioxide are generally used. This method can be utilised where the process machines are designed to employ it, but otherwise it may be impracticable. It has been successfully used in oil tankers to inert the cargo spaces during tank washing.

Earthing in large spaces

Flammable dust clouds can acquire a dangerous spatial charge which can spark over to an earthed electrode such as a measuring conductor, sampling tube or poking rod protruding into the container. The danger is a function of the volume and ignitions are only likely in large spaces such as large storage silos and hoppers. The dust clouds occur during filling when probes should not be allowed in the cavity to avoid ignitions, or alternatively, the space should be divided by earthed partitions, rods or wires to prevent the build-up of HV spatial charges. Earthed partitions are also advisable in large tanks containing flammable liquids to prevent HV charging of the contents.

Earthing of mobile containers

Mobile containers on non-conducting rubber-tyred wheels, such as aircraft, road tankers and industrial trolley tanks and vats, can acquire static charges and should be discharged by means of an earthing lead before their flammable contents are handled. The earthing lead should be connected by a clamp with an insulating handle, and the lead should next be earthed at a position remote from the flammable material. This practice serves the dual purpose of avoiding ignitions from an incendive spark and an unpleasant shock to the operator.

Operator training

Operators can initiate ignitions by unwittingly causing incendive sparks in flammable atmospheres. For example, if an operator who is well earthed by standing on a conducting floor or being in contact with earthed metalwork seeks to touch a charged surface, he can enable an incendive spark to jump the gap to his finger immediately before contact. Alternatively, an operator wearing insulating footwear or standing on an insulating floor can become charged, and an incendive spark is possible when he goes to touch an earthed object. In circumstances where the latter hazard is significant, operators should wear clothing made of natural, rather than man-made, fibres to reduce the generation of charge, and should wear anti-static footwear to allow charge to leak away to earth before it reaches a hazardous level. In this case, measures should be taken to ensure that the floor has semi-conducting properties – the charge will not relax to earth if the floor is non-conducting.

Operators in hazardous areas where static can occur need suitable training if ignitions are to be avoided.

Speed reduction

The generation of static charge is often related to the speed of the process, such as the speed at which a highly flammable liquid such as toluene is dispensed from a pipe into a vessel. A reduction of speed may be a sufficient precaution, in some cases, to reduce the static charge to a harmless level. This may be achieved, for example, by using larger bore pipes and thus decreasing the flow rate for the same output. In the case of webs and belts, a reduction in the linear speed or an increase in the diameter of rollers and pulleys will reduce the static charge.

Maintenance

To ensure safety in explosive atmosphere locations, maintenance staff should be suitably trained and competent to carry out the necessary inspection, testing and repair work. This means, for example, that an electrician has to be trained and qualified in electrical technology and also the techniques peculiar to hazardous atmosphere apparatus and wiring installations. There are industry schemes in place to provide electricians and technicians with the competencies they need for this type of work.

One such scheme is known as COMPEX. This is a national training and assessment scheme for electrical tradesmen who carry out work in potentially explosive atmospheres. The scheme was jointly developed by the Engineering Equipment and Materials Users Association (EEUMA), the training arm of the Electrical Contracting Industry (JTL) and the National Electrotechnical Training Organisation (NET, formerly EIEITO). On successful completion of practical and written assessments of each unit, candidates are awarded a Certificate of Core Competency which carries national recognition.

The scheme comprises a total of six units which are delivered via a network of approved centres and may be studied individually or in pairs covering installation and maintenance. The six training units which make up the scheme are as follows:

- Unit EX01: Preparation and Installation of Ex 'd', 'n', 'e' and 'p' Systems
- Unit EX02: Maintenance and Inspection of Ex 'd', 'n', 'e' and 'p' Systems
- Unit EX03: Preparation and Installation of Ex 'ia' and 'ib' Systems
- Unit EX04: Maintenance and Inspection of Ex 'ia' and 'ib' Systems
- Unit EX05: Preparation and Installation of Equipment Protected by Enclosure for Use in the Presence of Combustible Dusts
- Unit EX06: Maintenance and Inspection of Equipment Protected by Enclosure for Use in the Presence of Combustible Dusts

16

Tests and testing

Introduction

Electrical apparatus, equipment and installations are inspected and tested to prove that they are safe and fit for purpose and that they comply with relevant standards, specifications and codes of practice.

The need for testing, and its extent and type, varies through the life cycle of a particular item. During development, prototypes of apparatus and equipment need detailed and extensive testing to prove the design, both in terms of functional safety and inherent electrical safety; this testing is usually done in the laboratory. Once the design has been finalised and the item is put into production, less elaborate testing is needed during manufacture to ensure that each product is safe and meets the specified requirements. For more complex equipment, testing at various stages of assembly on the production line is common to detect faulty components and to minimise the failures of the completed product when it is tested before despatch.

Some testing may be required when equipment is being commissioned, often to provide the client with evidence that it is safe and functions correctly. Further testing is then needed during the in-service life of the equipment as part of routine preventive maintenance to ensure its continuing safety integrity; the topic of maintenance is covered in Chapter 18. Additional testing is often needed after the item has been subject to any repair, refurbishment or modification.

Some test facilities, such as those provided for high power short circuit testing, are specialised and expensive and are usually provided by testing authorities or companies to whom manufacturers submit their prototype products for testing in preference to providing the facility themselves. Where products have to be certified, e.g. explosion-protected apparatus, it is usually necessary to have the testing done by an approved laboratory, or notified body, which may issue compliance certificates itself or the certificates may be provided by a certifying authority for whom the laboratory works. Other test facilities used for routine testing of products are less complex, and the precautions to be taken against electrical injury are well-defined, not least in the standard BS EN 50191:2010 *Erection and operation of electrical test equipment.*

Electrical installations are usually inspected and tested on completion and before commissioning, but large installations may be tested and commissioned in sections as each part is completed, as explained in Chapter 13. Subsequently, they need periodic inspection and retesting as part of the maintenance procedure to detect incipient faults and to prevent the development of fault combinations that may lead to danger.

This chapter describes, firstly, the tests routinely carried out on low voltage installations; and, secondly, the testing of apparatus and equipment, and the precautions against electrical injury that should be taken in test facilities.

The law

The law covering electrical testing falls into three main camps.

Firstly, there is law covering the safety of products that are supplied to end users. This includes Section 6 of the Health and Safety at Work Act and product safety legislation such as the Electrical Equipment (Safety) Regulations and the Supply of Machinery (Safety) Regulations.

Secondly, there is law covering the maintenance of electrical equipment and installations and the need to ensure that it is maintained in a safe condition. This includes Section 2(a) of the Health and Safety at Work Act, Regulation 4(2) of the Electricity at Work Regulations, Regulation 5 of the Provision and Use of Work Equipment Regulations, and Regulation 6(7) of the Dangerous Substances and Explosive Atmospheres Regulations.

Thirdly, there is law covering the safety of the testing activity itself, especially where live testing is undertaken. This includes Section 2 of the Health and Safety at Work Act and Regulations 4(3) and 14 of the Electricity at Work Regulations.

Low voltage installation inspection and testing

As explained in Chapter 12, the recognised standard for extra-low and low voltage installations is BS 7671:2008. Part 6 of the standard provides recommendations for the inspection and testing of installations, with amplification published in the IET's Guidance Note No. 3 *Inspection and Testing*. The information in these publications has much in common with what follows in the next few pages. The standard stipulates that the results of inspections and tests, be that at initial verification or subsequent routine inspections and tests, should be recorded in inspection and test schedules which, together with the circuit diagram and equipment manufacturer's instructions, should form part of the operational manual and always be available to the installation user and whoever carries out the periodic inspections. Compliance with the recommendations in Part 6 and Guidance Note 3 is one way of demonstrating compliance with the Electricity at Work Regulations, Regulation 4(2), on maintenance.

Inspection

When an electrical installation is completed and before it is energised, a visual inspection should be carried out to ensure that the installation complies with the specification and with BS 7671 and that the connected apparatus and wiring are undamaged. Some inspections will need to be done while the installation is being erected, an example being a visual check that cables buried in walls are routed in the preferred zones, which can be done only before the wall covering is applied.

Strictly speaking, inspection does not fall under the heading of 'testing', but it is an essential prerequisite. Both BS 7671 and IET Guidance Note No. 3 provide a comprehensive checklist that should be used to ensure that nothing is missed. Some checklist items merit particular attention:

- In the connections of conductors, strands are not cut back to accommodate one or more conductors in an inadequately sized terminal. There are no loose terminal screws and wiring insulation is not cut back too far.
- The conductors are of adequate size not only for current rating but also for voltage drop.
- The conductors are labelled where necessary to facilitate subsequent maintenance.
- Cables, conduit and trunking which pass through walls or floors designated as fire barriers are sealed externally and internally.
- The wiring is not located where it may be adversely affected thermally, e.g. in the vicinity of radiators or covered by thermal insulation, unless its current rating and insulation is adequate and suitable.
- The wiring and apparatus are suitable for the environmental conditions, e.g. corrosive, polluting, wet or dusty atmospheres, which may occur in parts of the premises and which may not have been appreciated when the specification was prepared.
- Isolating and control devices, e.g. the positions of emergency stop push buttons for potentially dangerous machinery, are correctly located.
- The labelling of switches, fuses and circuits is suitable.
- The earthing terminals of Class 1 equipment are connected to the protective conductors.
- The installation is earthed and the main earthing conductor is the correct size.
- The main equipotential bonding conductors are present and of the correct size.
- Cable glands are properly made off; wire armouring and braiding is securely held by the gland and dust boots are correctly fitted.
- RCD protection is provided as required by BS 7671.
- Supplementary equipotential bonding is provided where necessary.
- Fuses are of the correct type and rating and excess current trips correctly set.
- Drainage holes are provided at low points in conduit systems.
- Flexible metallic conduit is not used as the protective conductor.
- Any electrical equipment installed in the hazardous zones of bathrooms and shower rooms are suitable for those locations.
- The wiring diagram or schedule is correct and the operational manual adequate and complete.

Whereas the foregoing checks are for pre-commissioning inspections, routine visual inspections of the installation covering the same issues should be an integral part of the in-service maintenance regime (see Chapter 18). They should form part of the periodic formal inspection and test.

In addition to the checks listed previously, any signs of overheating, damage, wear and tear or other distress of the components making up the installations should be used as an indication that something may be wrong and that a detailed investigation of the problem is warranted. Importantly, this includes the distributor's and supplier's apparatus (cable head, cut-out, meter, earthing equipment, meter tails) which should be checked to ensure that they are in good condition, undamaged, and not showing any signs of overheating.

When the person carrying out the inspection is satisfied with the results, the tests listed in the BS 7671 Section 612 should be carried out, following the recommended sequence to ensure that it is safe to energise the circuits for the final three live tests. The main tests are described in the following sections.

Continuity tests

Protective conductors

It is not necessary to measure the continuity of phase and neutral conductors in radial circuits as an open circuit is self-evident, but the continuity of protective conductors, including the main earthing conductor and the main and supplementary equipotential bonding conductors, has to be verified. For conductor sizes not exceeding 35 mm^2, their inductance, which is small, may be ignored and the testing can be done with a low resistance d.c. ohmmeter, which may form part of a combined instrument such as a multi-function tester. It should be capable of reading to two decimal places, and the results should be recorded for subsequent use in calculating the earth fault loop impedance and for comparison purposes when carrying out periodic inspection and testing. Where, however, conductor sizes exceed 35 mm^2 and the inductance contribution to the impedance cannot be ignored, an a.c. low impedance ohmmeter of similar resolution should be used to ensure that realistic impedance values are obtained. It should operate on SELV but be capable of providing a test current of 1.5 times the design current but need not exceed 25A.

Radial circuits are tested by connecting the phase and protective conductors together at the distribution board and then applying the test between the phase and earth terminals at each outlet point. Alternatively, the 'wandering lead' method can be used, whereby one end of a long length of wire is attached to the earth terminal and the test instrument is connected between the other end of the wire and various locations in the circuit under test. A higher than expected reading should be investigated and will probably be due to a defective connection.

To measure the continuity of the earthing conductor and the main and supplementary equipotential bonding conductors, their ends remote from the consumer's earthing terminal should be temporarily disconnected to avoid parallel paths and a test applied between these remote ends and the main earthing terminal.

Where there is an insulated protective conductor run in metal conduit, trunking, MIMS or armoured cables, it should be tested during erection and before the item supplied is in place so as to avoid a parallel path through the exposed conductive parts. Where, however, the exposed conductive parts are used as the protective conductor and are secured to the building steelwork or other extraneous conductive parts, there may well be parallel paths. In these cases, the impedance or resistance of the exposed conductive parts is likely to compare favourably with that of the phase conductor because of their greater cross-sectional area, provided their terminations are properly connected and the design is correct, so any parallel paths through extraneous metalwork is a bonus. However, a test should be done, and if the reading is less than the designed amount, it can be regarded as satisfactory. If there is any doubt about the integrity of the connections, the a.c. high current ohmmeter should be used for the test or a test equipment of the type shown in Figure 16.1 should be employed and the resistance calculated.

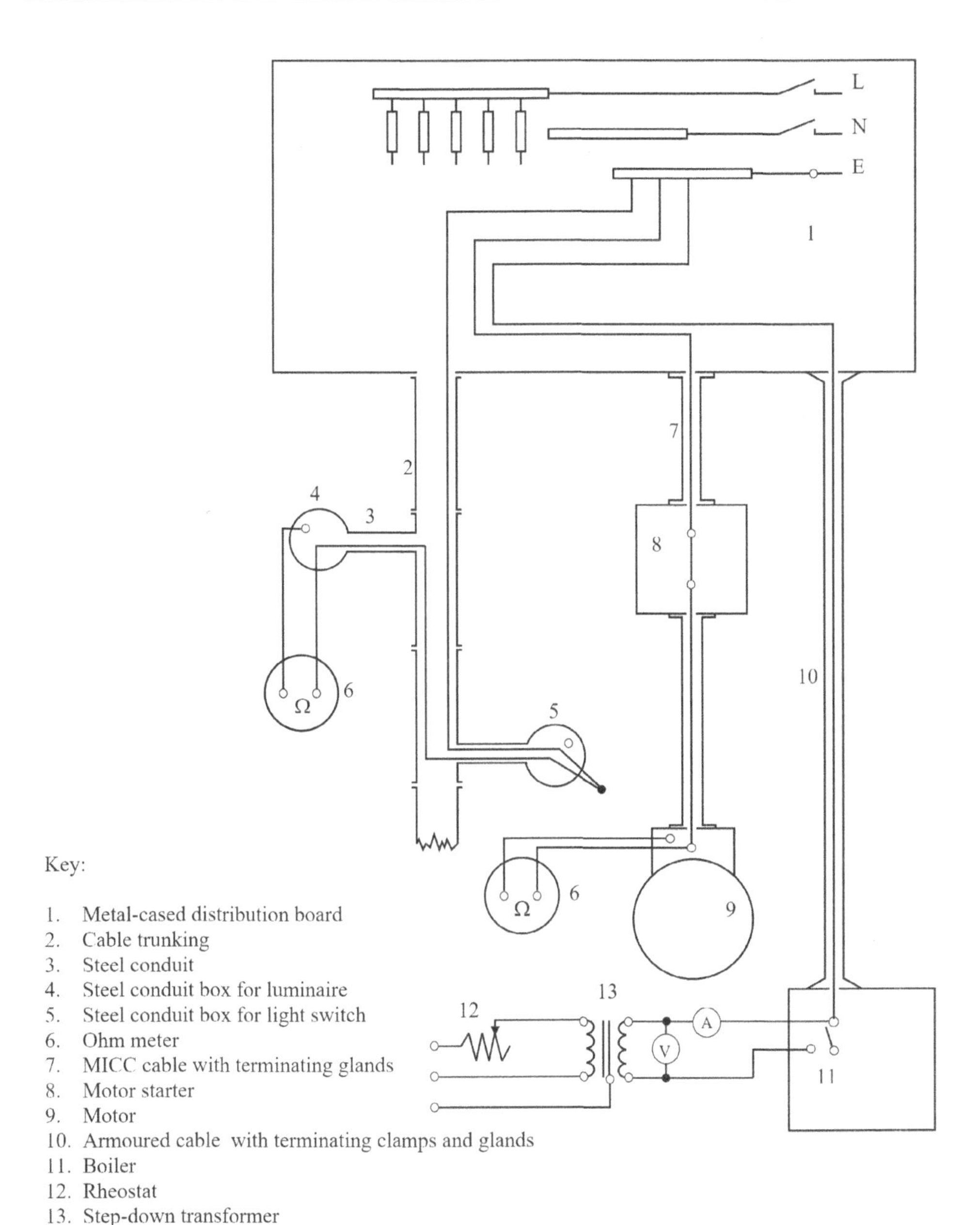

Figure 16.1 Measuring the earth continuity of exposed conductive parts

Continuity of ring final circuit conductors

The first test to be done consists of joining the phase and neutral conductors of opposite ends of the ring at the distribution board, as shown in Figure 16.2a, and measuring the resistance between the other two to prove continuity of the phase and neutral conductors. A high resistance reading indicates an open circuit or wrong connection. These other two

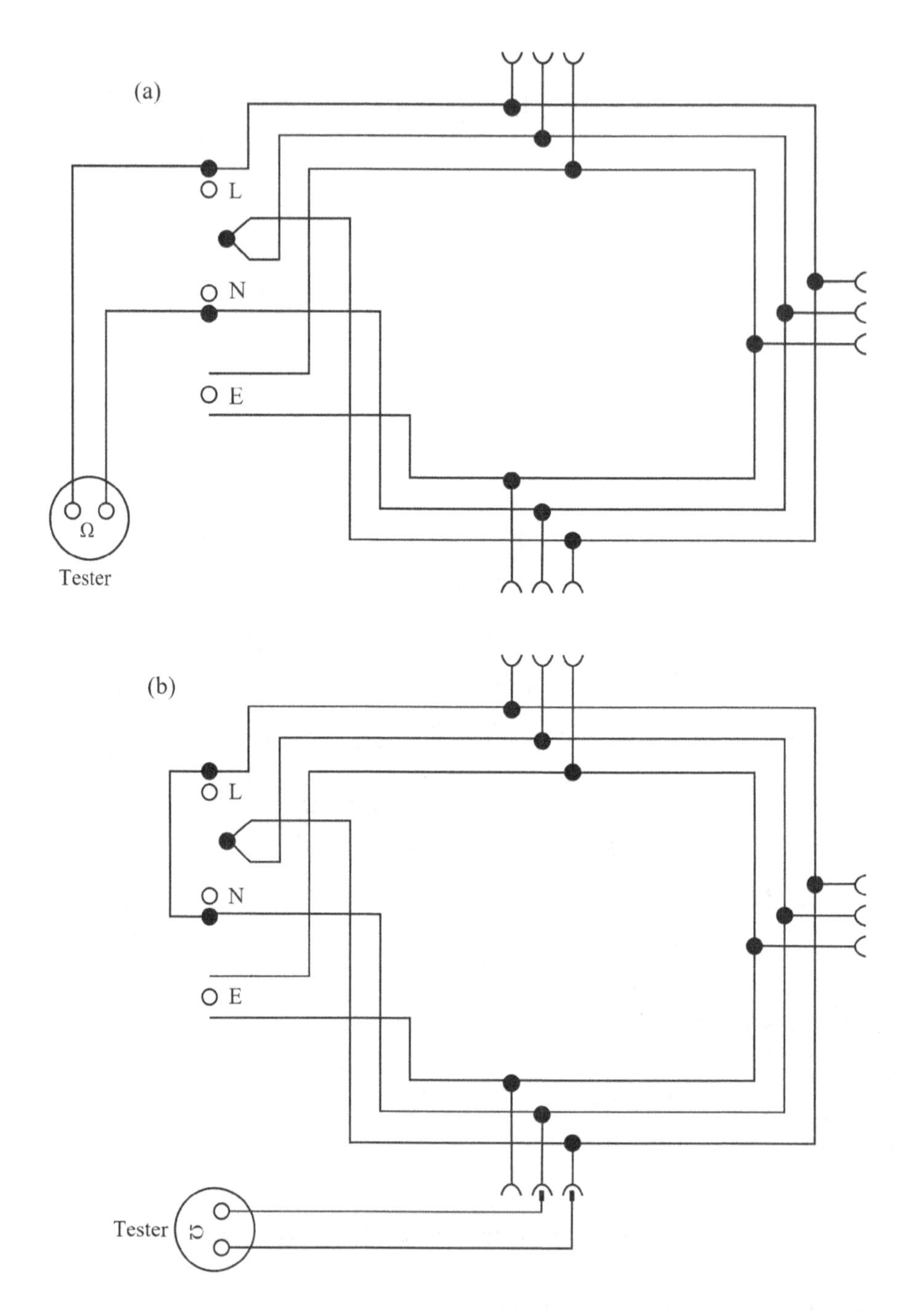

Figure 16.2 Measuring the continuity of the phase and neutral conductors in a ring circuit

are then connected together and the ohmmeter connected between the phase and neutral conductors at each socket-outlet in the ring (see Figure 16.2b). Each reading should be about the same provided that there are no interconnecting loops. The method is repeated, connecting the phase and protective conductors, as shown in Figure 16.3. The reading taken at the socket-outlet at the ring mid-point is the loop resistance to be used in calculating the earth loop impedance of the ring circuit.

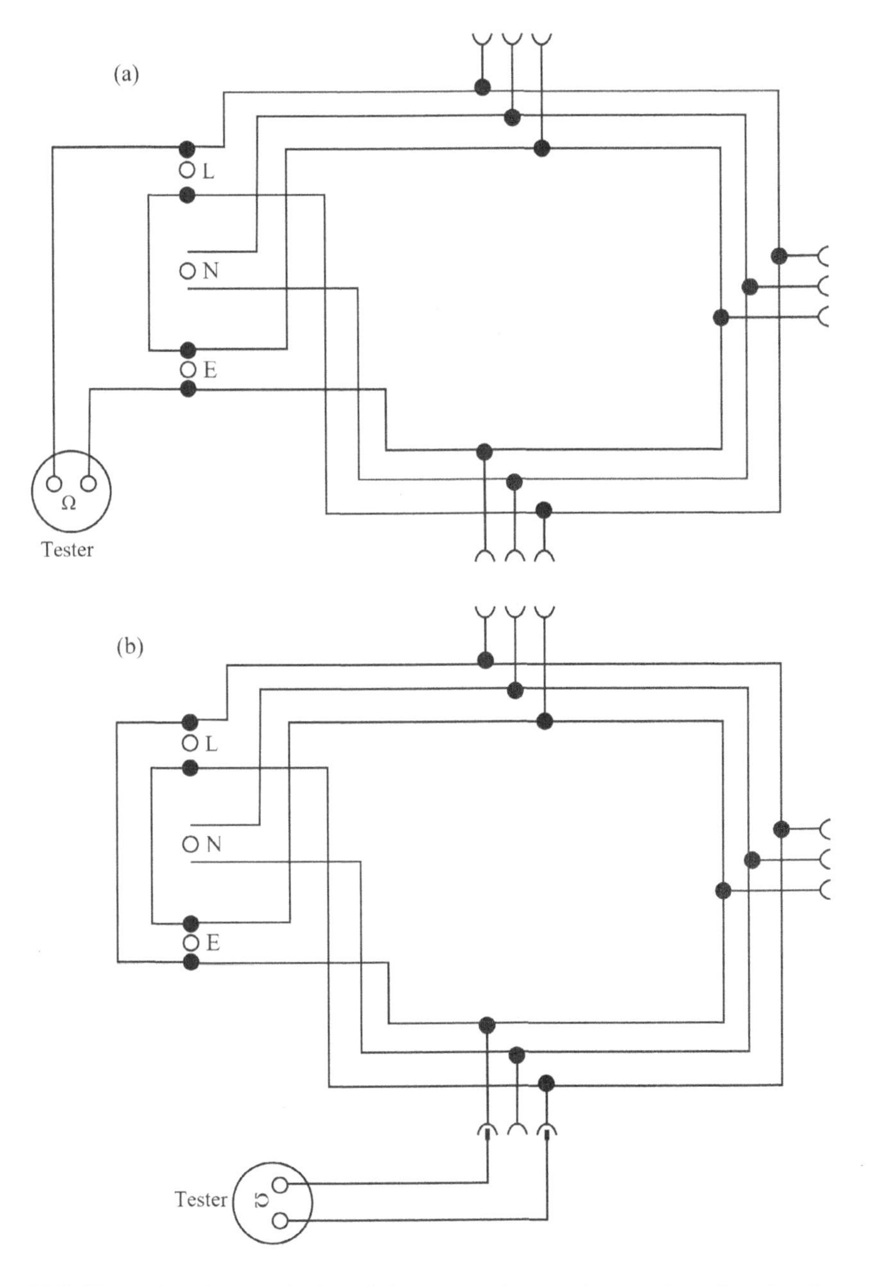

Figure 16.3 Measuring the continuity of the protective conductors in a ring circuit

Earth electrode resistance

The fall-of-potential method of measuring earth electrode resistance is indicated in Figure 16.4 and is appropriate if an electricity supply is available; alternatively, where a supply is not available, a dedicated earth electrode test instrument may be used, following the manufacturer's instructions. For this type of measurement, the earth electrode being tested should be disconnected from its installation to prevent any parallel paths interfering with the test results.

The current and voltage electrodes, which are usually 10–15 mm diameter steel rods, are pushed or driven into the ground by means of a vibrating hammer to a depth of preferably about 1 m; if the contact resistance between the ground and electrode is too high, it should be driven deeper and/or watered with brine. The current electrode has to be outside the resistance area of the electrode under test. A distance of 30 m apart should ensure this in most cases. The voltage electrode is first located midway between the other two at position 1 in Figure 16.4 and the current and voltage read. Readings are then obtained with the voltage electrode relocated at positions 2 and 3; for accurate results, the voltmeter should be a high resistance type so that its operating current is a negligible proportion of the current between the earth and current electrodes. Provided the readings are about the same, the earth electrode resistance is the mean voltage divided by the current.

Figures 16.4 and 16.5 show the voltage readings plotted against distances, where it will be seen that the maximum voltage gradient occurs near to the earth and current electrodes. Where the resistance areas do not overlap, the curve has a horizontal section and the three voltage readings are sufficient to establish the earth electrode resistance. Where the conductivity of the ground changes between the electrodes and/or their resistance values are significantly different, the resistance areas of the electrodes will vary and perhaps overlap, resulting in the shape of the curve varying. In this case the separation distance between the earth and current electrodes should be increased and the test repeated. The horizontal section of the curve may not be midway between them, so additional voltage readings should be taken if necessary nearer to each electrode to establish its location. If the readings fluctuate, it is due to the presence of stray currents of the same frequency in the ground. In this case, the results will be unreliable and the test should be carried out at a different frequency. This is best done by using a portable purpose-made earth electrode tester.

IET Guidance Note No. 3 describes two other measurement techniques. One involves the use of a test instrument that does not require the earthing conductor to be disconnected or extra electrodes to be placed in the ground. The other technique is given for RCD-protected TT installations. This consists of connecting an earth loop impedance tester between the earth electrode and the phase conductor of the installation, at the intake. Before doing this test, all equipotential bonding conductors are disconnected temporarily to ensure that all the test current passes through the electrode. The earth loop impedance so obtained is treated as the electrode resistance for calculation purposes. If the measured impedance is lower than 1667 Ω, the required disconnection time can be achieved using an RCD with a rated residual operating current of 30 mA, although the Guidance Note advises that earth electrode resistances higher than 200 Ω may indicate unstable conditions.

Insulation resistance tests

Insulation resistance tests are carried out on isolated systems with the purpose of detecting any degradation in insulation and short circuits between live conductors and between

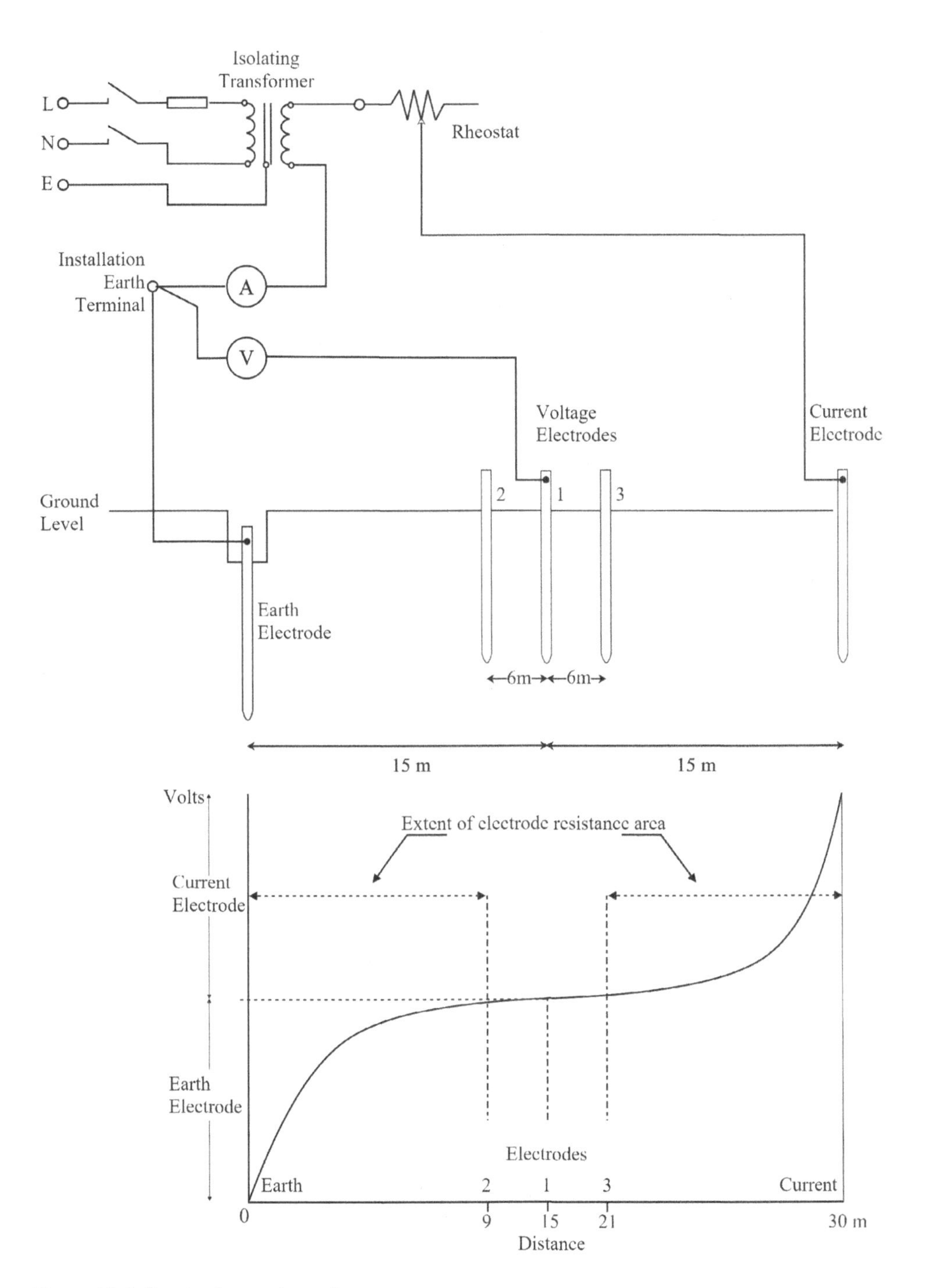

Figure 16.4 Earth electrode resistance measurement

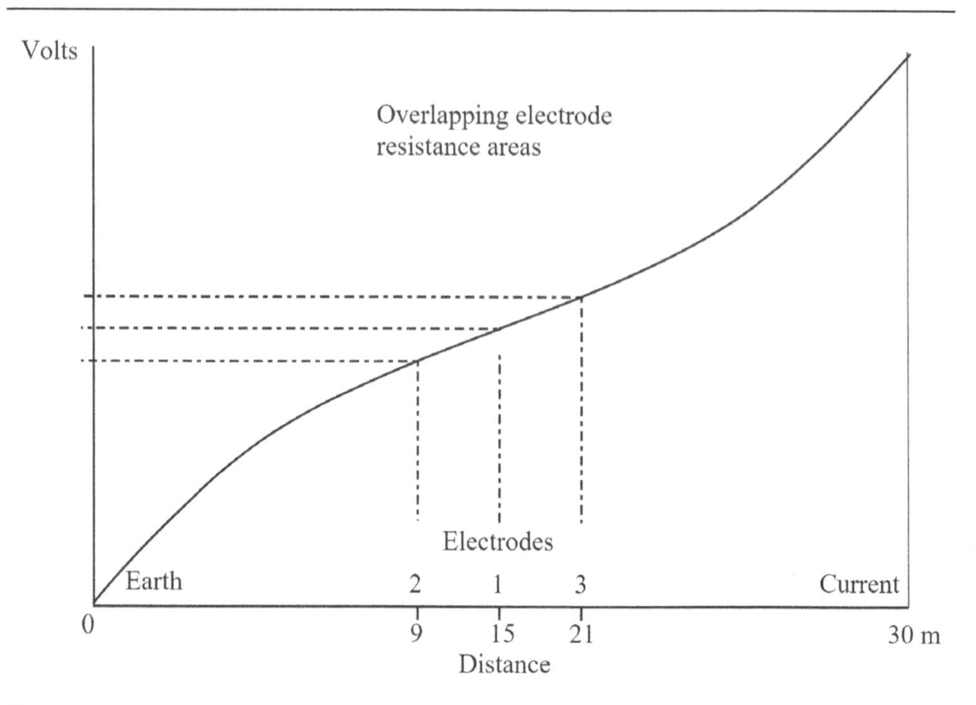

Figure 16.5 Voltage/distance curve where the electrode resistance areas overlap

live conductors and circuit protective conductors connected to the installation's earthing arrangements.

BS 7671 specifies that d.c. test voltages should be used for insulation resistance tests, and that the minimum acceptable insulation resistance values should be in accordance with Table 16.1. The instruments used should be capable of providing an output of not less than 1 mA. For installations designed to operate at voltages up to 500V, e.g. those connected to the 400/230V public supply and the 110V systems supplying portable apparatus, a 500V d.c. insulation tester is adequate and suitable for the prescribed test. A 1000V tester should be used to test systems over 500 V up to 1000 V. Large installations should be tested in sections, each of not more than fifty outlets. An outlet, for example, consists of a lighting point, a switch or a socket-outlet.

As the object is to test the installation and not the connected apparatus, which should have been tested by the maker, it should be prepared by switching off the main switch, removing lamps, and, for discharge lighting, unplugging or disconnecting one pole of capacitors or starting devices and also one pole of bell transformers, indicator lamps, or anything else connected in parallel with the supply. Disconnect connected apparatus, such as a cooker, by opening the double pole switch in the control unit and, for fixed apparatus controlled by a single pole switch, open it and disconnect the neutral. Voltage-sensitive equipment such as dimmers and any other susceptible electronic gear should be protected from damage by connecting the terminals together.

Table 16.1 BS 7671 advocated d.c. test voltages and minimum acceptable insulation resistance values

Circuit (Volts)	*Test (Volts d.c)*	*Minimum insulation resistance (MΩ)*
Extra-low, i.e. not exceeding 50 V a.c. or 120 V d.c. (SELV and PELV)	250	0.5
Over extra-low and up to 500 V, including FELV	500	1.00
LV over 500 V but not exceeding 1000 V a.c. or 1500 V d.c.	1000	1.00

Insulation resistance between current-carrying conductors and earth

The first test is between the current-carrying conductors and earth and is done by connecting the phase(s) and neutral conductors together at the supply point, as shown in Figure 16.6 and measuring the insulation between them and the consumer's earthing terminal, which must be connected to the means of earthing. The requirement that the protective conductors is connected through to the installation's earth was a new requirement introduced in BS 7671:2008. Both 'on' positions in two-way switched circuits should be tested. This test is not, of course, applicable to TN-C installations.

Insulation resistance between current-carrying conductors

The second test, illustrated in Figure 16.7, is to prove the insulation between the current-carrying conductors and is carried out by testing between each phase/neutral and the other conductors connected together at the supply point.

Suspect insulation

In most cases, the insulation readings will far exceed the minimum tabulated values but, where a reading is fewer than 2 MΩ, each circuit should be separately tested and if any is fewer than 2 MΩ it should be checked for a latent defect.

Site-applied insulation

Where equipment is fabricated or repaired on site and the work involves the insulation of live conductors, the insulation standard should be comparable with that of similar factory-built apparatus and be subjected to the same HV testing. In the case of LV switchgear and controlgear to BS EN 61439–1:2011 *Low-voltage switchgear and controlgear assemblies. General rules*, the dielectric test voltage, between live parts of different polarity and between live parts and earth, varies according to the system voltage and is 2120V d.c. or 1500V a.c. for a 230V system and 2670V d.c. or 1890V a.c. for a 400V system, applied for 5 seconds. Table 16.2 lists the voltages at which lower voltage control circuits are tested.

There is also an impulse voltage test which is dependent on the main circuit voltage and reference should be made to the British Standard for detailed information and guidance.

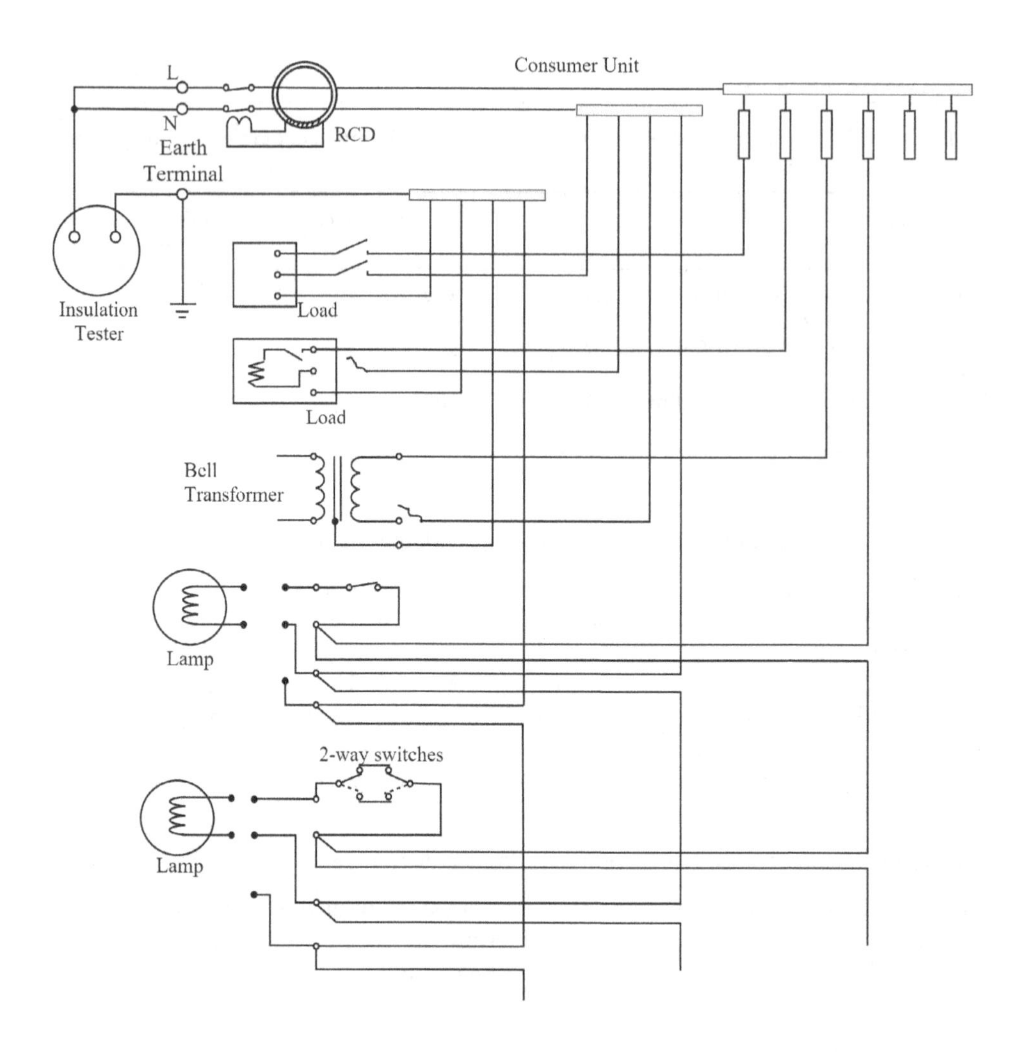

Figure 16.6 Insulation testing between phase(s) and neutral conductors and earth

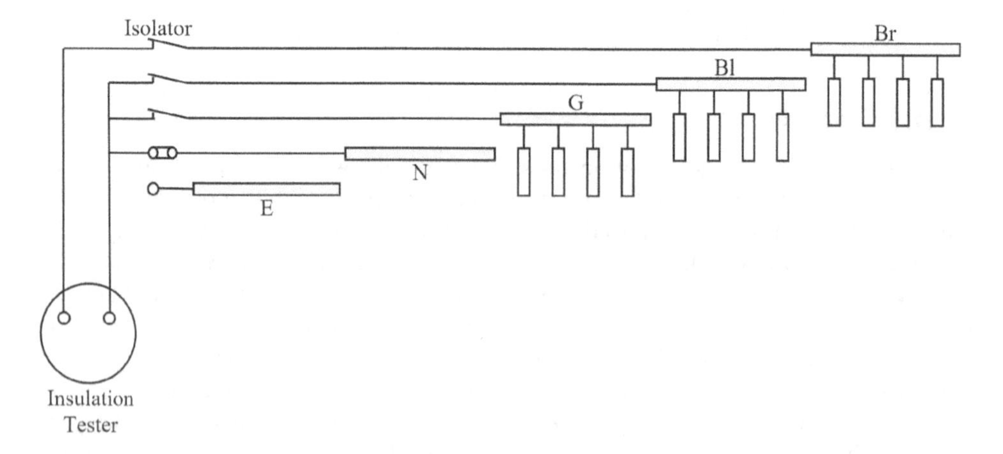

Figure 16.7 Insulation test between each phase or neutral conductor and the other conductors connected together

Table 16.2 Test voltages for lower voltage control circuits

Circuit volts U_i	*Test voltage (a.c. rms)*
Up to 12	250
12 to 60	500
Over 60	$2\ U_i$ + 1000, minimum 1500

SELV and PELV circuits

If the installation includes SELV circuits, for battery charging or lighting, for example, it is essential to prove that there is no possibility of the SELV circuits being energised at a higher potential because the live parts of such circuits may be handled. This possibility exists where the power source is a transformer and/or where the SELV cables are not separated from cables operating at a higher voltage. The IET recommended tests comprise a 250 V d.c. insulation test between the input and output circuits, the insulation resistance to be not fewer than 0.5 MΩ, and an insulation test at 250 V d.c. between the SELV circuit and the protective conductor of the supply input circuit, with the same minimum acceptable resistance. Where the SELV conductors are separated by just insulation, the test voltage should be 500 V d.c. with a minimum resistance value of 1 MΩ.

PELV circuits should have their basic insulation between line conductors and all other circuits tested at 250 V d.c. with a minimum resistance value of 0.5 MΩ.

Walls and floor in non-conducting location

If the installation includes any special facilities such as a test bench where it is necessary to establish and maintain a non-conducting location, the walls and floors must be made of non-conducting material. Their insulation value has to be proved by measuring the leakage current between earth and at least three points on each relevant surface. So a test bench against a wall would require a test at not fewer than three points on the adjacent wall and three on the floor. If there is an extraneous conductive part in the location, such as a metal window frame or service pipe, one of the test points should be not fewer than 1 m or more than 1.2 m from it, with the other two being further away. Check that the extraneous conductive part is separated from any exposed conductive part so that they cannot be touched simultaneously. If they can, the section of the extraneous part, within reach, has to be insulated to withstand a 2 kV a.c. test with a leakage current not exceeding 1 mA, which should be done after a 500 V d.c. test with a 1 MΩ resistance value; see Figure 16.8.

Although BS 7671 gives no guidance on the test probes to be used, which should simulate the contact of hands and/or feet with the test surfaces, IET Guidance Note No. 3 describes two types of electrode, one square and the other triangular. The square one has 250 mm sides and is used with a square of damp paper or cloth between it and the test surface. The other is a tripod with three circular electrodes, spaced at 180 mm, each having a contact area of 900 mm. The electrodes are in a flexible material such as conducting rubber. They are applied to the test surface which is either moistened or covered with a damp cloth. In both cases, the electrodes are pressed against the test surface with a force of 250 N for walls and 750 N for floors. Where the test surfaces are uneven, a more resilient type of electrode with a contact area of 250 × 250 mm, such as that illustrated in Figure 16.8, may prove to be more suitable.

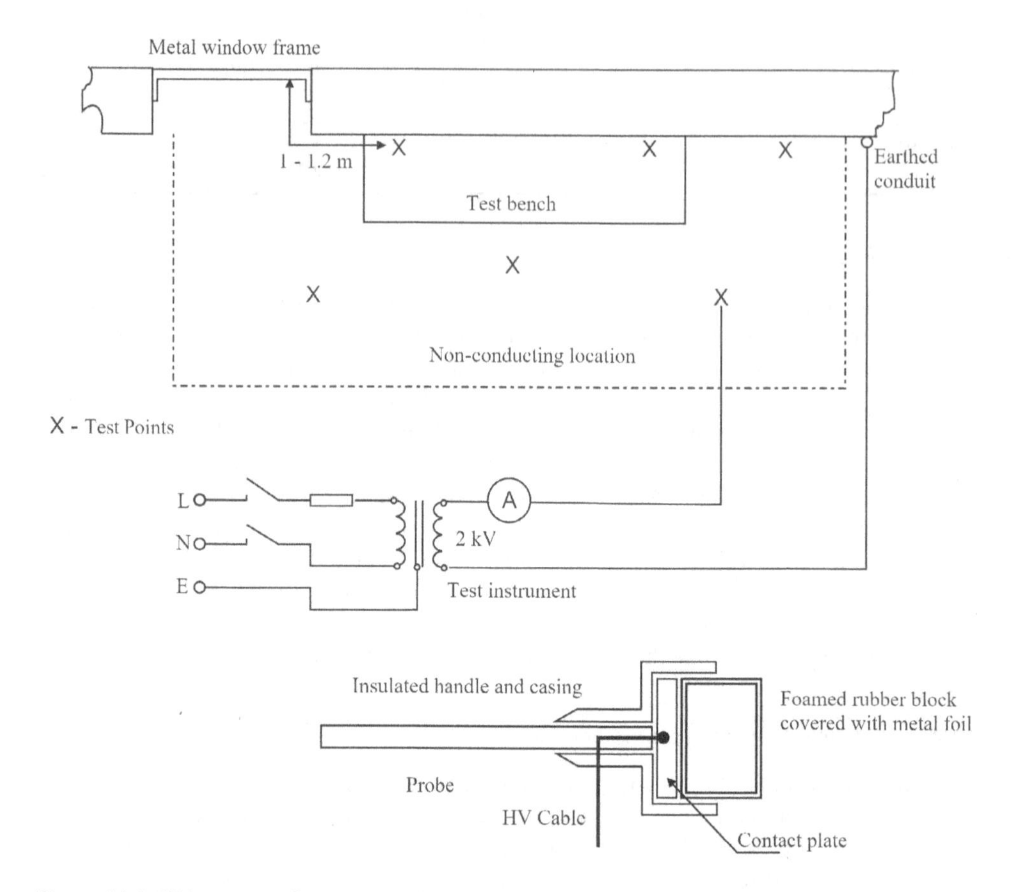

Figure 16.8 HV testing of a non-conducting location

One test lead should be connected to the probe and the other to the nearest protective conductor. The test voltage should be at least 500 V d.c. or 1000 V d.c. if the installation's nominal voltage exceeds 500 V. If the measured resistance exceeds 50 kΩ, the floors and walls may be regarded as being non-conducting.

Polarity test

There have been cases where the phase and neutral of single phase TN-S supplies have been interchanged at the intake so, for a new installation, it is worth checking that this has not happened. If the polarities are correct, a voltmeter connected between the conductor labelled 'phase' and anything earthed will display the mains voltage.

So far as the installation is concerned, the object of the polarity test is to prove that the fuses and single pole control devices are in the phase conductors only and that socket-outlets and lampholders are correctly wired. The continuity tests, already carried out, have confirmed that the socket-outlet ring circuit connections are correct so no retest is needed. For the remaining radial circuits, IET Guidance Note No. 3 recommends using the same test method employed for continuity testing, connecting the phase conductor to the protective conductor at the distribution board and then test between the phase and earth terminals at each outlet.

Loop impedance and prospective short circuit current testing

Loop impedance tests are carried out to determine the loop impedance between the power source(s) and the point in the installation where the test is done and are an essential component of checking the effectiveness of fault protection using automatic disconnection. The test equipment employed measures the current which passes through a resistor and displays the result in ohms. It is used to determine the loop impedance between phases, phase to neutral or any phase to earth. Some instruments incorporate a transformer to enable the neutral/earth loop impedance to be measured. From Ohm's law these readings can then be expressed not just as impedances but also in terms of prospective short circuit fault currents.

The object of measuring the phase/phase or phase/neutral loop is to check that the short circuit rating of the protective devices is adequate. It should not be necessary to carry out this test on a new installation where the short circuit rating of the selected apparatus should be suitable for the calculated values of short circuit current, but the measurement is a useful check. The measurement, however, can only indicate the value when the reading is taken. Subsequent alterations in the supply network and connected apparatus can affect the value, so there should be a substantial safety margin between this ascertained value and the short circuit rating of the protection to allow for reductions in the loop impedance and a consequent increase in the prospective short circuit current.

The earth fault loop impedance should be measured so as to enable the prospective fault currents to be determined. These currents have to be sufficient to operate the protective fuses or circuit breakers within the specified time in the event of an earth fault so as to meet the required standards for automatic disconnection; the assumption is made that the earth fault itself has zero impedance. Again, the readings obtained indicate the present position, which may alter subsequently. In most cases load growth tends to increase the prospective fault current, but in areas subject to dereliction the load may decrease and the consequent network alterations may result in a fall in the prospective fault current.

Tests should be carried out at the origin of the installation, at each distribution board, at all fixed equipment, at all socket-outlets, at the furthest point of every radial circuit, and at a sample of lighting outlets (typically 10% at fittings farthest from the supply). They are carried out after the main bonding conductors have been disconnected so as to prevent parallel paths affecting the measurements. The test should be repeated at least once to allow for the effect of transient variations in the supply voltage.

In a socket-outlet ring main they are taken at the socket-outlet nearest the mid-point of the ring and at the end of non-fused spurs. The readings are then converted into prospective fault currents by Ohm's law, i.e. $I = E/Z$, where E is the phase-to-earth voltage; many instruments do this automatically and display the results. These currents are then used to ascertain the corresponding protective device's operating time from the time/current curves of the protective device. BS 7671 incorporates tables of earth fault loop impedance values for different ratings of the most commonly used protective devices so as to negate the need for calculation.

The measurements should be taken between a phase and earth, not neutral and earth, and at a current exceeding 10 A. Instruments which are readily available operate at 20 to 25 A where the supply system is TN-S or TN-C-S and the earth loop impedance is low, but usually at a much lower current on TT systems where high earth loop impedances are common. In these cases the operating current may be under 10 A, and if the installation has a protective conductor which is mainly steel conduit, the reading will not be accurate. A second test is made, therefore, of the earth loop impedance external to the installation by measuring between one of the incoming phase terminals and the main earthing terminal with the main equipotential bonding conductors disconnected. The effective value of the earth loop impedance is then taken as twice the first reading less the second.

In circuits protected by RCDs and where the current exceeds the RCD tripping current, carrying out the test may trip the RCD. Some test instruments are designed to prevent this happening by, for example, first passing d.c. current to saturate the magnetic core of the core balance transformer in the RCD. Where such an instrument is not available, or it fails to prevent the RCD tripping, the measurements can be taken after defeating the RCD's tripping mechanism or bridging the phase terminals. It may not be practicable to do this safely on some RCDs built into consumer units. In this case, the earth loop impedance is measured at the incoming mains location as described previously, and the total earth loop impedance for any point in the installation is taken as the summation of this value and the resistance of the phase/earth loop as measured in the continuity tests.

Bridging the phase terminals of an RCD will prevent its inductance being included when the earth loop impedance is measured, and the calculated resultant fault current will be slightly higher than the true value. However, this is immaterial, as in the event of an earth fault the RCD will trip anyway within the required time. It is, of course, axiomatic that any bridging conductors are removed once the tests have been completed.

Figure 16.9 shows a domestic installation connected to a TT system of supply and an earth loop impedance tester connected to measure the earth loop impedance external to the installation. It will be seen that the loop consists of the phase conductor from the incoming terminals of the consumer unit via the metering and supply company's fuse to the HV/LV distribution transformer's secondary winding, then via the transformer neutral and earth electrode, through the ground to the consumer's earth terminal. If the RCD trip can be defeated or bridged, the earth loop impedance at the socket-outlets shown can be measured by connecting the instrument at these points. Otherwise, the calculation method described previously will have to be used.

When carrying out phase/earth loop impedance tests, the protective conductors, the metalwork of Class I apparatus and bonded extraneous metalwork are energised and there will be a potential difference between them and the ground. For high earth loop impedances, this would be a substantial proportion of mains voltage, so to avoid any shock hazard no one should touch the protective conductors or metalwork during the test.

RCD testing

BS 7671 stipulates that RCDs should be tested with their loads disconnected by pressing the test button and also by supplying an earth leakage current and checking the tripping time. Figure 16.10 shows a small installation where the RCD is a circuit breaker which

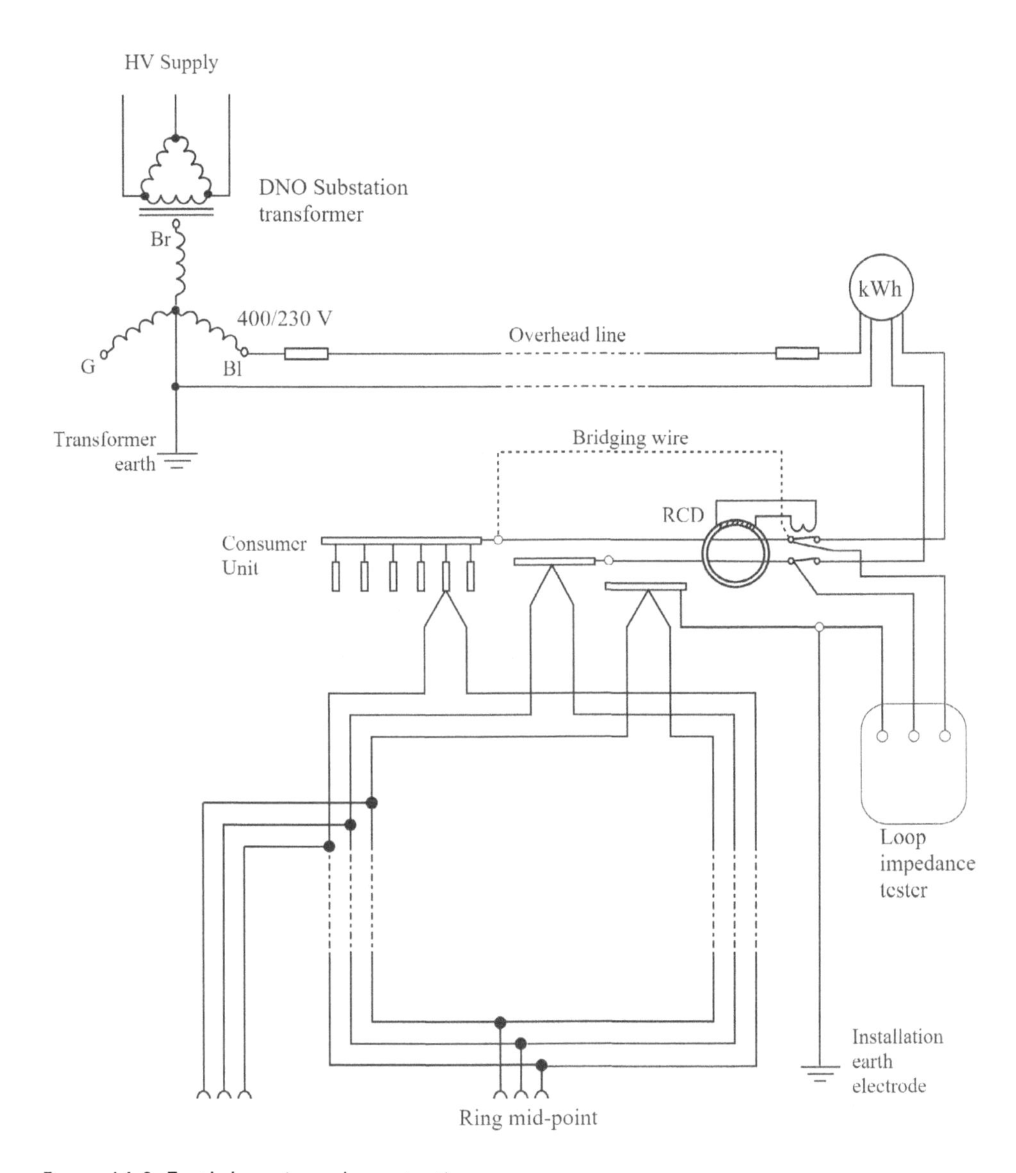

Figure 16.9 Earth loop impedance testing

also serves as an isolating switch. The second test can be safely carried out by the following procedure:

(i) Open the RCD.
(ii) Remove the fuses.
(iii) Connect the tester between the incoming side of a fuse carrier and the earth busbar.
(iv) Close the RCD.
(v) Operate the tester.

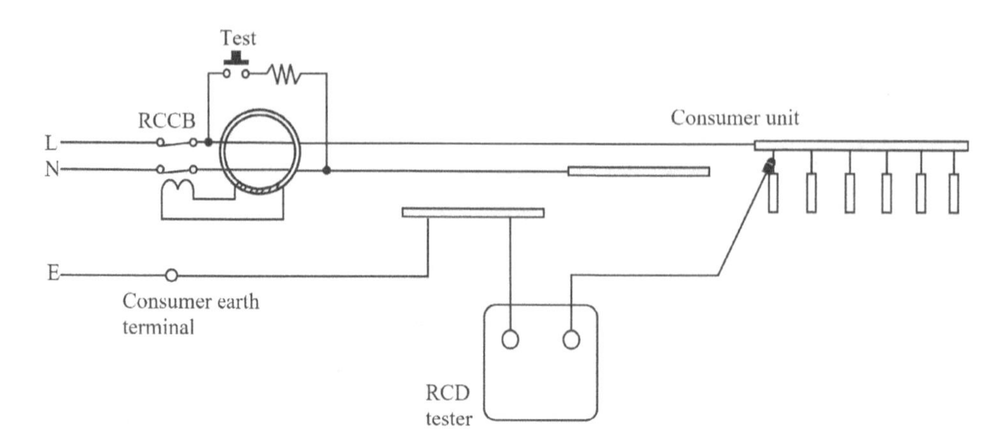

Figure 16.10 Checking the tripping time of a residual current circuit breaker

The specified requirement is that the rated tripping current should cause the circuit breaker to operate within 0.2 seconds, or 0.3 seconds for devices to BS 61008 and BS 61009, or at the delay time declared by the device maker. Where the RCD is providing additional protection and/or fault protection, it should operate within 40 milliseconds for a residual current of 150 mA (5 × $I_{\Delta N}$). On larger installations, where there is a switchboard feeding a number of distribution fuse-boards, for example, earth leakage protection may be provided on the incoming and/or outgoing circuit breakers as well as on the final sub-circuits. All the devices should be tested and any designed discrimination between them verified.

The operation of the tester will cause the protective conductors, Class 1 apparatus metalwork and bonded extraneous conductive parts to become live with respect to the ground at a voltage dependent on the earth loop impedance, so during the tests no one should touch such metalwork.

Excess current protection testing

Apart from functionally checking the actuating switch of circuit breakers, BS 7671 does not require the testing of excess current protection, because circuit breakers should have been tested by the maker. However, where excess current and the operating time of the trip are set on site, it is advisable to verify the performance using primary or secondary current injection apparatus. Such testing is essential on high voltage systems, where excess current protection is usually provided by relays fed from current transformers. Incidentally, such apparatus can conveniently be used to test earth leakage protection where the leakage trip current and time are outside the limits of the normal RCD testers.

Sampling during periodic inspections and tests

In the context of routine periodic inspections and tests, it is relatively straightforward to fully inspect and test small electrical installations in, for example, private dwellings. On the other hand, inspecting and testing 100% of the installations in larger more complex premises such as hospitals, factories, universities and so on is a very time consuming, complex and expensive task. Because of this, it has become custom and practice to inspect and test only a representative sample of larger installations, with the sample size being increased if problems are found that would justify the increase.

The job of determining the sample size falls to the engineer carrying out the periodic inspection and test in consultation with the client. The factors to be taken into account include the age and general condition of the installation, the extent to which has been maintained over the years, the history of faults and failures, the environmental conditions, and the results of previous periodic inspections and tests.

Records of the sample of circuits and accessories that have been inspected and tested should be retained to ensure that the same sample is not inspected and tested at the next maintenance interval.

As far as inspections are concerned, IET Guidance Note 3 suggests that the external condition of all main switchgear and the presence and tightness of all earthing and main bonding conductors should be inspected. However, all other parts of the installation, including the internal parts of main switchgear, final distribution boards, and accessories such as socket-outlets and light switches, may be sampled with a sample size greater than 25% for final distribution boards and greater than 10% for all other items.

IET Guidance Note 3 applies a similar principle to the tests to be carried out on an installation, suggesting that the sampling should be at the discretion of the test engineer but should be no less than 10%.

Electrical apparatus and equipment testing

There are very many circumstances in which electrical apparatus and equipment need to be tested; some examples are

- during fault finding on electrical installations and equipment;
- appliances and equipment being tested for electrical safety on a production line;
- motors being tested after having been rewound;
- computer monitors and televisions after having been repaired; and
- portable, transportable and fixed equipment, as well as installations, being tested as part of a routine preventive maintenance regime. (The legal duties relating to maintenance of equipment were set out in Chapter 6 and the general topic of maintenance is covered in Chapter 18.)

The tests will be a combination of functional tests and electrical safety tests, the latter being aimed at ensuring that the equipment will not be electrically dangerous when energised. This will principally involve testing the insulation resistance characteristics of the device and ensuring the integrity of the earthing arrangements on Class I appliances, although many other types of tests are carried out depending upon the application (e.g. measurement of magnetising losses, energy efficiency, thermal effects, in-rush currents, harmonics, electrical and acoustic noise levels, radiation levels etc.).

The risks of electrical injury during testing can be inherently high because conductors energised at dangerous voltages are often exposed to touch. It is therefore essential that measures are taken to ensure that the persons who are undertaking these tests are not exposed to unacceptable levels of risk, and the following text describes the main techniques for achieving this. The HSE has published guidance on the topic in INDG354 *Safety in electrical testing at work*.

The relevant legal requirements are the Electricity at Work Regulations, mainly Regulations 4(3) and 14. It should be borne in mind that the legal duty is that the work should always be done dead, unless it is unreasonable for the conductors to be dead, the level of risk is acceptable and suitable precautions are taken against injury. A very common misconception is that testing does not constitute a work activity and therefore does not come within the remit of the regulations. This is, of course, not the case – testing is a work activity so the full panoply of the law applies.

In those situations where testing must be done with the conductors live, the most important requirement is that the tester is protected against direct contact with those conductors energised at dangerous potentials. There are two main ways in which this can be achieved:

(i) the use of interlocked jigs and enclosures to keep people away from the live conductors during testing; and
(ii) the use of techniques such as limitation of energy, electrical separation, and earth leakage protection, coupled with precautionary measures such as the use of barriers, warning indicators, and insulated tools, and the wearing of PPE such as insulated gloves and footwear.

Jigs and enclosures

Testing during the course of manufacture on an assembly line or before and after the repair of components can often be done using interlocked jigs and enclosures. The general principle is that the item being tested is placed inside the enclosure, perhaps after test probes and instrumentation have been attached to the item, and the enclosure door is then closed. Only once the door is closed can energy be applied to the item under test and the action of opening the door would lead to the supply being disconnected. One such type of interlocked jig, and the associated interlocking device, is described in Chapter 17.

Enclosures used for this purpose should generally be constructed with an ingress protection rating of IP3X, unless access to the live parts is otherwise safeguarded by, for example, the use of electrosensitive protective equipment. Moreover, the interlocking system should be designed so as not to fail to danger in the event of a single fault occurring, meaning that the safety-related parts of the control system should be risk assessed and designed in accordance with the principles of BS EN ISO 13849–1:2015 *Safety of machinery. Safety-related parts of control systems. General principles for design* (see Chapter 17).

In high volume manufacturing, for example in the production of white goods such as washing machines and dishwashers, the testing procedure is often automated. Partly assembled products are tested on a moving production line after test cables have been attached. The appliances are safeguarded by fixed and interlocked guards while they are energised, making them inaccessible to personnel while they are energised, and measurements are taken and recorded using automated test equipment, frequently under the control of programmable control systems such as Programmable Logic Controllers (PLC).

Live testing

Live testing of equipment, apparatus and installations is a common cause of electrical injury and requires careful consideration and planning of the precautions needed to reduce the risks to an acceptable level. It should be carried out only if there is no alternative, such as fault tracing using continuity testing rather than measuring voltage. An example of a fatal accident will illustrate the point.

Case study

An electrician employed by a catering equipment supply company was electrocuted while fault finding on a 3-phase commercial metal-clad dishwasher located in the small kitchen of a café. The café had reported that the automatic detergent dosing system was not working, although they could continue to use the dishwasher by manually pouring detergent into it during each wash cycle.

The electrician arrived at the café during a busy lunchtime period and proceeded to remove the machine's steel panels to gain access to the control circuitry, including the 2-wire 230 V a.c. circuit supplying the detergent peristaltic dosing pump, which he must have suspected to be the cause of the problem. He did this even though there were a number of kitchen and waiting staff working in the immediate vicinity, passing close by him as they made their way between the kitchen and the café. He was kneeling on the kitchen floor at the side of the dishwasher.

He initiated a wash cycle with the intention of using a voltmeter to measure the voltage on the supply to the pump, which was meant to become energised for a short period during the wash cycle when the pump was meant to dispense detergent into the wash tank. It is presumed that he did this to confirm that there was a supply to the pump, in which case if the pump failed to work he could conclude that it was faulty and needed to be replaced. To do this, he disconnected the pump's two supply wires and held the spade terminal that was attached to the end of the live wire in his bare hand, touching one probe of his meter against it and pressing the other probe against some earthed metalwork. Unfortunately, the insulating shroud that should have been surrounding the spade terminal had been removed during some previous maintenance work.

When the wire became live during the wash cycle, he suffered a fatal electric shock between his right hand holding the terminal and his legs which were touching the kitchen floor. A customer in the café, on hearing a commotion in the kitchen, ran into the kitchen and had the presence of mind to use a wooden spoon to knock the still-live wire out of the electrician's hand, but it was too late.

One of the features of this accident assessed during the investigation was whether or not he could have attempted to locate the fault by continuity testing on the isolated machine rather than by testing for voltage on the energised machine. The answer was that he could have done so if he had had access to the machine's electrical schematics, whilst recognising that this method is more complicated and time consuming. Moreover, there was no rush for the machine to be repaired, and the fault finding work could have been delayed until the café was not busy and employees could be excluded from the kitchen, making it easier to carry out the work in safety and in a less pressured environment.

General precautions

In all cases it is essential that the person carrying out live testing is competent, having received adequate training and instruction on the risks and the techniques to be used to avoid injury. There should always be a sufficiently comprehensive risk assessment carried out, with a clear and unambiguous explanation of the safety precautions. The main techniques that can be employed, most often in combination, are as follows:

(i) The use of SELV supplies to the items under test
(ii) The use of isolating transformers, coupled with the removal of conducting paths to earth in the vicinity of the test area
(iii) Where isolating transformers cannot be used, shrouding and screening off adjacent live and earthed conductors and providing sensitive earth leakage protection where possible (such as an RCD with a residual operating current of no greater than 30 mA)
(iv) Limiting to a safe level the amount of current that can be supplied into a load representative of the human body. In general this means that the current should be limited to a maximum of 5 mA. As an example, this limitation is commonly applied to the probes of insulation resistance test units which may be energised in excess of 1000 V but which can be handled without undue risk because the output current is reliably limited to a nominally safe level.
(v) Carrying out the work in an 'area set aside', delineated by barriers and screens, to prevent both unauthorised access and the tester being disturbed. In some circumstances, such as high voltage test areas, the barriers should be interlocked to prevent access while the unit is energised. People at different work stations on a test bench should be separated by insulating screens.
(vi) The use of warning signs and indicators to highlight the danger
(vii) The provision of emergency switching devices to ensure the supplies to the test area can be switched off quickly in an emergency
(viii) Accompaniment of the tester in high risk situations such as high voltage testing of switchgear and other equipment
(ix) The use of tools and test equipment suitable for the application
(x) The wearing of PPE where necessary – such as antiflash clothing where there is a significant risk of high energy flashovers, and/or insulated gloves to prevent direct contact

Some examples of these techniques follow.

Earth-free areas

Some types of product, a photocopier perhaps, require adjustments to be made to various components in proximity to exposed live parts as part of the test procedure. If the potential shock voltage exceeds extra-low voltage, the danger should be minimised by avoiding the possibility of a shock between a live part and earthed metal or other conducting materials. This is done by carrying out the work in an area that has been made as earth-free as possible and/or by using an isolating transformer to provide an unreferenced earth-free supply. This type of testing should always be preceded by insulation tests to ensure that there is no earth fault which could make the exposed metalwork live. It is also advisable to exclude conducting materials from the test area which might become live from a fortuitous contact with live parts of the product. Thus, wooden or plastic benches are better than metal ones and insulated rather than metal-cased instruments are preferable.

In addition, insulating sheeting such as polythene and neoprene can be used to shroud off conducting paths to earth.

Earthed areas

Some types of electrical apparatus have to be tested and adjusted with their metalwork earthed, so an earth has to be introduced into the test area. This would increase the potential shock risk considerably if the normal mains supply, with earthed neutral, was to be employed. One way of avoiding this is to use an isolating transformer and earth each pole of the output winding instead of the neutral through a high resistance in the earthing conductor so that an operator, touching either pole and the earthed metalwork, would experience a potential shock voltage of less than 115 V instead of 230 V and a shock current that is limited by the resistance to a harmless value. The circuit diagram is shown in Figure 16.11. The relay should be a high impedance sensitive type with an operating current of not more than 1 mA.

Batch product test

For products that are batch rather than mass produced and supplied in a range of sizes, it is often convenient to establish a test facility between the assembly and packing and despatch departments. Figures 16.12 and 16.13 show an example of a test facility for the

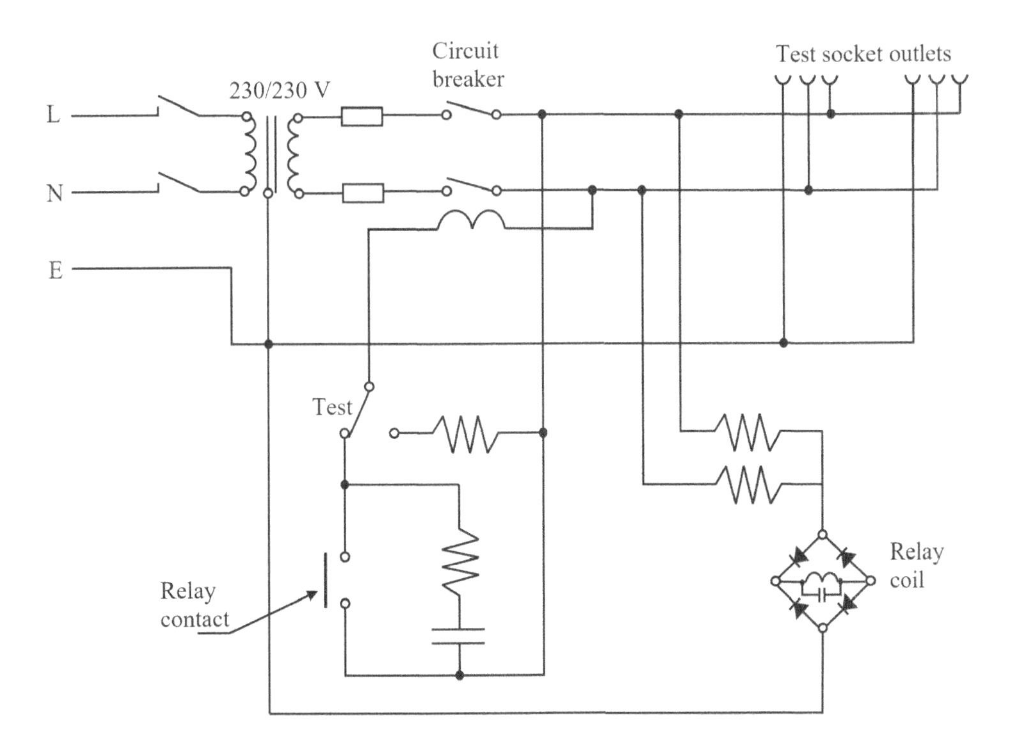

Figure 16.11 Isolated supply circuit for Class I electronic product testing

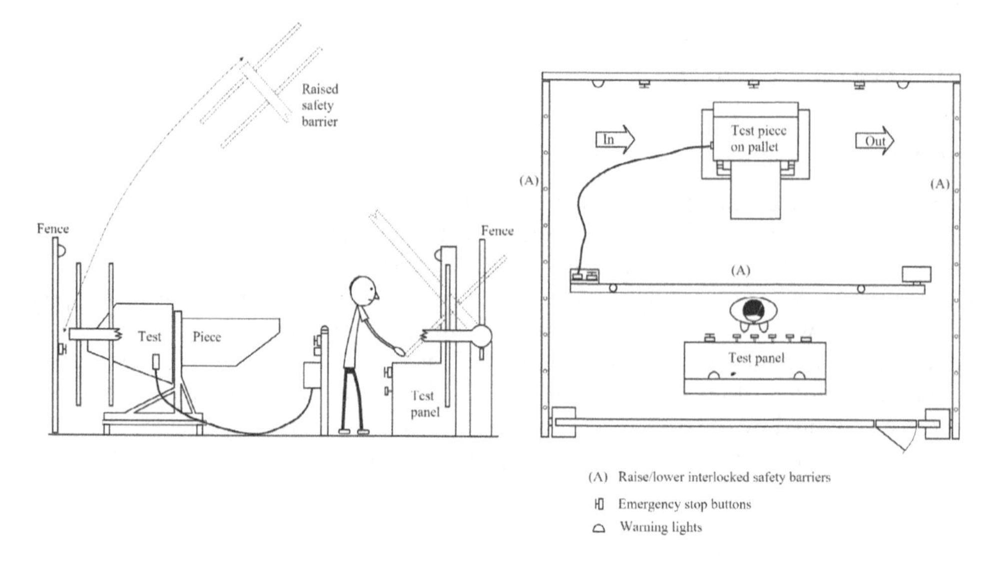

Figure 16.12 **Layout of test facility for batch produced refrigeration units**

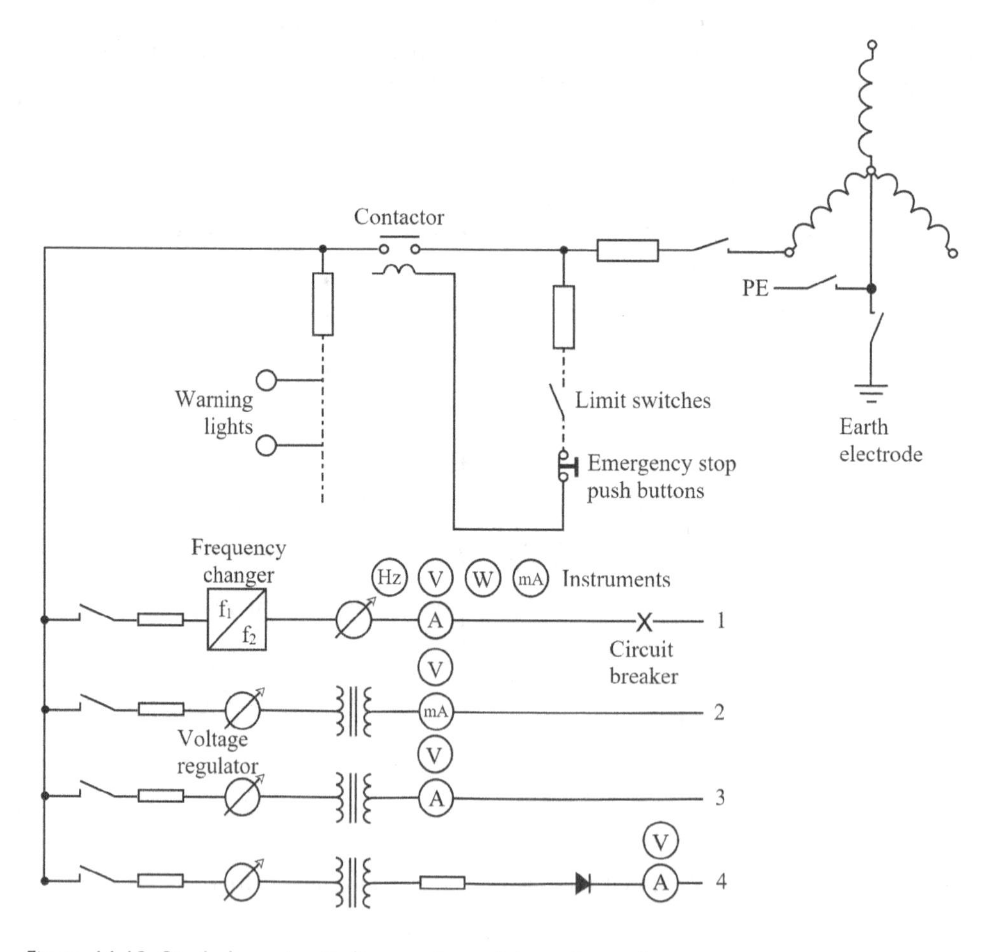

Figure 16.13 **Single line circuit diagram for refrigeration units test**

transportable refrigeration units used on refrigerated lorries and containers. These units usually have a diesel engine to drive the compressor when in transit by road and a mains voltage electric motor drive for use when parked. The control circuits are usually 12 V or 24 V d.c.

In the factory the units are moved on pallets by forklift trucks. Two sides of the test enclosure comprise push button controlled barriers which can be raised to permit truck ingress and exit. Limit switches on the barrier pedestals ensure the supply is off when either barrier is raised. There is also an internal safety barrier between the test panel and test piece which has a limit switch on its pedestal to switch off the supply when the tester raises the barrier for access to the test piece. Other safety features include connecting the test cables from terminal boxes near the test piece to minimise their length and avoid a tripping hazard at the test panel. The connections between the test panel and the terminal boxes are in floor ducts. The terminal boxes are near the side barriers so that the test cables can be run alongside them and minimise the lengths adjacent to the machine and the tripping hazard. Short lengths of moulded rubber floor mats, with a recess for the cables, could be used over these exposed lengths.

There are a number of manually resettable emergency stop buttons, suitably located, to enable the supply to be cut off and red warning lamps which are illuminated when the supply is 'on'. To protect the operator against shocks between live parts and earth, the floor of the test area should he non-conducting and the test supply unearthed but with a facility for earthing it if a test specification should require it. When carrying out a.c. tests at voltages exceeding SELV with the supply earthed and the metalwork earthed via the protective conductor, sensitive and rapid operating earth leakage protection may be employed to minimise the shock hazard.

Manually applied test electrodes for HV testing should preferably be used only where the HV current is restricted to a maximum of 5 mA, a potential shock current which is

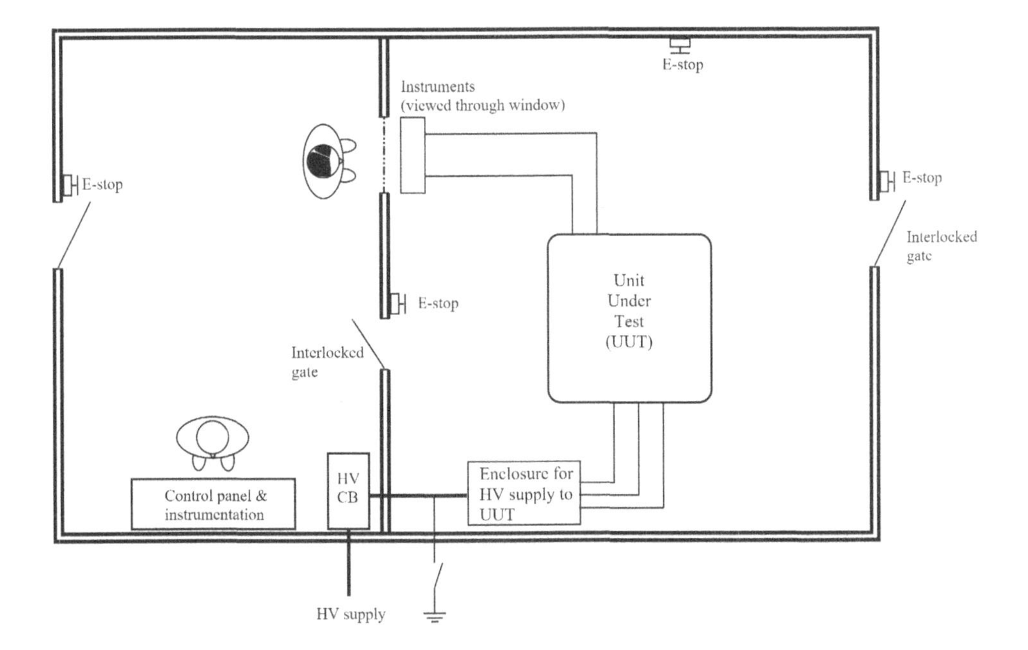

Figure 16.14 High voltage test area

tolerable. However, where it is necessary to burn out faults, for example, and greater currents are needed, it is safer to effect the HV connections to the work piece manually when dead and then control the test at the test panel. As an additional precaution against handling live connections, the operator can be provided with two probes connected together by a short length of flexible conductor and with long insulated handles. By applying these probes to terminals of opposite polarity or between a potentially live terminal and the apparatus metalwork, the operator can verify that the supply is 'off' and the circuit discharged before touching the terminals.

Displayed on the test panel there should be a circuit diagram for the test installation and the test safety rules.

This test layout was devised for a particular range of electromechanical products where bought-in electrical components, which had already been tested by their makers, were incorporated in the product. The testing, therefore, was mainly confined to checking the wiring and the performance of the machine and was different to that required in an electrical manufacturer's factory. For such a manufacturer, of say electric motors and generators, the test layout would have to be suitably modified, but similar safety precautions would be relevant.

High voltage test area

High voltage testing of motors, electrical distribution gear such as circuit breakers and transformers, and generator sets is a specialised activity that carries significant risks. However, the general principles are not much different to those set out for low voltage systems. Figure 16.14 illustrates a high voltage test area of the type commonly found in manufacturers' premises. The fundamental safety principle is that under no circumstances should anybody be inside the test area while the unit under test is energised and there are uninsulated live conductors that can be touched. It is commonly argued that engineers must be inside the area to read instruments, but this is only permissible if there is no possibility of the person coming into contact with live uninsulated conductors. Moreover, it is well within the state of the art nowadays for instruments to provide remote outputs via data links, without the necessity for people to enter the hazardous area to read them.

There should be a positive indication that the high voltage has been removed from the unit under test once the testing is complete and it is advisable to provide automatic earthing facilities on high voltage gear to ensure that all charge is discharged to earth before the people can enter the enclosure.

17

Safety-related electrotechnical control systems

Introduction

This chapter is concerned with electrotechnical control systems that perform safety functions; such systems are commonly known as safety-related control systems or electrical / electronic / programmable electronic (E/E/PE) safety-related systems. In this context, the term 'safety function' refers to a control action that needs to be taken to ensure that the risk associated with a particular hazard is reduced to a level that is acceptable or tolerable. For example, an interlocking circuit associated with an interlocked guard protecting access to live conductors inside a test rig may work to ensure that the conductors are made dead when the guard is opened and that the conductors remain dead while the guard remains open. This interlocking facility performs a safety function.

Whilst the subject does not strictly come within the scope of 'electrical safety', the chapter is included because many electrical engineers, technicians and electricians are required or expected to have an understanding of control system technology. The purpose of the chapter, therefore, is to provide an overview of, and insight into, the main devices used and the principles that underpin the achievement of adequate levels of safety.

So as to limit the scope of the chapter to a manageable level, only applications associated with machinery safeguarding and the safeguarding of electrical hazards are addressed in detail. This means that highly specialised areas such as safety-related systems for emergency shut down and instrumented protective functions in petrochemical and chemical plants are not considered, although the principles will be the same as those for the machinery sector. The fact that the chapter addresses electrotechnical control systems means that the design of systems that are not based on electrical, electronic and programmable electronic technologies is not covered.

A generic electrotechnical control system used in machinery and other applications is depicted in Figure 17.1. This shows that at the heart of the system is the 'processing unit' (often called a 'logic solver' or 'logic unit'), the subsystem that processes data and signals from input devices and sends signals to output devices. The processing unit may be implemented in a variety of technologies, ranging from electromechanical devices such as relays and discrete timers, to electronic systems, to programmable electronic systems such as Programmable Logic Controllers (PLCs). The inputs to the processing unit will typically emanate from

- control buttons and switches for operator-initiated functions such as 'start', 'stop', 'mode change', 'inch' and 'pause';

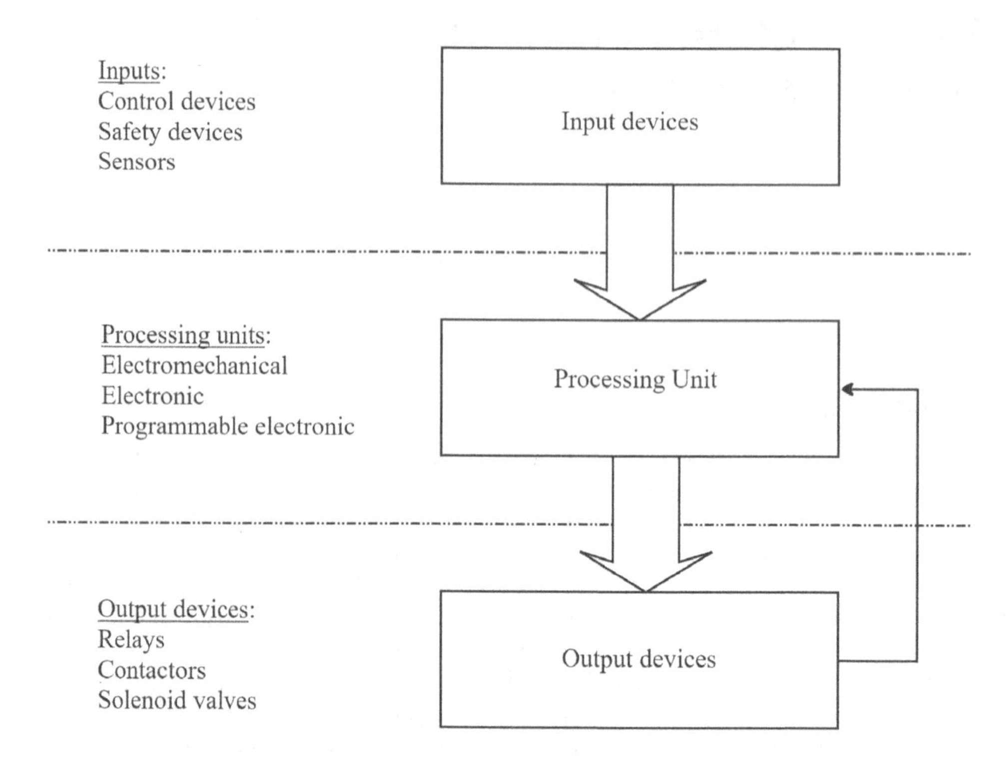

Figure 17.1 Generic block diagram of a control system

- sensors on a machine which detect the position and rate of change of position (speed) of parts of the machine (including guards) and the product being handled by the machine, such as limit switches, photoelectric sensors, capacitance switches, magnetically operated switches and so forth;
- sensors that detect parameters such as temperature, pressure, level and flow; and
- other control systems on linked machinery.

The outputs from the processing unit will typically interface with electromechanical contactors, relays, variable speed drives, solenoid-operated valves and indicating light bulbs.

The chapter describes the essential elements of electrotechnical safety-related control systems and identifies the principle CEN and CENELEC 'A' and 'B' standards that cover them. It also briefly describes three standards that have an important influence in the field: BS EN ISO 13849–1:2015 *Safety of machinery. Safety-related parts of control systems. General principles for design* and BS EN 61508–1:2010 *Functional safety of electrical/electronic/programmable electronic safety-related systems. General requirements* with its machinery sector implementation BS EN 62061:2005+A2:2015 *Safety of machinery. Functional safety of safety-related electrical, electronic and programmable electronic control systems.*

Accidents and incidents

A minority of accidents in the machinery sector are the result of control systems failures. Most are the result of machinery operators and maintenance staff entering hazardous areas by removing or bypassing guards and other safeguarding devices without having isolated power from the actuators that cause dangerous movement; safe isolation procedures are just as important in machinery safety as they are in electrical safety. For example, operators clearing blockages on machinery frequently intervene in the machine with the power still applied, clear the blockage and then get hurt when the machine moves. And maintenance staff carrying out fault finding work with the power still applied may cause sensors to change state while they are working in the hazardous zone and then get hurt when the machine responds by moving unexpectedly.

However, whereas unsafe systems of work dominate the machinery accident statistics, it is important that the safeguarding systems, and the safety-related control systems associated with them, are properly specified, designed, built and maintained so as to maximise safety. The HSE's publication *Out of Control, Why control systems go wrong and how to prevent failure* claims that some 44% of control system failures are a result of incorrect or inadequate specifications.

Basic principles

The electrotechnical safety-related systems being considered generally aim to safeguard against exposure to electrical hazards (shock and burn injuries and the effects of explosions and fires that have an electrical origin); mechanical hazards (crushing, shearing, cutting, puncturing and similar injuries); thermal hazards associated with parts and liquids or gases that are at a high or very low temperature; and other hazards that may occur during the operation, setting and maintenance of the equipment under control.

Many of the systems are referred to as interlocking systems because they provide an interlocking function, usually between a guard and a hazard, such that when a guard is opened the hazard protected by the guard is removed or, alternatively, the hazard must be removed before the guard can be opened. However, interlocking functions are not, by any means, the only safety functions that need to be addressed. Overspeed control on paper and steel mill machinery, limitation of temperature in thermoforming machines, emergency stop systems and anticollision systems on roller coasters are all examples of safety functions that would not necessarily be categorised as 'interlocks'.

The legislation and standards covering machinery safety stipulate a hierarchy of measures that should be taken to protect against the hazards identified here. Initial efforts must be made to design them out. If this cannot be done, they should be safeguarded using fixed guards and barriers. Next in the hierarchy is the use of interlocking and trip devices, which make use of the type of safety-related control systems being considered here to ensure that where people need to be near dangerous parts they cannot approach them to the point where they might be injured. Finally, and least acceptable, is the provision of instructions and information to people who may be exposed to those hazards that remain.

As a general principle, safety-related control systems such as interlocks should be well designed, be of simple construction, use proven components for which there is reliability data, and be sufficiently robust for the application. Furthermore, it should not be possible to easily defeat interlocking devices, yet where appropriate it should be possible for authorised persons to override them for tests or other necessary purposes.

A particularly important requirement for such systems is that they must have a safety integrity that is matched to the amount of risk reduction that the system is aiming to achieve, where the term 'safety integrity' is usually understood to be a 'goodness' or 'effectiveness' factor that combines the concepts of reliability and fault tolerance. The complexity, and hence cost, of the safety systems should be proportionate to the amount of risk reduction they are required to achieve; the aim is to reduce risks so far as is reasonably practicable, which as explained earlier in this book is not the same as saying they must be eliminated altogether.

This may seem a difficult concept at first, and its realisation is often far from easy, but the principle is straightforward. It simply says that the higher the contribution that a safety-related system makes to safety, the better must be the system's safety performance. This means that the risk assessment process that designers have to carry out and document is of fundamental importance in determining the control system's required risk reduction performance and hence the specification for its safety integrity; this specification determines the design requirements for fault tolerance and reliability. Safety performance of such systems is considered later in the chapter when standards such as BS EN ISO 13849 and BS EN 62061 are discussed, after the main elements shown in Figure 17.1 are described in more detail.

Legal requirements

The general provisions of the Health and Safety at Work etc Act apply to the safety of control systems, both in terms of their supply and their use. Whereas the act is goal-setting and non-prescriptive, there are regulations that provide more specific legal requirements.

In the context of safety-related control systems, the Essential Health and Safety Requirements of the Supply of Machinery (Safety) Regulations 2008 lay down generic requirements that must be considered by suppliers for the safety and reliability of control systems, control devices, starting and stopping devices, mode selection, failure of the power supply and the control circuit, software, and movable guards.

Users of machinery and other work equipment need to be familiar with the Provision and Use of Work Equipment Regulations 1998 and the associated Approved Code of Practice. As explained in Chapter 8, these regulations have a number of provisions relating to control systems that tend to mirror those in the Essential Health and Safety Requirements of the Supply of Machinery (Safety) Regulations. Indeed, the regulations impose a duty to provide and maintain suitable work equipment that conforms to standards specified in other legislation implementing European Directives, including the Machinery Directive (this is the type of requirement that might need to be amended after the UK's exit from the EU). The relevant regulations are

- Regulation 11: Dangerous parts of machinery;
- Regulation 14: Controls for starting or making a significant change in operating conditions;
- Regulation 15: Stop controls;
- Regulation 16: Emergency stop controls;
- Regulation 17: Controls; and
- Regulation 18: Control systems.

Main standards

The overarching standard for machinery safeguarding is BS EN ISO 12100:2010 *Safety of machinery. General principles for design. Risk assessment and risk reduction*, which is one of the small number of CEN 'A' standards covering machinery safety (see Chapter 8's section on the Supply of Machinery [Safety] Regulations for a description of Type 'A', 'B' and 'C' standards). This standard sets out the general principles for safeguarding machinery.

Another useful source of general guidance is the excellent machinery safeguarding standard, BS 5304, originally produced by BSI in 1984 and which is now published by BSI as Published Document PD 5304:2014 *Guidance on safe use of machinery*. It remains a first class reference document, if only because the illustrations in it, and the descriptive nature of its text make it far more user friendly than any of the newer European harmonised standards.

In the context of this book, the following 'B' standards address the main electrical and control systems issues on, and systems used in, machinery safeguarding (in addition to the previously mentioned BS EN ISO 13849–1:2015 and BS EN 62061:2005+A2:2015):

- BS EN 60204–1 Electrical parts of machinery;
- BS EN ISO 14119 Interlocking devices associated with guards;
- BS EN 61496 Electrosensitive protective equipment;
- BS EN ISO 13850 Emergency stop devices;
- BS EN 1037:1995+A1:2008 Prevention of unexpected start-up; and
- BS EN 574:1996+A1:2008 Two-hand control devices. Functional aspects.

The techniques described in these standards are aimed at machinery safeguarding, but the principles also apply to control systems, particularly interlocking systems, which are used to safeguard electrical hazards.

Input, interlocking, trip and emergency stop devices

There are many types of devices used in safety-related systems that provide inputs to the processing unit, with the main ones described in the following paragraphs. Some of these act as interlocking devices, usually associated with guards. These devices – limit switches, key-actuated switches, guard locking devices, proximity and magnetic switches – are described in BS EN ISO 14119. Other devices act as presence sensing and trip devices – photoelectric light curtains being an obvious example – and are generally covered by B standards that are specific to the type of device, an example being BS EN 61496.

Other devices measure parameters such as speed, acceleration, pressure, level, flow, and temperature. These types of devices are not described in this book but the main technologies are

- Flow measurement: flow meters using ultrasonics, electromagnetics, turbines, and pressure differential measurement
- Temperature measurement: thermometers, thermistors, thermocouples, radiation pyrometers and bimetallic strips
- Force measurement: spring balances, strain gauges, piezoelectric transducers
- Pressure measurement: force-balance and elastic element systems, and solid state pressure transducers
- Velocity measurement: tachogenerators
- Acceleration measurement: seismic-mass, strain-gauge, potentiometric, piezoelectric and servo accelerometers

Limit switches

The use of limit switches is widespread in interlocking systems, and they are also used to indicate the position of moving parts. A typical limit switch used in machinery safeguarding applications is depicted in Figure 17.2, which shows a switch with a roller plunger actuator; another common type of actuator is a lever. The figure also shows an example of a typical contact arrangement in these switches. In this case, the switch has two normally open (NO) and two normally closed contacts (NC), although manufacturers offer many other combinations of contacts.

The contacts in limit switches can be wired into different parts of control circuits, but it is usual to wire normally open contacts into primary safety circuits. This is because safety functions are frequently best implemented on deenergisation to ensure, for example, that open circuit faults do not lead to dangerous failures and that the system moves to a safe state in the event of a power failure.

In the context of safety applications, there are two generic types of limit switch, known as positive mode and negative mode switches. These are shown in Figure 17.3, which illustrates the switch's plunger being moved by a rotating cam.

The positive mode switch is safer than the negative mode switch because its contacts are positively forced apart by the rotating cam whereas the negative mode switch relies on spring pressure to open the contacts. The negative mode switch could fail to danger if the spring were to break, or if the contacts were to weld together, or if the plunger were to become stuck. Moreover, it can easily be defeated by taping or wedging the plunger down.

Limit switches, and other interlocking devices, can either be inserted in the power supply circuit (called power interlocking) or in the control circuit (called control interlocking), depending upon the size of the current to be interrupted and the rating of the switch. In general terms, power interlocking is safer than control interlocking because the former acts directly on the supply, whereas the latter relies for successful operation on interposing components such as relays and contactors and therefore has the potential to be less reliable; for this reason, power interlocking is frequently used in high risk applications

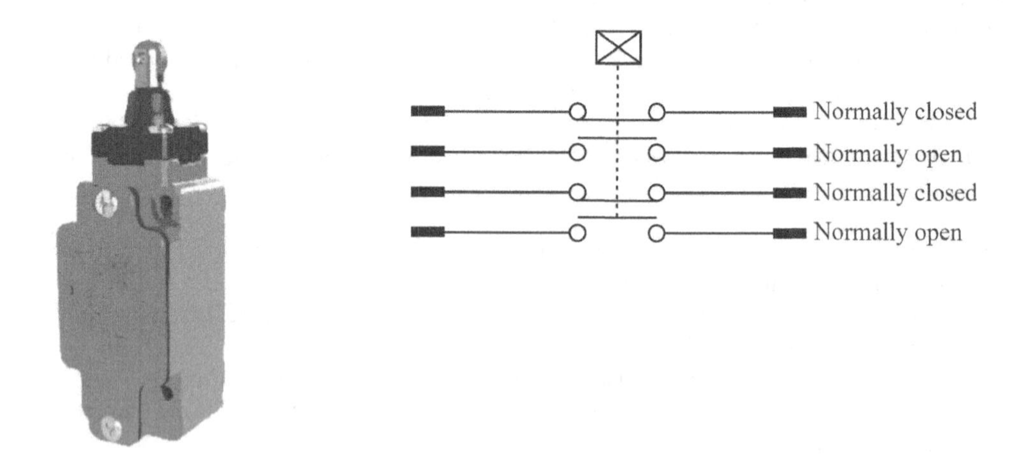

Figure 17.2 Typical limit switch and contact configuration

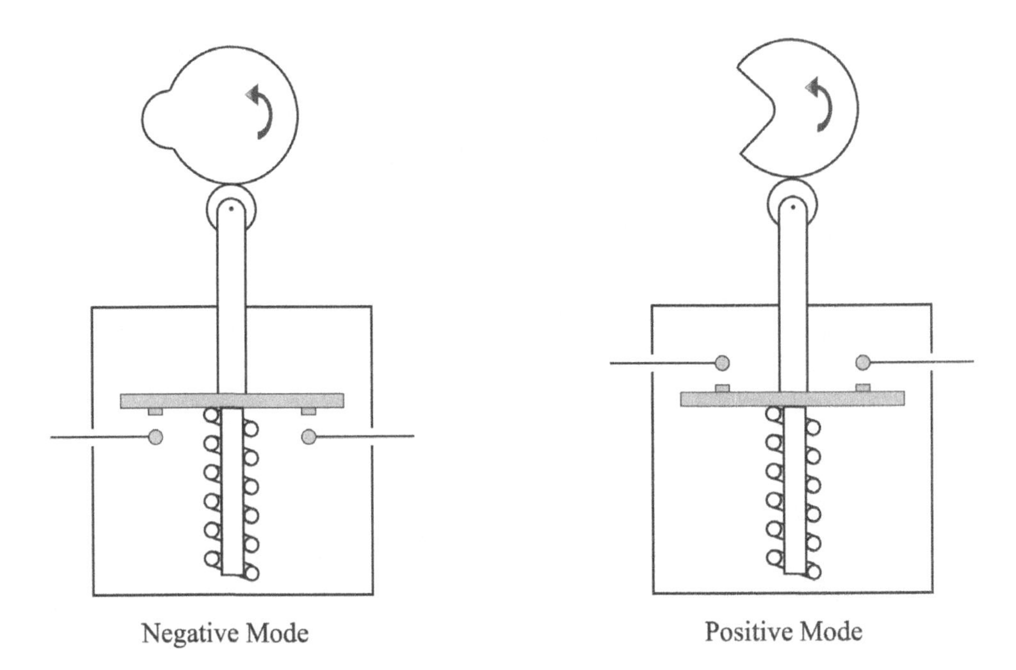

Figure 17.3 Positive and negative mode operation

such as plastic thermoforming machines and power presses where opening a gate or guard leads directly to disconnection of the power source. However, control interlocking is more commonly used than power interlocking, because in many applications it is undesirable simply to switch off the supply to the machine and because of the current breaking limitations of the interlocking switch.

In general terms, interlocks must not be easily defeatable. Figure 17.4 shows a limit switch used in the interlocked cover of a test jig in which a component to be tested is first inserted in the jig and becomes automatically connected to the power supply via the test contacts. Closing the lid allows the limit switch to complete the control circuit, which energises the coil of the contactor leading to the contactor's contacts closing and the test contacts becoming live. Opening the lid causes the limit switch contacts to be forced apart by positive action, thereby deenergising the contactor coil, causing the contacts to open and the test contacts to become not live. This means that the test contacts will only be live when the lid is closed. Moreover, the limit switch is not easily defeated so as to switch the test contacts on while the lid is open.

A possible dangerous failure mode in this type of design is the failure of the contactor with its contacts in the closed position, meaning that the test contacts would remain live with the lid open. A possible failure mechanism would be the contacts welding together if excessive current is passed through them. This can be prevented by ensuring that the contactor is of good quality, is built to a recognised standard, and is properly rated for the duty, and by inserting a fuse or a circuit breaker in the supply to provide adequate overcurrent protection.

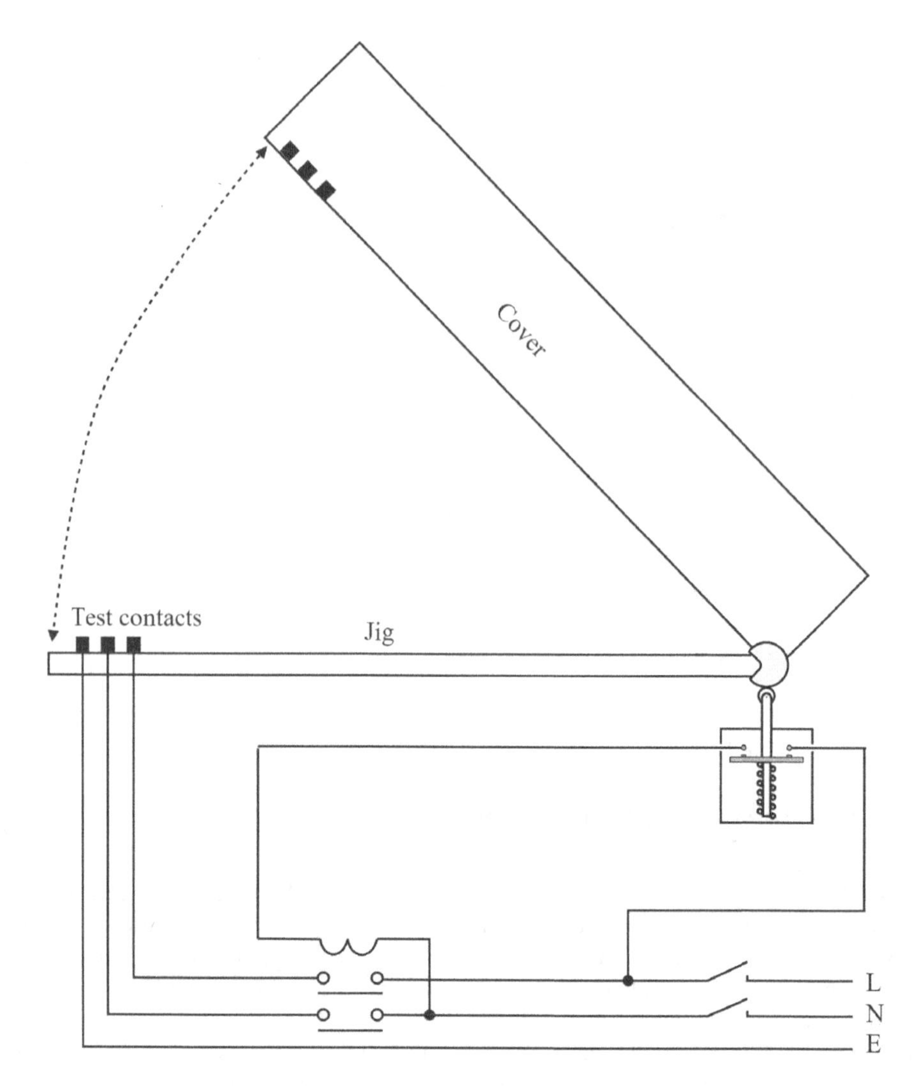

Figure 17.4 Test jig with interlocked enclosure

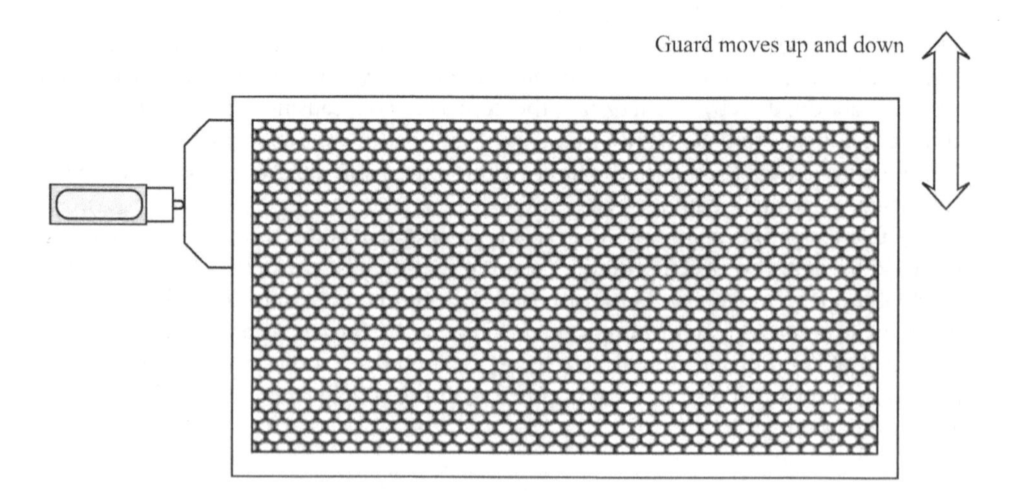

Figure 17.5 Limit switch operated by ramp on sliding guard

Although the discussion so far has concentrated on safeguarding access to live parts, limit switches have important applications in machinery safeguarding. The cam-operated switch configuration shown in Figure 17.4, for example, is commonly used on hinged guards that prevent access to mechanical hazards on machinery.

Roller plunger limit switches are often used in association with sliding guards, as depicted in Figure 17.5. Movement of the guard causes the shaped ramp to push open the positive mode contacts of the switch, opening the supply to a contactor and thereby switching off the prime mover such as an electric motor.

Positive and negative mode switches are often used together in safeguarding applications to prevent common mode failure caused by, for example, the displacement of a guard. An example of this configuration is shown in Figure 17.6, which depicts a sliding guard with a ramp attached; the ramp operates the plungers of the two switches. If the guard were to become displaced, it is most likely that one of the two switches would operate. Provided that the system is properly maintained, a single fault is unlikely to cause a failure to danger of the interlocking system.

Another example of the application of limit switches is in lift safety circuits. These employ hardwired components connected together in series in such a way that they must all be closed before the lift can operate, as shown in Figure 17.7. If any one of the components were to become open circuit, the safety circuit would cause the main up or down contactors to open and the lift to stop. The ultimate top and bottom switches in

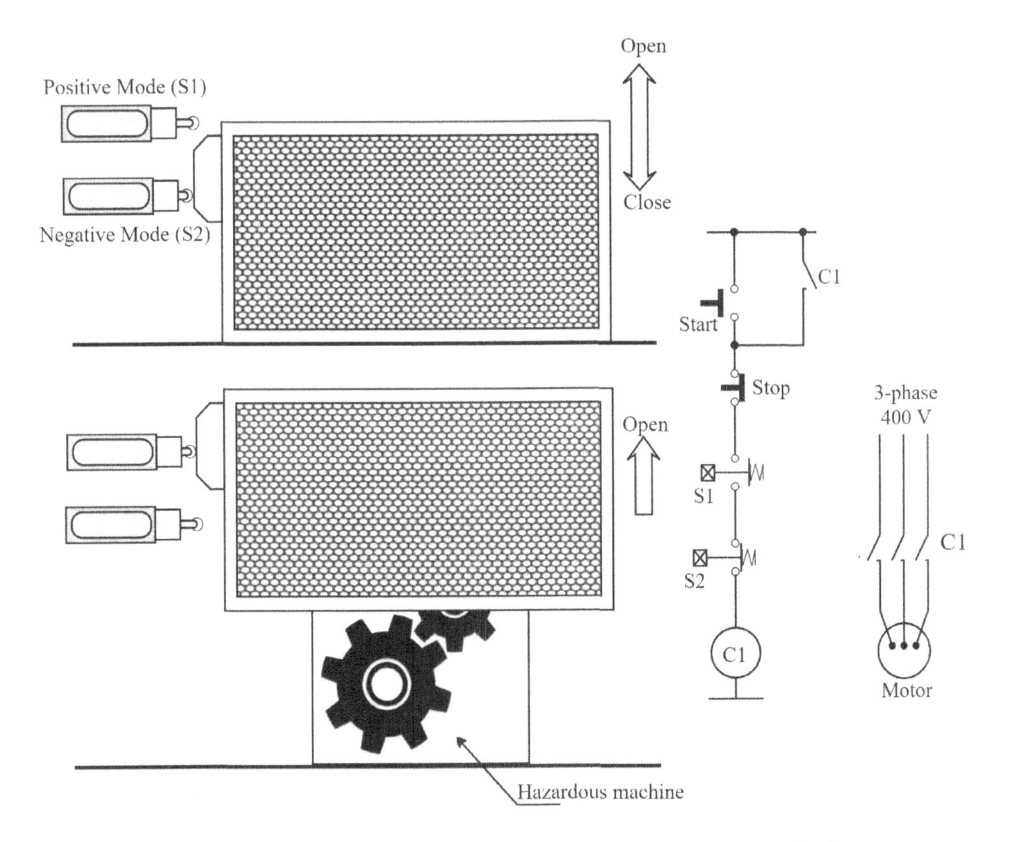

Figure 17.6 Positive and negative mode limit switches used in guard interlocking

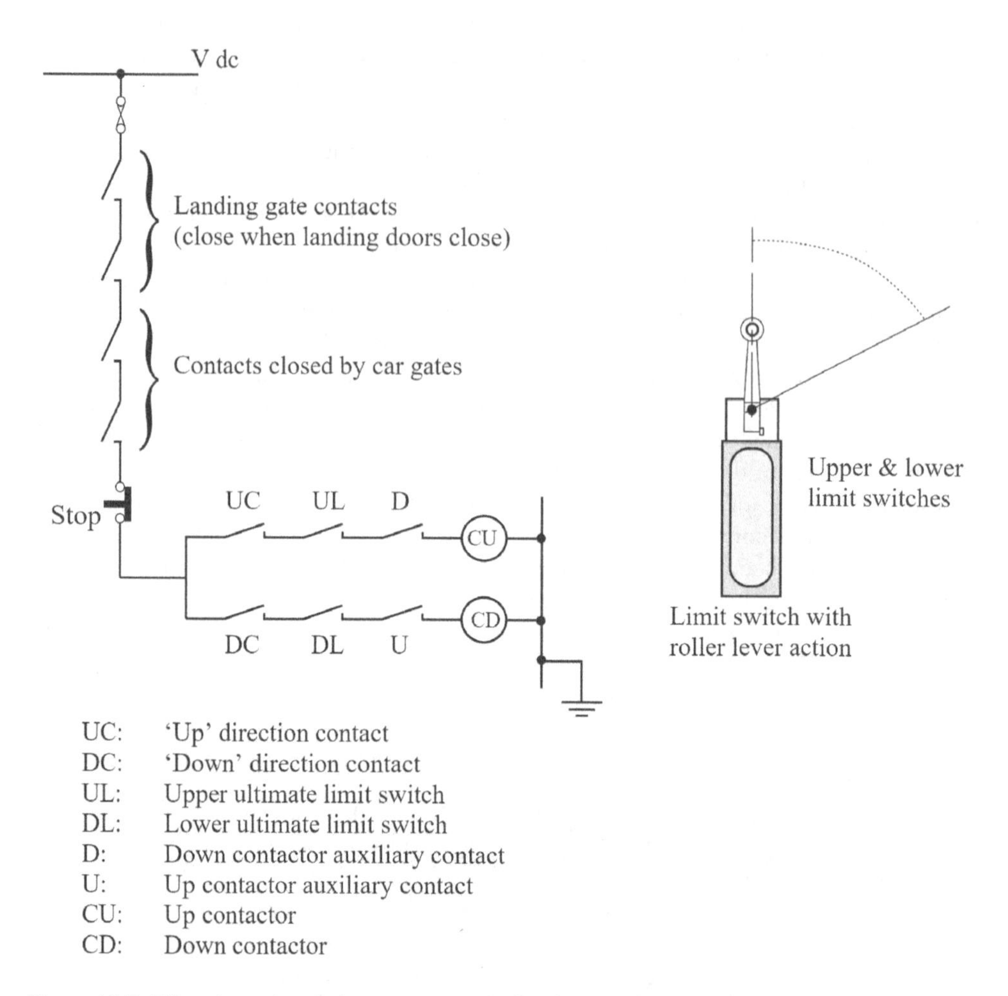

Figure 17.7 Lift safety circuit, incorporating roller lever switches for upper and lower ultimate limit switches

the lift shaft are limit switches, which have contacts that open if the lift overshoots by a predetermined distance the top or bottom floors. The safety circuit is usually independent of the much more complex control system that controls the movement of the lift car and the car and landing doors.

Key- or tongue-operated switches

A special type of limit switch in common use in machinery safety applications is the key-operated switch, also called tongue-operated, an example of which is illustrated in Figure 17.8. In this case, a shaped metal key that enters or leaves an aperture in the body of the switch positively operates the internal contacts both when being inserted into and withdrawn from the switch, thereby minimising the chance that the switch contacts could stick together. The key would be attached, for example, to a moveable guard and

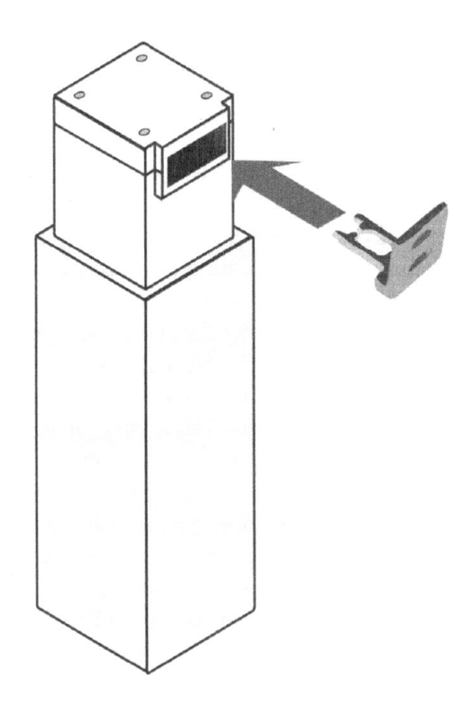

Figure 17.8 Key-operated limit switch

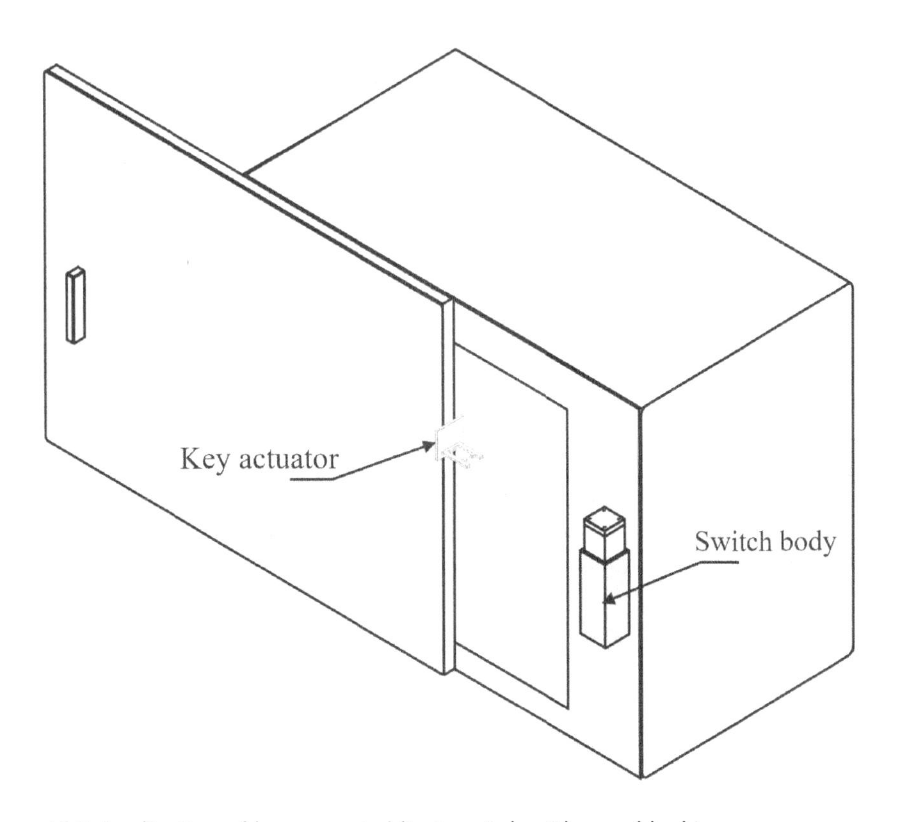

Figure 17.9 Application of key-operated limit switch with guard locking

the body of the switch would be attached to the frame of the machine, as depicted in Figure 17.9.

It is possible to defeat this type of switch by inserting a spare actuating key, so any spare keys should be kept locked away or thrown out to prevent this happening. An alternative means of defeat is to remove the key from the machine and to leave it inserted in the switch. To prevent this happening, the key should be securely attached to the machine and should only be capable of being removed by using a special tool – in this context, a 'special tool' does not include common screwdrivers, Allen keys, wrenches and so on.

An advantage of the key-operated switch is that the switch mechanism can be designed to lock the key in place once it is inserted into the body of the switch. This is commonly achieved using a spring-applied catch that is released by a solenoid-operated bolt.

The key actuator is inserted in the switch as normal when the guard to which it is attached is closed, positively operating a set of normally open and normally closed contacts that are wired into the machinery control circuit. The key is automatically locked in place by a mechanical latch so that the guard cannot be opened; this is known as an interlocking guard with guard locking. The key can only be unlatched by energising a solenoid inside the switch – this controls a plunger which acts on the latching mechanism. The solenoid will normally be energised by the machinery control system when a safe state has been achieved, such as when hazardous rotating parts have stopped or when power has been removed from normally live parts.

Some guard locking interlocks contain a separate lock and key in the body of the switch. When somebody enters the hazardous area the key switch can be turned to the 'lock' position and the key removed and retained by the individual while working inside the hazardous area. This prevents the switch from being operated again until the key is replaced and turned to the 'unlock' position, thereby providing the person entering the hazardous area with control over the interlocking system.

Time delay switches

One of the main purposes of the interlocking switch with guard locking is to prevent a guard being opened until the hazard safeguarded by it has been removed. Another technique that achieves the same end in the case of rotating machinery is a time delay switch, an example of which is illustrated in Figure 17.10 which shows the switch attached to a centrifuge. The time delay switch has a long threaded bolt with a hand-operated knob at the end. To release the guard, the operator has to unscrew the bolt. During the first few turns, the plunger of the limit switch is forced down as it moves out of the indentation in the bolt and the switch contacts are opened, causing the control circuit to become deenergised. The bolt must then continue to be unscrewed to allow the hinged lid to be released, and this takes sufficient time to ensure that the machinery has come to a halt and the hazard removed before the lid can be lifted.

Non-contact switches

There are a variety of techniques that can be used to produce switches that have a non-contact mode of operation. By far the most common are devices that employ magnetic coupling.

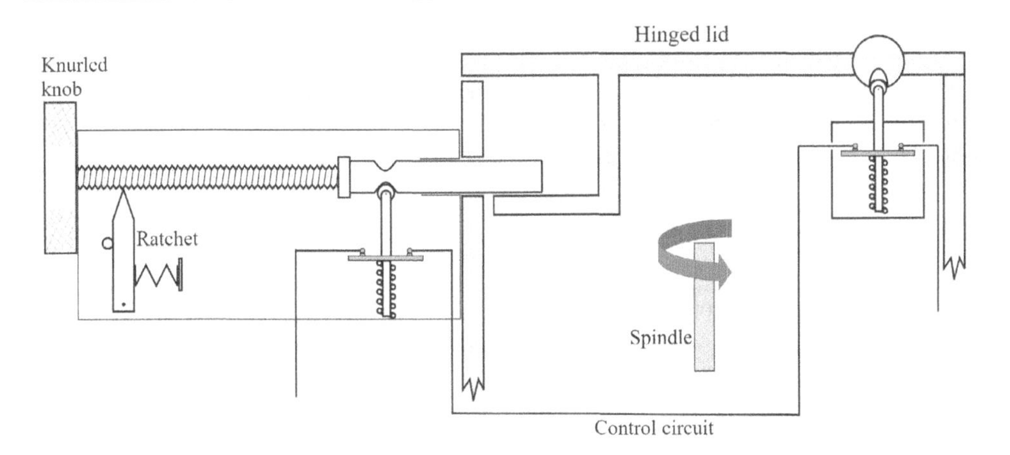

Figure 17.10 Centrifuge with two interlocking switches closed in the run mode

Magnetic switches have two components: a unit containing a magnet and a unit containing a switch, typically a reed switch, that is switched on and off by the magnet. In safeguarding applications, it is normal to mount the magnet unit on a movable guard and the switch unit on the fixed part of the machine.

For correct operation, the magnet must be correctly aligned with the switch unit. Bringing the two close together, typically to within 5–10 mm, will cause the switch to close. If it is connected into a control circuit, this can then be used to enable machinery movement.

Only magnetic switches that are specifically designed for safety applications should be used – they are designed to be resistant to vibration and with overcurrent protection to prevent the switch contacts welding together.

Bringing a magnet up to the switch unit could defeat it, so it is important that the switch unit is mounted so that it is not easily accessible when the guard is open. In most modern designs, the magnetic coupling is designed to reduce the likelihood of the switch being defeated in this way. In one design, the magnet has to be inserted into the body of the switch for it to operate. The hole in the switch is made a complex shape so that only a magnet of matching shape can enter, meaning that a simple bar magnet cannot be used to defeat the interlock. In another design, the system has two channels and features an anti-tamper code that ensures that it can only be operated by its own magnetic actuator unit.

In another design of non-contact safety switch, a coded signal is transmitted between the two heads (one on the fixed part of the machine and one attached to the guard) when they are in close proximity. The code is unique to the particular unit, making it very difficult to defeat.

One of the advantages of non-contact safety switches is that the assemblies are hermetically sealed against the ingress of moisture and particles, providing them with a very high IP rating. This makes them especially useful for applications where hygiene may be an important issue, such as in food and pharmaceutical processing machines where particles may be able to enter and decay inside the actuator and plunger openings of other types of forced contact switches.

Their main disadvantage is that the switch contacts are not physically forced apart, so they are not suitable for use on their own in applications that demand a high level of safety integrity, as will be discussed further in a later section of this chapter. However, the level of integrity can be improved by using redundancy and monitoring techniques, and devices suitable for this are supplied by many of the safety component manufacturers.

Captive-key and trapped-key devices

A captive-key system comprises a switch and integral lock, typically fitted to a machine or equipment enclosure. The switch is operated by inserting a key into the lock, with the key being secured to the moving guard or enclosure door, usually as an integral part of a handle. When the guard or enclosure door is closed, the key enters the lock. Turning the handle then turns the key that operates the switch. The switch can therefore only be operated by closing the guard or door and turning the handle, locking the guard or door in its closed position.

A trapped-key system has a single key that is transferred between a lock on a guard or enclosure door and a key-operated switch incorporating a lock on a control panel. The key is trapped in either the lock on the guard or the lock on the control panel. The key can only be released from the guard when the guard has been closed and locked. Once it is released from the locked guard, the key can then be inserted in the control panel's lock and then turned to switch on the control function. At this point, the key becomes trapped in the lock. The key must then be turned before it can be released to open the guard. This has the effect of turning off the control system, removing the hazard covered by the guard. There are many variations on this theme, some of which involve a multiplicity of keys and locks to ensure that all guards are closed and/or all power sources removed before guards can be opened.

An example of an electrical system that uses more than one key is shown in Figure 17.11. The system is used to prevent the paralleling of transformers feeding a low voltage switchboard, which is divided by bus-section switches. In normal operation the bus-section switches are open and each switchboard section is fed from its own transformer, so the feeder switches are closed. All switches are provided with locks, and there are three keys that are free when the switches are locked open but trapped when the relevant switches are closed. Key X will only fit its feeder switch, and the adjoining bus-section switch XY and key Z will similarly fit only its feeder switch and the adjoining bus-section switch YZ, but key Y will fit both its own feeder switch and either of the bus-section switches XY and YZ. This arrangement permits the following operational choices:

(i) Three feeder operation: Bus-section switches are open and the switchboard is fed from its own transformer.
(ii) Two feeder operation: Bus-section XY is closed, bus-section YZ is open and sections X and Y are fed from either X or Y. Section Z is fed from feeder Z.
(iii) Two feeder operation: Bus-section XY is open, bus-section YZ is closed and sections Y and Z are fed from either Y or Z. Section X is fed from feeder X.
(iv) One feeder operation: Both bus-sections are closed and the whole switchboard is fed from only one of the three feeders.

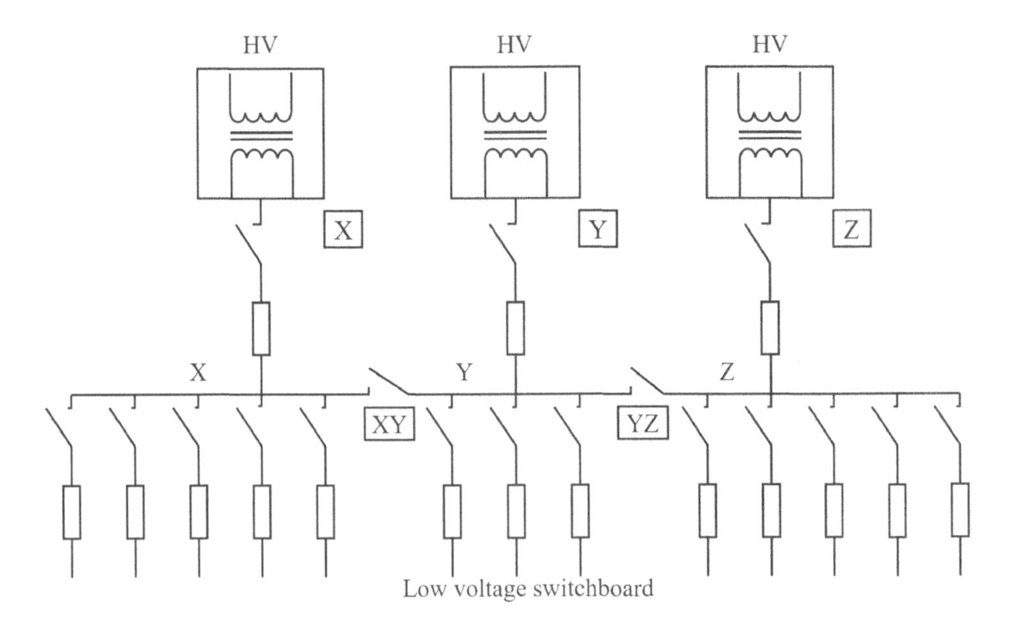

Figure 17.11 Trapped-key interlock system on a feeder switchboard to prevent paralleling of the transformers

Two-hand control devices

A two-hand control device requires activation of two control buttons to initiate a control action. Most devices are synchronous two-hand control, meaning that the design of the system must ensure that the buttons need to be pushed within 0.5 seconds of each other, so that they can only be actuated by a single person. Operation of the two buttons by a single hand is prevented by, for example, spacing the buttons at least 550 mm apart and/or by shielding them.

The principle is that these features will result in the operator's hand being kept away from the danger zone. It does not, of course, prevent another operator accessing the danger zones, which means that the machine must be built and laid out so as to prevent this happening.

The requirements are set out in BS EN 574:1996+A1:2008.

Trip devices

Some safeguarding devices are used to detect the presence of people and to trip the machines they safeguard. The main type described here is Electrosensitive Protective Equipment (ESPE), which includes photoelectric safety systems and laser scanners, both of which are types of Active Optoelectronic Protective Devices (AOPD). Pressure sensitive devices such as pressure mats, and trip bars, are other types of trip devices.

Photoelectric safety systems generally comprise a transmitter unit and a receiver unit between which a light beam, set of beams, or curtain is established. Breaking the light curtain or beams causes the control unit to open the switch or switches, known as Output Signal Switching Devices, on its output. These switches are connected into the control system of whatever machine the light curtain is protecting, so that opening the switches causes the machine to stop or some other action to be taken.

The detection capability of light curtains must be specified by the manufacturer so as to describe the minimum size of object that will cause actuation of the system. In many applications, the light curtains must be actuated by objects with a size in the same order as that of a human hand, although in some applications larger object detection capability may be appropriate.

Light curtains have many safety applications, but they are commonly used to safeguard dangerous parts where both frequent and infrequent operator access to hazardous areas is necessary, such as the blades of paper-cutting guillotines and the tools of power presses. In these types of applications, the detection capability of the device must be objects the equivalent size of a human hand or even smaller. They are also used to protect apertures in perimeter fencing where products have to be fed in or taken out, examples being palletisers and depalletisers, manufacturing robots, and packaging machines. In these latter applications, the detection capability may be enlarged because it is whole- or part-body access that is being prevented. The two types of application are depicted in Figure 17.12, where the guillotine has frequent access through the light curtain and the other machine has infrequent access.

Photoelectric devices can also be used to safeguard electrical hazards, but it is not common; an example can be found in high voltage meat stimulators. These devices, depicted in Figure 17.13, are used to tenderise meat by applying a pulsed voltage of about 800 volts to the carcasses of animals such as pigs, cattle and sheep shortly after slaughter. Essentially, the carcasses are suspended from a chain conveyor that forms one pole of the electrical supply. As they move through the enclosure, the carcasses rub against an electrode that is energised at the stimulating voltage. The current that flows through the carcass from the rubbing electrode to the conveyor makes the meat more tender than it otherwise would be, a fact that is used as a selling point.

The physiological effects that lead to the perceived improvement in the quality of the meat will not be considered here. What is of relevance is the potential for serious electrical injury to anybody who enters the stimulator enclosure while the rubbing electrode is live. Given that suspended animals as large as cattle need to be moved into and out of the enclosure, it can be appreciated that the entry and exit apertures are large enough for people to walk in and out with ease. The risk is enhanced by the fact that, as in all abattoirs, the environment tends to be wet.

There are a number of ways in which the risk can be controlled but the most common is the use of light curtains to scan the floor across the entry and exit apertures. The light curtains are wired into the control system in such a way that the main contactor providing electrical power to the stimulator is switched off if the light beams are interrupted. Some designs have light curtains that scan the whole of the floor area inside the stimulator to ensure that the device cannot be energised while somebody is inside the enclosure. One of the potential problems is that any fluids falling from a carcass may trip a light curtain and cause nuisance tripping of the production line, but experience has shown that this is not a significant issue.

Photoelectric safety devices are 'safety components', as defined in the Supply of Machinery (Safety) Regulations. A Notified Body, who will use BS EN 61496 as the

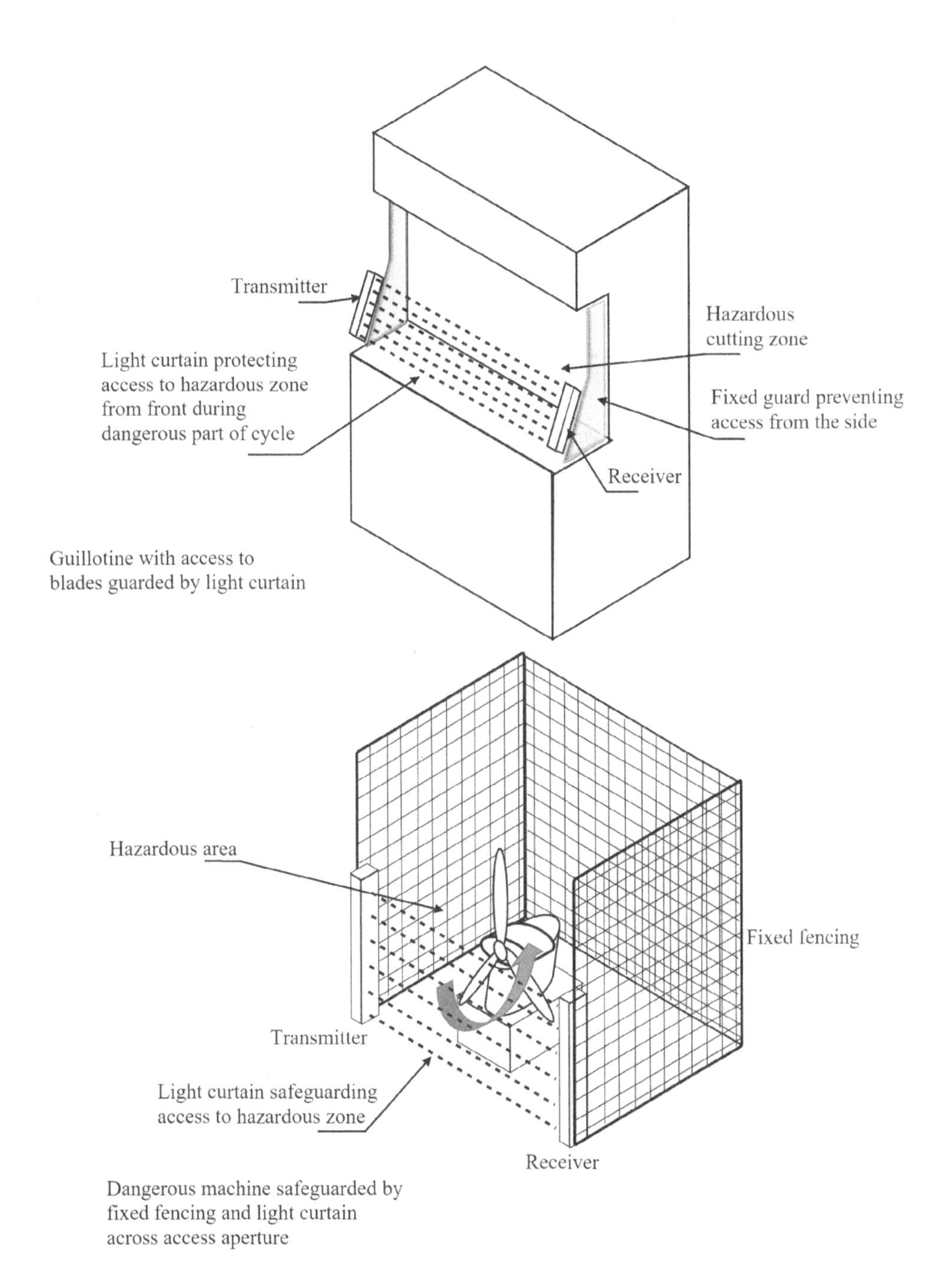

Figure 17.12 Typical safeguarding applications of photoelectric systems

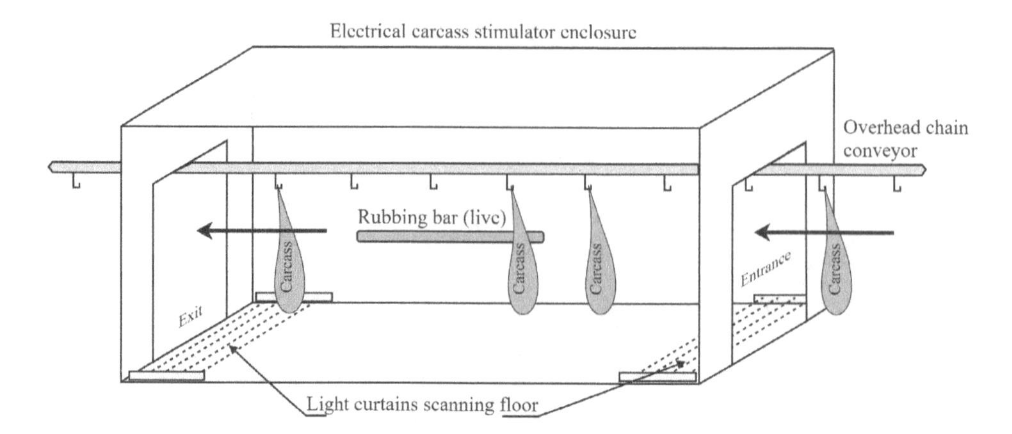

Figure 17.13 **Carcass stimulator with entry and exit apertures guarded by floor scanning light curtains**

baseline standard, must therefore check their conformity with the regulations. This standard has three parts:

(i) Part 1 of the standard lays down general requirements for the design, construction and testing of electrosensitive protective equipment.
(ii) Part 2 specifies particular requirements for photoelectric systems such as single-beam switches and light curtains that have transmitter and receiver elements; these are more properly known as Active Optoelectronic Protective Devices (AOPD).
(iii) Part 3 sets out requirements for Active Optoelectronic Protective Devices responsive to Diffuse Reflection (AOPDDR), which are devices such as laser scanners. These are optical devices that scan infrared laser beams to monitor a hazardous area near a machine. The scanner detects light reflected off objects and determines their exact position within the area and compares them with preprogrammed object locations, tripping the device if an object is detected within a protected area.

The standard specifies three types of devices according to their safety performance: Types 2, 3 and 4. Types 2 and 4 relate to AOPDs and Type 3 relates to AOPDDRs:

(i) Type 2 AOPD devices can be manufactured with a single Output Signal Switching Device or equivalent safety-related data interface but must have a periodic test to reveal any failures to danger, which means that the device can fail to danger if the fault occurs in the interval between tests. Such failures must be detected immediately, or as a result of the periodic test occurring after the failure occurs, or at the next actuation of the light curtain. They are suitable in applications where Safety Integrity Level (SIL) 1 of BS EN 62061 or Performance Level (PL) c of BS EN 13849 are specified – these terms are described later in this chapter.
(ii) Type 3 AOPDDR devices are designed such that the output circuit of at least two output switching devices go to the OFF-state when the sensing device is actuated,

or when the power is removed from the device. Any single fault that will cause the loss of the detection performance must cause the device to go to a lock-out condition. They are suitable in applications where Safety Integrity Level (SIL) 2 of BS EN 62061 or Performance Level (PL) d of BS EN 13849 are specified.

(iii) Type 4 AOPD devices must be designed so that a single fault that adversely affects their detection capability will be detected and cause the device to lock-out. Type 4 devices must have at least two Output Signal Switching Devices or equivalent safety-related data interfaces. They are suitable in applications where Safety Integrity Level (SIL) 3 of BS EN 62061 or Performance Level (PL) e of BS EN 13849 are specified.

The standard defines a series of optional extras. For example, there is an option for a stopping performance monitor, which allows for monitoring of the time taken for the machine to reach standstill after the beam has been interrupted. As another example, there is an option for external device monitoring which allows, for example, auxiliary contacts on contactors to be fed back to a light curtain controller and for the controller to enter a lock-out condition if there were to be a detected disparity between the desired and actual condition of the contactor. The standard also specifies the optical requirements for the devices, including their immunity to optical interference, reflections and so on.

The interfacing arrangement between the electrosensitive protective equipment and the machinery control system is a particularly important facet of the safety system's design. It is the overall control system incorporating both the AOPD and the machine's safety-related control system that determines the level of safety that can be achieved. Close attention therefore needs to be paid to the machine's control system and the way in which the photoelectric equipment's final switching contacts are connected into it.

Another important issue is the separation between a light curtain and the hazardous parts protected by it. This is because the hazard must be removed by the time the light curtain and the machinery control system and any moving parts respond to the actuation of the light curtain by, say, an operator reaching through the light curtain. This matter is addressed in BS EN ISO 13855:2010 *Safety of machinery. Positioning of safeguards with respect to the approach speeds of parts of the human body*. The standard specifies that the minimum distance, S, from the danger zone to the detection point shall be calculated according to the formula:

$$S = (K \times T) + C$$

where

K is the approach speed of the part of the human body (e.g. 2000 mm/s for a hand/arm approach).

T is the overall system stopping performance, which depends on factors such as the stopping time of the machine, the response time of the safety-related control, the response time of the protective device (ESPE), and additions according to the detection capability of the ESPE, the protective field height and/or the type of approach.

C is an additional distance in millimetres, based on intrusion towards the danger zone prior to actuation of the light curtain, which will typically be linked to the detection capability of the device.

In some applications, such as packaging machinery, products such as loaded pallets have to pass through a safety-related light curtain from, for example, the palletising machine into a despatch area. The light curtain guards the aperture in the machine's perimeter fencing to prevent humans entering the danger zones. This means that the light curtain must be muted, or effectively switched off, while the pallet load is passing through it; otherwise the pallet load would trip the machine's control system and then be reinstated as soon as the load has passed through. The layout of the system should be such that the pallet load fills the aperture while it is passing through it, so as to prevent people passing though while the light curtain is muted. Sensors that detect the presence of the pallet load as it approaches the aperture normally activate the muting system. It is important that these sensors cannot be activated by anybody trying to defeat the light curtain.

Emergency stops

Emergency stopping facilities must be provided where they will contribute to risk reduction. Therefore, they are not needed where they will not lessen the risk of injury or in hand-held portable and hand-guided machines. The particular requirements for the equipment are laid out in BS EN ISO 13850:2015 *Safety of machinery. Emergency stop function. Principles for design.*

Emergency stops may be initiated by a number of different means, including actuators such as mushroom-type push buttons, wires, ropes, bars, handles and footpedals that do not have protective covers. The devices must be designed such that they latch-in when actuated to generate a stop command, with the emergency stop command being maintained until the device is reset. The actuators must be coloured red and, where a background exists, it should be coloured yellow as far as practicable.

Most machines supplied nowadays have a mushroom-headed emergency stop button fitted on the control panel. On larger machines, there will normally be more than one button, with all the buttons (or groups of buttons) wired in series.

The pull wire device is particularly suitable for conveyors and other long machines that have a hazard along their length. It consists of a long wire running alongside the machine and connected at its extremities to two limit switches connected in series in the control circuit, as depicted in Figure 17.14. Pulling the wire operates one or both of the switches and causes the control circuit to respond, usually by deenergising a main contactor to switch off prime movers such as motors, hydraulic rams, pneumatic cylinders and so on. The device should latch in the off position and require a deliberate reset to allow power to be restored to the machine. Figure 17.15 shows the system redesigned so that the switches would operate if the wire were to break. In this configuration the wire must be tensioned to ensure that both switches are 'on' at the same time.

Emergency stops must function as either a category 0 or category 1 stop, as defined in BS EN 60204–1. The former initiates immediate removal of power to the machine actuators and braking where necessary. The latter is a controlled stop with power being retained to achieve the stop and then removed when the stop has been achieved, such as the stopping of a paper making line under the control of a PLC, with power being removed via contactors and similar devices when the high-inertia paper reels have been brought to a controlled stop. The common feature of the two categories is the eventual removal of the power source to the machine's prime movers.

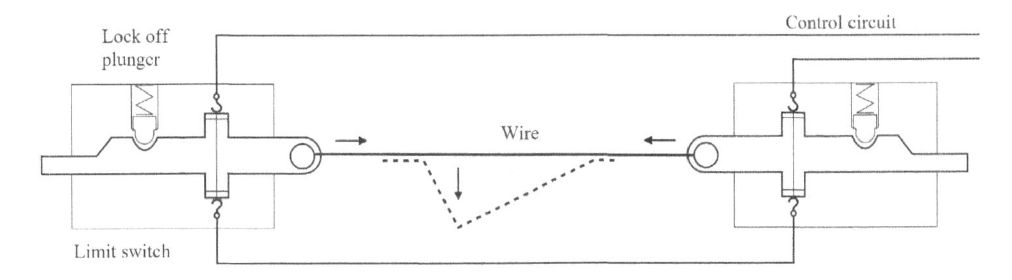

Figure 17.14 Basic 'pull wire' emergency stopping system

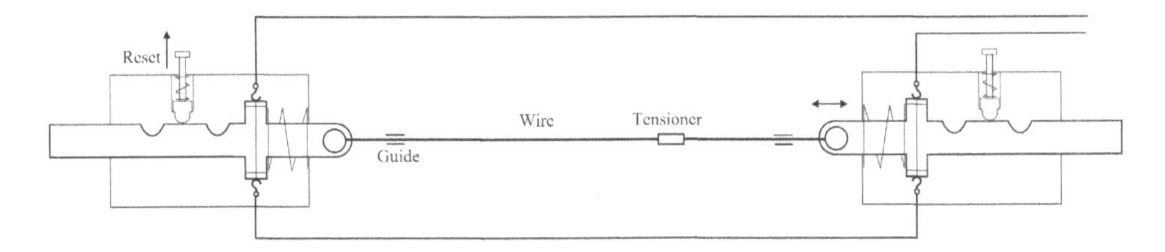

Figure 17.15 Improved 'pull wire' emergency stopping system

Switches, push buttons etc

Although the discussion so far has concentrated on devices that perform interlocking and trip functions, inputs to the control system processing unit are commonly in the form of signals from selection switches, push buttons and other devices. There is a huge range of such devices and, as far as safety is concerned, it will simply be noted that these devices are commonly used to implement start, stop and mode selection functions.

The processing unit

As previously noted, the 'processing unit' (also known as the 'logic unit' and 'logic solver') lies at the heart of the control system, processing the signals received from the input devices and providing actuating signals to the output devices. In its very simplest form, it may be a single circuit, as shown in Figure 17.16, that directly connects an input device, a switch, to an output device, a contactor coil, whose contacts switch a motor on and off. It is more likely, however, that the processing unit will be somewhat more complex than this.

For many years, processing units in electrotechnical control systems were implemented using electromechanical devices such as relays, switches, push buttons and so on. This discrete component technology remains in widespread use today. One of its disadvantages, particularly in more complex applications, is that as the component count increases so does the amount of cabling needed to interconnect them all, with concomitant increases in cost, unreliability, and problems with maintainability.

Increasingly since the late 1970s, control systems have used programmable devices, principally programmable logic controllers (PLCs) and embedded microcontrollers, to

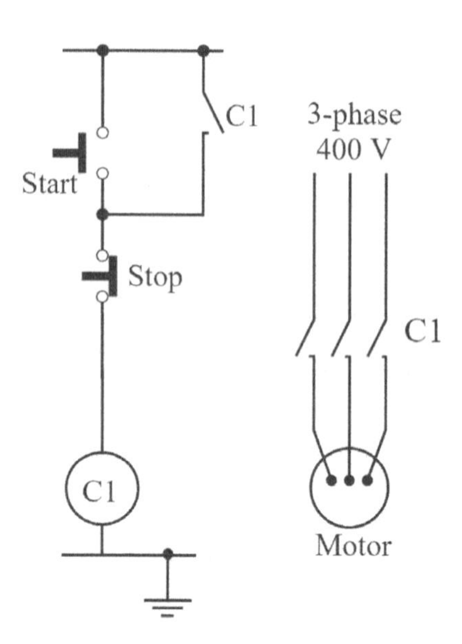

Figure 17.16 Simple configuration processing unit

implement control strategies in software. This type of technology provides designers with a considerably increased degree of flexibility and functionality, with increased processing speed and decreased wiring/component (and cost) overheads. PLCs can also be connected onto data buses to provide a network of systems capable of controlling large and complex manufacturing installations.

The ubiquitous PLC is essentially a microprocessor that takes inputs from a number of field devices and provides outputs to a number of output ports. The fundamental configurations are shown in Figures 17.17 and 17.18. The former is a high-level schematic of a basic system structure with a photograph of a basic PLC installed in an industrial cabinet. Figure 17.18 shows in a simplified way that the PLC scans data from a number of inputs, processes the data according to the program stored in its internal memory, and then outputs the processed data to a set of outputs.

PLCs were, and in many cases still are, programmed using a graphical programming technique known as 'ladder logic' because it provided a representation analogous to the traditional hardwired systems. This eased the transition for many designers from discrete component technology to programmable technology. As an example of this, Fig 17.19 shows how an element of a very simple hardware control system maps into the equivalent ladder logic program.

More modern devices have the facility to use languages, both graphical and text-based, other than ladder diagrams, largely because the complex functions that they offer are not amenable to the simplistic ladder logic representations and because of their ability to incorporate branching statements such as IF-THEN-ELSE conditional statements. The languages are defined in an IEC standard, IEC 61131–3, which is the international

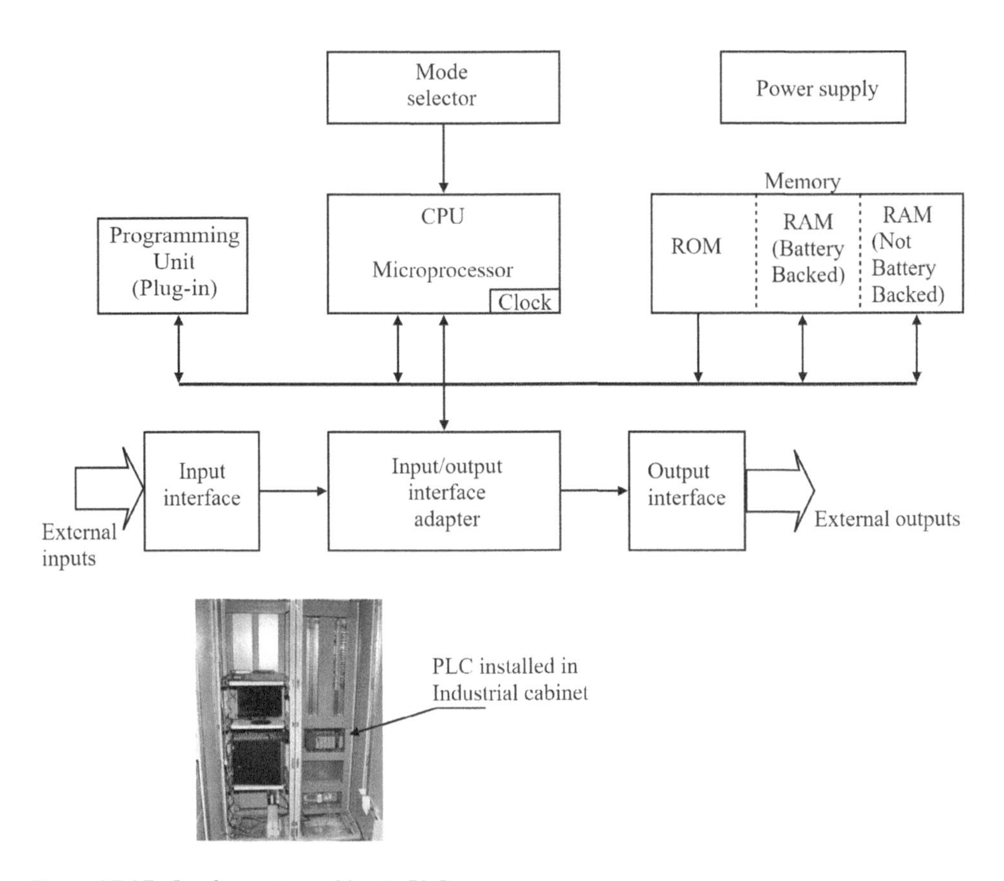

Figure 17.17 Configuration of basic PLC

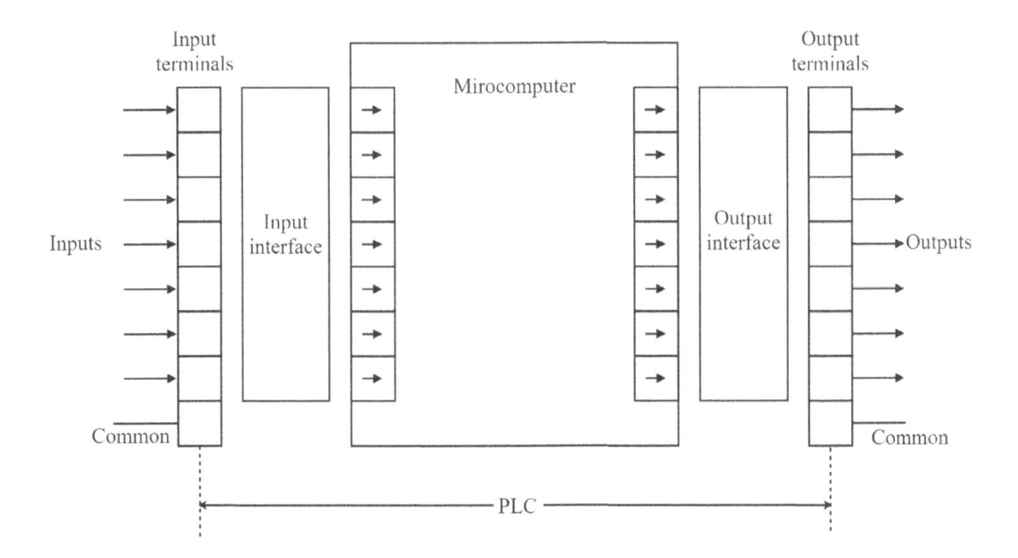

Figure 17.18 Scanning data into and out of PLC

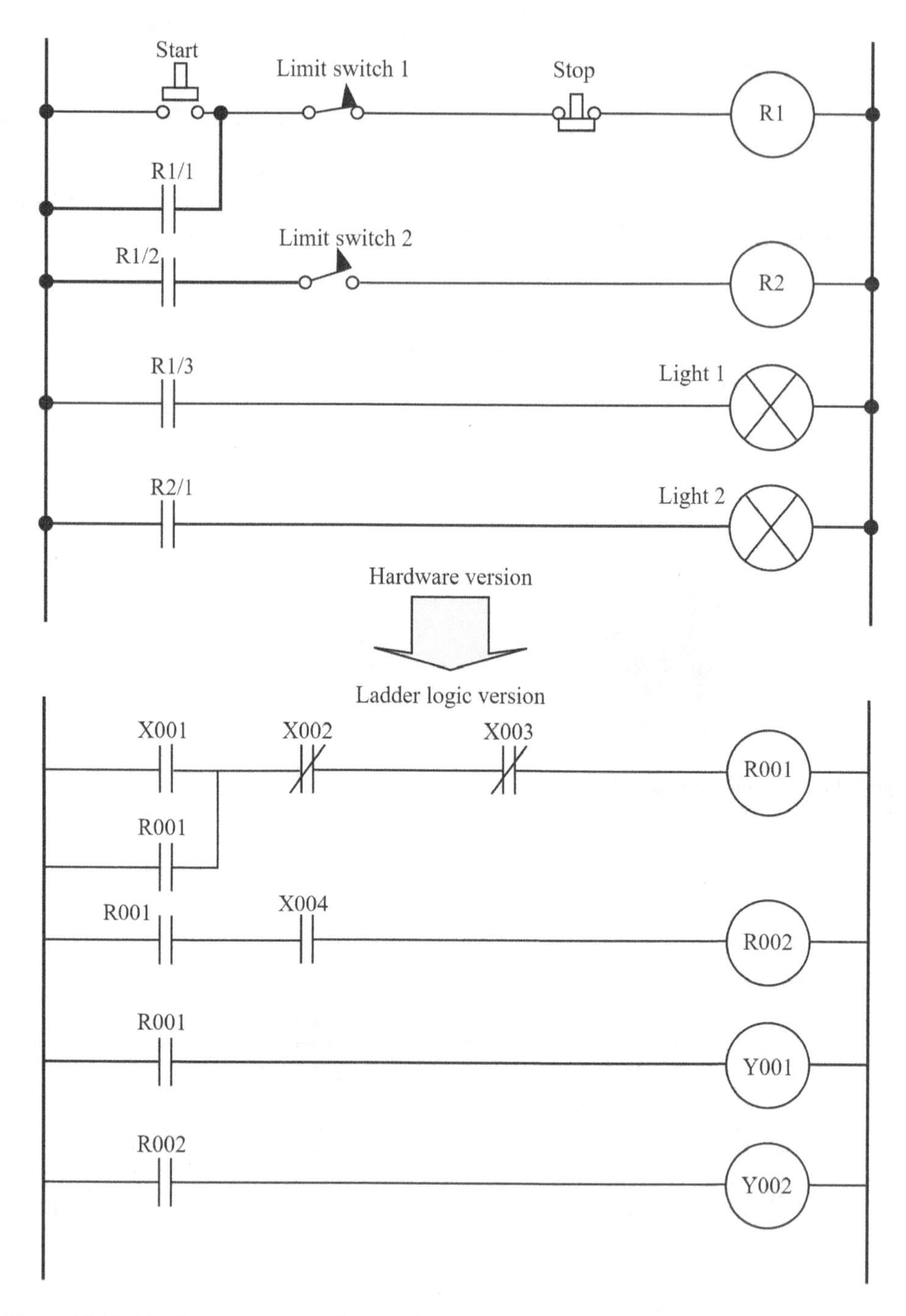

Figure 17.19 Hardware circuit and equivalent ladder program

standard for programmable controller programming languages. It specifies the syntax, semantics and display of five types of PLC programming languages:

- Ladder diagram (LD);
- Sequential Function Charts (SFC);
- Function Block Diagram (FBD);
- Structured Text (ST); and
- Instruction List (IL).

One of the main benefits of the standard is that it allows multiple languages to be used within the same PLC. This allows the program developer to select the language best suited to each particular task. It also facilitates the production of modular and structured programs, which in turn reduces errors and increases programming efficiency. These are significant advantages where software is used in safety applications.

Devices such as PLCs are used to control the operational functions of machines – to ensure that the machine does what it is meant to do. However, in the machinery safety field there has been a long-held concern about the integrity of programmable devices used to implement safety functions. The concerns stemmed from the difficulty experienced in testing software with sufficient rigour to provide confidence that it is free of systematic faults – faults or 'bugs' designed into the software, both at the level of the operating system and at the application level. Whereas electromechanical control systems can have their reliability and failure modes predicted because there is a substantial amount of relevant data available, it is considerably less easy to predict the reliability and failure modes of programmable systems.

Another concern has been the relative ease with which the software in PLCs and other programmable electronic devices can be altered. This raises the spectre of users of programmable systems that perform safety functions altering the code in a haphazard way without assessing the consequences on the safety performance of the system and, perhaps, losing control of the configuration management of the software. This latter point means that unauthorised or uncontrolled modifications, without the modifications being documented, can result quite quickly in the users not having an understanding of the software and how it is configured. Ultimately, this can lead to chaos and danger.

Imagine the situation, for example, in a 'fly-by-wire' passenger aircraft in which all the control surfaces (elevators, rudder, ailerons and so on) are controlled by computers which work to maintain the aircraft in steady and stable flight and which respond to inputs from the pilot and on-board sensors. It is unlikely that the flying public or the regulatory authorities would be pleased if they felt that the safety of the aircraft was being compromised by engineers or aircrew making modifications to the software code in the computers to see what effects they could have on aircraft performance or to find out if they could improve fuel consumption. The need for tight configuration management and control of modification procedures in programmable systems that perform safety functions is of paramount importance. Whereas users in the high risk industries (nuclear, petrochemical, avionics and so on), where safety-related programmable systems are quite common, generally do have very good and comprehensive processes and procedures, the same is not true in the machinery sector.

These issues led the standards makers who wrote editions of EN 60204–1 *Safety of machinery. Electrical equipment of machines. General requirements* and EN 954–1 *Safety of machinery. Safety related parts of control systems. General principles for design* published in the 1990s and early 2000s to insert notes in the standards to warn against the use of single channels of programmable electronics in safety applications.

An example of a design incorporating hardwired safety circuits in a system with a PLC as its main control element is shown in Figure 17.20. This shows how the outputs of the interlocking switch and the emergency stop actuator are routed not only to the input of the PLC controlling the machine's actuators but also to the output side of the PLC, effectively bypassing the PLC. The aim of this configuration is to ensure that actuation of the interlocking or emergency stop devices will lead to the appropriate outputs being switched to the correct state by hardwired means, without reliance on the programmable elements of the control system. It also ensures that the program correctly represents the current state of the machine. Although the diagram shows a contact being used to switch just one of the PLC's outputs, a more common configuration is for the contact to be placed in the power circuit to the PLC's output common module. This arrangement means that all the PLC's outputs would be switched off in the event of the contact opening.

In many modern applications, certainly those designed since the middle of the 1990s, the hardwired safety functions are routed through so-called safety relays and controllers manufactured and supplied by companies such as PILZ, Telemecanique, and Rockwell Automation. These are components that, in general, provide an in-built dual channel cross-monitoring facility aimed at providing a means of implementing high integrity control systems capable of meeting the fault tolerance requirements of the harmonised standards; the concepts are described later in this chapter.

The situation has changed somewhat after the publication of standards such as BS EN ISO 13849–1, BS EN 61508 and BS EN 62061, which have enabled manufacturers of safety-related control devices to design, certify and place on the market PLCs and other types of programmable safety controllers that perform safety functions, including

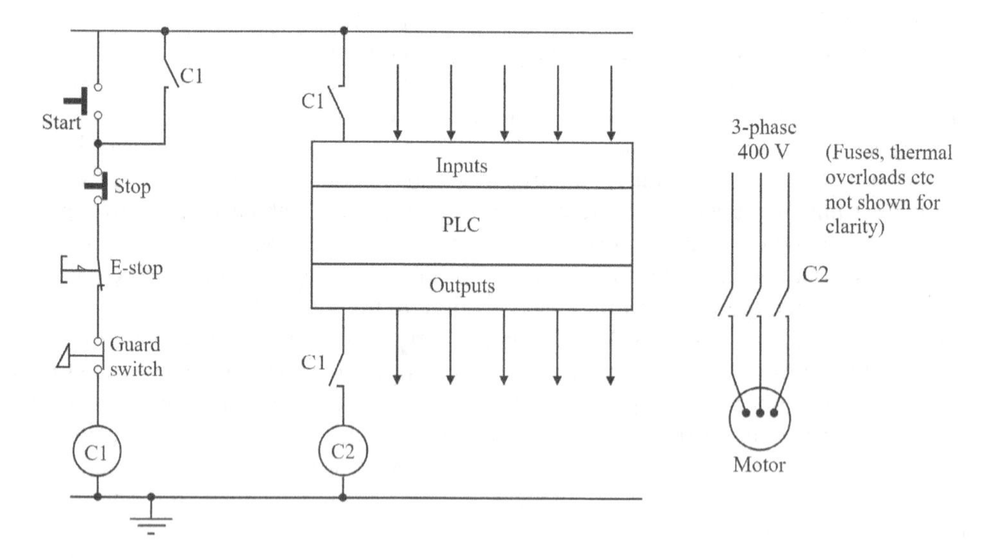

Figure 17.20 Hardwiring the safety circuit

programmable controllers that integrate the normal operational control modes and safety functionality into one platform. Safety PLCs use multiple microprocessors with internal comparators, diagnostics and watchdogs, together with certified software function blocks for specific safety applications such as e-stop and gate interlock, and are certified for use in applications with specific safety integrity requirements.

They are also amenable to connection into communication networks that allow safety applications (controllers, sensors and actuators) to access independent safety communication networks and standard fieldbuses such Ethernet/IP and DeviceNet using safety-related communications protocols. Such safety networks are covered by BS EN 61784–3:2010 *Industrial communication networks. Profiles. Functional safety fieldbuses. General rules and profile definitions* and include networks such as SafetyNet p, Profisafe, and CIP (Common Industrial Protocol) Safety. They have the significant advantage of reducing cable costs, allowing safety functions and devices to be distributed around machines using a single communications cable, a feature that has obvious advantages in the case of large distributed production lines.

Allied to this has been the developments in recent decades of SCADA systems for use in machinery control applications, where SCADA stands for Supervisory Control and Data Acquisition. In broad terms the SCADA system sits above the machinery control system's PLCs to provide operators with a human-machine interface (HMI) that allows them to monitor and control the machinery; the SCADA system also gathers data and creates records of events that are held in a file that can be called up for analysis. SCADA systems appear in many industrial sectors, including in power distribution systems where DNOs control their networks using centralised SCADA-based systems.

However, there is still a tendency, especially in the small machine sector, for safety functions to be hardwired to make them independent of the electronic and programmable electronic elements of control systems. At present, the more complex, and costly, programmable safety-related systems and SCADA systems tend to be used in larger integrated manufacturing systems. An increasing concern in this regard are the risks associated with cybersecurity and the potential for these complex software-based networked systems to be hacked and disrupted in a fashion that has the potential to compromise safety and the integrity of manufacturing facilities and national infrastructure. There is no shortage of stories of vulnerable systems being disrupted in this way, and dealing with it is an essential aspect of effective risk management.

Output devices

In most applications the ultimate aim of the processing unit is to cause the machine's prime movers – principally electric linear and rotary motors, braking systems, bolts and plungers, and hydraulic and pneumatic rams – to move in the required direction at the appropriate speed and acceleration, or to cause devices such as heaters and laser cutters to be switched on and off. This is normally achieved via interposing components that are directly switched and controlled by the signals output by the processing unit, although some devices are directly switched by the processing unit. The most common of these components are relays, contactors, solenoid-operated valves and speed controllers.

In safety applications on machinery, the most common output reaction to a safety-related input such as an interlocked guard being opened or the emergency stop button being pushed is for a machine to be placed in a safe state and a stop function to be implemented by power being removed and/or brakes being applied and for them to remain applied until a reset signal is provided. In electrical-only systems, the reaction is to have power disconnected to remove the risk of electrical injury.

Output devices used to implement safety functions must be adequately rated for their duty and have an appropriate level of reliability and fault tolerance. They should also be designed so that, wherever possible, faults are revealed and do not lead to the loss of the safety function; for example, electromechanical brakes should be designed to be held-off by being energised so that a loss of supply fault will lead to the brake being applied.

The stop function

The stop function is clearly of considerable importance in safety applications; it can be initiated by, for example, an emergency stop button being pressed, an interlocked gate being opened, trip devices being activated, or an operator pressing a stop button. BS EN 60204–1 defines three categories of stop, categories 0, 1 and 2:

- Category 0 stop involves the immediate removal of power to the machine drives such as motors and valves, which is why it is often known as an uncontrolled stop. Unless motors are braked by mechanical means, they will tend to continue to rotate until they stop naturally.
- Category 1 stop is a controlled stop with power remaining available to the machine drives to achieve the stop, at which point power is removed. This category of stop allows hazardous parts to be moved to a safe state and for powered braking to be applied. Guards must not be capable of being opened until the safe state has been achieved.
- Category 2 stop is a controlled stop with power left available to the machine drives. A normal production stop is considered a category 2 stop.

Within this hierarchy, a category 0 stop takes priority over category 1 and 2 stops.

The emergency stop function must be either a category 0 or a category 1 stop; a category 2 stop must not be used for the emergency stop function.

The stop function will often initiate some form of braking action. Braking systems can be quite complex in practice, but their principles are quite straightforward. The main techniques are described in the sections that follow.

Mechanical brakes

Mechanical brakes use friction linings to stop mechanical parts when the brake is applied. The brakes are normally held off under electrical, pneumatic or hydraulic power and are spring-applied.

Electrodynamic brakes

There is a variety of types of electrodynamic braking systems: d.c. injection, capacitive, plugging and regeneration being the most common. Note that all of these types of braking system require current to flow for the system to function and may therefore not be available under category 0 stops – they may also all fail to danger under certain fault conditions and this needs to be considered by carrying out a risk assessment during the design stages.

D.C. injection braking

In this type of system the a.c. electric motor is first disconnected from its supply and a d.c. current injected into its windings for sufficiently long to stop the motor. The circuit diagram in Figure 17.21 shows the basic features of this type of system. Pressing the 'start' button energises relay R and closes the direct-on-line starter contacts to supply 3-phase power to the induction motor, which will then accelerate to its working speed. The starter is held on via contacts R1.

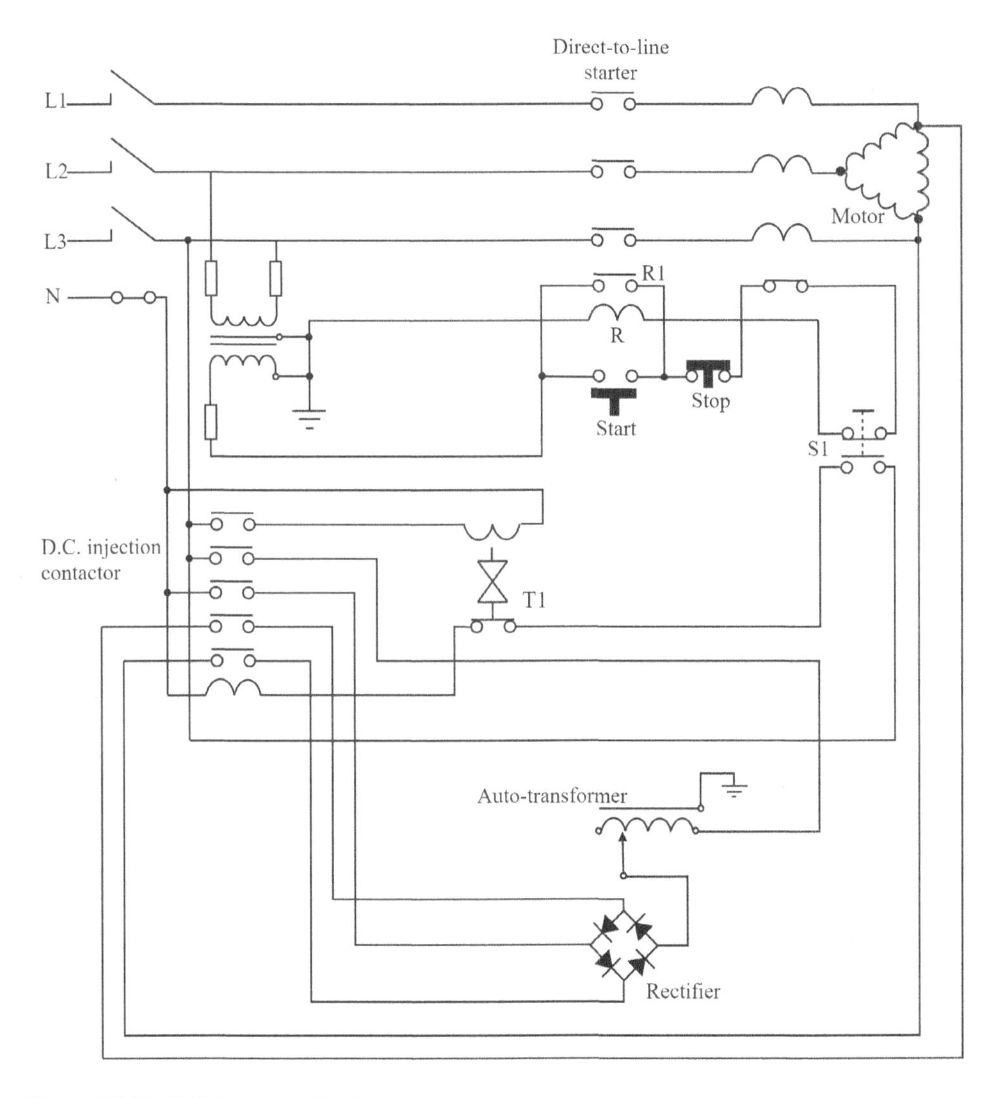

Figure 17.21 D.C. injection braking system

If the processing unit receives a safety signal such as an emergency stop it responds by toggling switch S1. This breaks the supply to relay R, tripping off the direct-on-line starter contacts and disconnecting the motor. Switch S1 also closes the circuit containing the coil for the d.c. injection contactor, closing the contactor. This provides a supply to the rectifier and leads to a d.c. current being injected into the coils of the motor, bringing the motor to a stop. After a period of time, the timer operates to break contacts T1, deenergising the d.c. injection contactor and removing the d.c. supply to the motor. The size and duration of the d.c. injection current must be set to the correct values for the system to work correctly.

Capacitor braking

In capacitor braking, the motor is disconnected from the supply, then connected to a capacitor for a preset time and then short circuited during the final deceleration stage to stop the motor.

Plugging

Plugging involves a reversing contactor that, when operated, swaps two of the phases on the supply to the motor, the motor is then said to be 'plugged'. This reverses the electromagnetic field that drives the motor and causes the motor to decelerate. If the plugging is not removed when the motor stops, the motor will then reverse its direction, which may be hazardous. Care must therefore be taken in the design of this type of braking system.

Regeneration braking

In regeneration braking, the power supply is removed from a motor, which then acts as a generator. This causes the motor to decelerate.

Safety integrity issues

Thus far the components that make up a control system have been described without considering in detail their architecture and how they are integrated to ensure that the systems have sufficient robustness, or safety integrity, for the safety application in which they will be used. Safety integrity refers to the reliability with which the system will perform the safety functions it has been designed to perform, as well as the extent to which it will continue to perform its safety functions in the event of faults occurring. We are therefore interested in system performance in terms of reliability and fault tolerance.

It would seem, intuitively, that the greater the contribution that a control system makes to safety the more reliable it should be and the more resistant it should be to faults that would adversely affect the safety function. These are important concepts covered in the three standards mentioned at the start of this chapter: BS EN ISO 13849–1:2015 *Safety of machinery. Safety-related parts of control systems. General principles for design*, BS EN 61508–1:2010 *Functional safety of electrical/electronic/programmable electronic safety-related systems. General requirements*, and its machinery sector implementation BS EN 62061:2005+A2:2015 *Safety of machinery. Functional safety of safety-related electrical, electronic and programmable electronic control systems*. As a general statement of principle, BS EN ISO 13849 is applicable to non-electrical, electromechanical, and simple programmable systems, and BS EN 62061 is applicable to more complex programmable systems, although it can also be applied to the more simple systems.

There are many other national and international standards that cover system safety issues but which tend to be targeted at very specific applications, an example being defence and military standards (DEF-STANs and MIL-SPECs) covering avionic systems in aircraft. These standards are not considered in this book.

BS EN ISO 13849–1:2015 Safety of machinery. Safety-related parts of control systems. General principles for design

BS EN ISO 13849–1 is the successor to BS EN 954–1, a standard that dominated the safety integrity topic in the machinery sector for a decade but which was unable to take account of developments in technology, particularly the advent of programmable systems used to implement safety functions. It is a Type B1 standard which is meant to be used by CEN/CENELEC Technical Committees preparing Type B2 and Type C European standards.

The standard provides guidance on the design of safety-related parts of control systems on machinery and is applicable to all types of technologies employed: electrical, electronic, programmable electronic, hydraulic, pneumatic and mechanical. This would typically be guard interlocking circuits, stop and emergency stop circuits, and trip devices. The standard is targeted at the machinery sector, but it has broad application to safety-related systems.

The aim is to ensure that those parts of a control system that perform safety functions have adequate safety performance for the application. Safety performance, as previously noted for safety integrity, relates to the reliability and fault tolerance of the safety-related parts, and their ability to provide the required level of risk reduction. It is therefore axiomatic that the process of satisfying the requirements of the standard for any particular machine starts with a risk assessment of that machine.

As explained earlier, in the hierarchy of measures to control risk it is preferred that risks are designed out. Those that remain will then need to be safeguarded using a variety of risk reduction techniques such as the use of fixed guards. Some of the techniques may make use of electrotechnical control systems. The risk assessment should therefore establish the safety functions of the control systems which will then allow the designer to identify the safety-related parts and to specify their contribution to risk reduction. The extent of this contribution will influence the design measures that must be taken. In the very large majority of machinery applications, the risk assessment and measurement of risk reduction is essentially a qualitative assessment, with some quantitative analysis associated with matters such as diagnostic coverage and reliability. BS EN ISO 12100:2010 provides guidance on how the risk assessment may be conducted, and BS EN ISO 13849 explains how this should be mapped into safety functionality of control systems.

BS EN 954–1 classified the design measures into five categories – B, 1, 2, 3 and 4 – on the basis of reliability and tolerance to faults, and BS EN ISO 13849 builds on these in terms of required Performance Levels (PL) for the safety functions of the safety-related control system. The main features of the five categories are as follows:

- Category B. The safety-related parts are to be designed, constructed, selected, assembled and combined in accordance with relevant standards so they can withstand the expected influence. The design is such that the occurrence of a fault can lead to the loss of the safety function.

- Category 1. The requirements of B apply, plus the use of well-tried components and well-tried safety principles. This type of system can suffer the loss of the safety function on the occurrence of a fault, but the event is less likely than for a category B system.
- Category 2. The requirements of category 1 apply, plus the safety function must be checked at suitable intervals by the control system. In this type of system the occurrence of a fault can lead to the loss of the safety function but the loss is detected at the next check.
- Category 3. The requirements of category 1 apply, plus the design must ensure that no single fault in the safety-related parts leads to the loss of the safety function, and the fault is detected whenever reasonably practicable. In this type of system, the safety function will be performed even when a single fault exists, and the fault may be detected, but might not be, and an accumulation of faults will lead to a dangerous situation.
- Category 4. The requirements of category 1 apply; the design must ensure that a single fault in the safety-related parts does not lead to the loss of the safety function, and the single fault is detected at or before the next demand upon the safety function. If this is not possible, then an accumulation of faults must not lead to the loss of the safety function. This means that the safety function will be performed in the event of faults occurring.

It should be noted that the categories are not a measure of the risk on the machine, as has been commonly and mistakenly assumed. They are simply a means of classifying the required safety performance of the safety-related parts of the control system on a machine and are used largely to determine the system architecture. In general terms, category B and 1 control systems employ single channel architectures while category 2, 3 and 4 control systems use dual channel architectures with redundancy and, possibly, diversity. The fault detection requirements imply the use of cross-monitoring between the redundant channels. In systems where sophisticated diagnostic capabilities exist for the detection of dangerous failures in single channel architectures, these generalisations may not apply.

BS EN ISO 13849–1 uses the BS EN 954–1 categories, which it refers to as designated architecture categories, but adds additional requirements relating to matters such as diagnostic coverage, reliability, common cause failures and, for programmable elements of the system, software-specific systematic requirements that system designers use to ensure that the required performance levels for the individual safety functions have been achieved. The standard specifies five performance levels: PL a, PL b, PL c, PL d and PL e, with the required PLs for any system being generated from the machine's risk assessment. The standard provides a risk graph that can be used to generate a required PL for each risk identified. The risk graph is shown in Figure 17.22, where it will be seen that PL a has the lowest level of safety integrity and PL e the highest.

Manufacturers of subsystems, such as actuators and safety relays, allocate PLs to their devices. A system designer will then use the subsystem data to calculate the overall PL of the system, taking due account of the designated architecture category, diagnostic coverage, mean time to dangerous failure of component parts, common cause failures, and

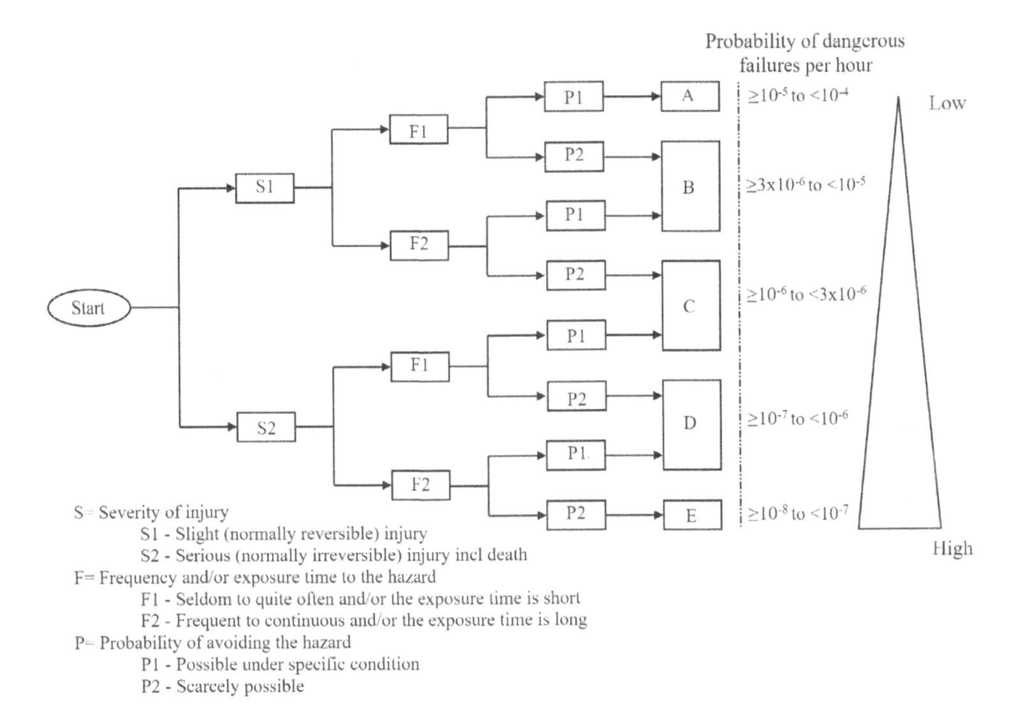

Figure 17.22 **Determination of required performance level from risk graph from BS EN ISO 13849**

certification of any programmable elements. The aim is to ensure that the achieved PL meets the PL specified in the risk assessment process.

Diagnostic coverage is a measure of the fractional decrease in the probability of dangerous hardware failures resulting from the use of automatic diagnostic tests. It is calculated as the ratio of the sum of the dangerous failure rates for each safety-related component to the sum of the dangerous detected failure rates for each component, and is a measure of the effectiveness of any automatic diagnostic capability designed into the system. The standard presents information on the relationship between PLs and designated architecture categories with varying degrees of diagnostic coverage and reliability. For example, a safety function implemented with a designated architecture category of 2, a medium diagnostic coverage, and high mean time to dangerous failure would have a PL of d.

Safety relays

It is helpful at this stage to have a look at 'safety relays' and see how they contribute to safety integrity and achievement of PLs. By way of explanation, the basic concept of these relays is depicted in Figure 17.23, which shows the basic circuit of a classic first generation safety relay. When power is first applied to the device, relay R3 is energised, which causes relays R1 and R2 to become energised. The output contacts of R1 and R2 close, and the output contacts of R3 end up in the closed position as relay R3 deenergises. If the

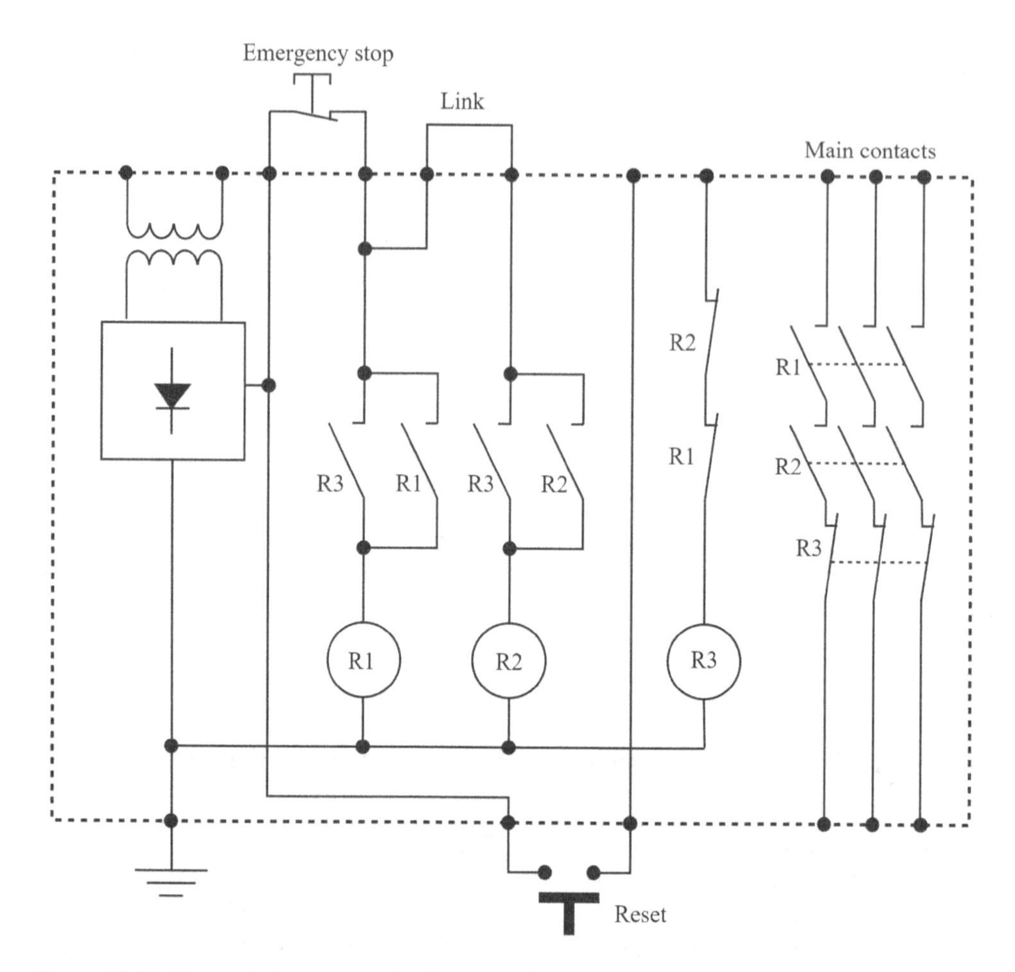

Figure 17.23 **Basic configuration of safety relay**

emergency stop button is pressed, relays R1 and R2 deenergise, leading to the opening of the output contacts, thereby removing power from connected devices such as contactors. If the contacts of either relay R1 or R2 were to fail in the closed position, the output would still be switched off by the other relay's contacts. If this fault were to occur, the relay could not be reset because relay R3 could not be energised, thereby revealing the fault. This technique for achieving single fault tolerance and failure detection is known as dual channel cross-monitoring. The technique has been in use for many years using discrete components – the safety relays encapsulate the function within a single component

These safety relays provide a facility for fault-tolerant and monitored control circuits, which can be used to meet the requirements of designated architecture categories 2, 3 and 4. In that sense, they offer a significant contribution to the achievement of safe machinery control systems. An example of a system with a designated architecture category of 3 using a safety relay is shown in Figure 17.24.

This shows a dual channel architecture with two contactors controlling power to the motor, the contactors being switched by the main contacts in the safety relay. There are

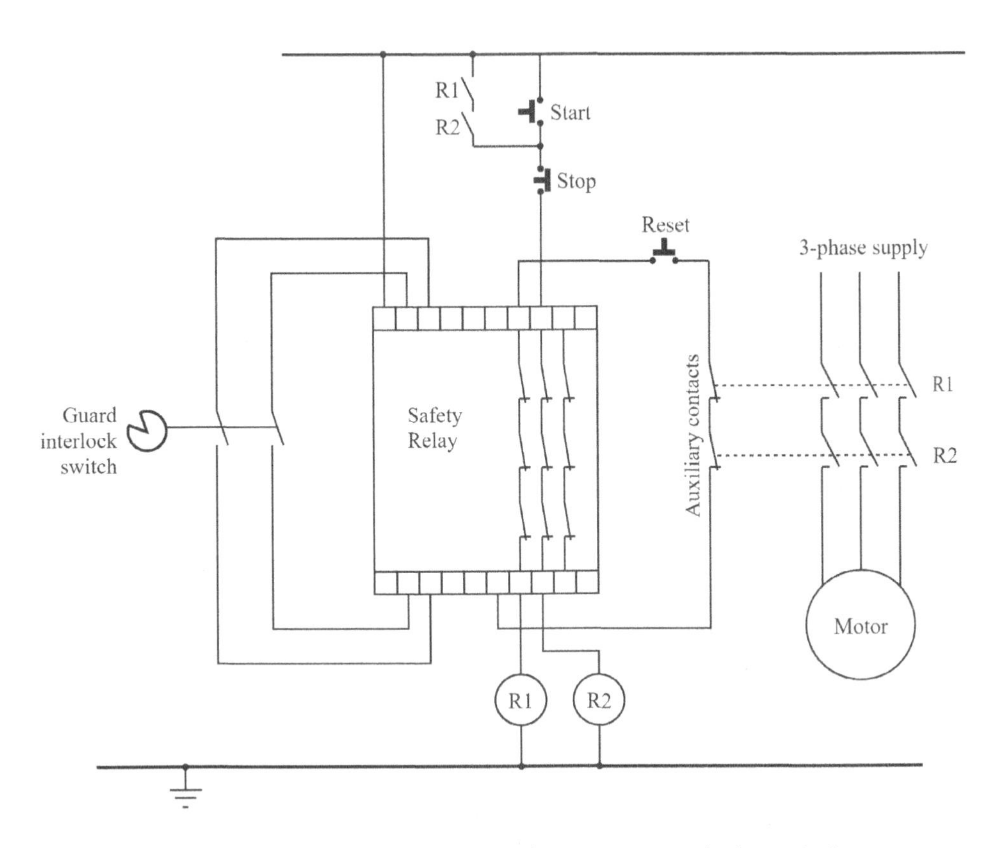

Figure 17.24 Designated architecture category 3 system using a 'safety relay'

auxiliary contacts on the contactors that are incorporated in the reset circuit of the safety relay, thereby monitoring the contactors. Any single fault occurring on this system will not cause the loss of the safety function, the safety function being the disconnection of the motor when the guard is open. Moreover, the single fault will be detected. This type of architecture is appropriate where the interlocking circuit is making a significant contribution to risk reduction. An example would be the interlocked access door on a bottling machine, through which an operator has to go once an hour to replace labels and pick up debris and where the hazard is loss of an arm if the machine were to move unexpectedly.

The manufacturers of these safety relays and similar devices, including programmable devices with safety-related function blocks, advise that when used for the applicable safety functions using the dual channel architecture shown, and components with suitable levels of reliability, the safety function is capable of having a PL of d, or even e.

BS EN 61508 Functional safety of electrical/electronic/programmable electronic safety-related systems

The international standard IEC 61508, published in the UK as BS EN 61508, provides guidance on mechanisms that can be used to ensure that safety-related electrotechnical systems implemented in electrical, electronic and programmable electronic technologies are safe. It comprises seven parts, the first of which were finalised and published in 2000, some 15 years after the IEC initiated work on the drafting of the standard. The seven parts have the following titles:

Part 1. General requirements
Part 2. Requirements for electrical/electronic/programmable electronic systems
Part 3. Software requirements
Part 4. Definitions and abbreviations
Part 5. Examples of methods for the determination of safety integrity levels
Part 6. Guidelines on the application of IEC 61508 Part 2 and Part 3
Part 7. Overview of techniques and measures

The standard is large and complex, and its contents are not easily absorbed. Whereas this may not be a particular issue for companies developing safety-related systems for the likes of major petrochemical companies, it acts as an impediment to its adoption by small and medium size machinery manufacturers. Moreover, it is generic in nature, meaning that it is not targeted at any particular applications, although the thrust of it is more appropriate for complex safety-related control systems in the process, nuclear, railway and similar industries than for simple non-complex machinery control systems, for which BS EN ISO 13849 is more suited.

However, in the context of the discussion in this chapter, the standard has the significant advantage over BS EN ISO 13849 that it describes how the functional safety of complex electronic and programmable electronic control systems can be assured. In recognition of this, a machinery sector implementation of the standard has been produced – BS EN 62061. This is described in the next section after this brief overview of BS EN 61508.

At BS EN 61508's core is the definition and explanation of a safety life cycle. The overall safety life cycle for safety-related control systems is depicted in Figure 17.25. The important attribute of this concept is that it imposes a formal structure on the planning and management of the specification, development, installation, use, maintenance, modification and final disposal phases of the life of safety-related systems.

The specification of safety requirements is an essential step, with the requirements based on the contribution the system is making to risk reduction, taking into account the contributions made by other technologies such as pressure-relief valves and any other external risk reduction measures. The concept is fundamentally the same as that set out in BS EN ISO 13849, although the realisation of it is entirely different.

The specification will lead to the identification of safety functions, again as recommended in BS EN ISO 13849. BS EN 61508's safety functions are defined in terms of their functionality and their safety integrity. The safety integrity of each safety function is roughly equivalent to the concept of the Performance Levels of BS EN ISO 13849. BS EN 61508's safety integrity levels are tightly defined in terms of their target failure measure. There are four levels of safety integrity, ranging from Safety Integrity Level (SIL) 1 to SIL 4, with SIL 4 being the most safe.

There are two types of safety integrity level, those for safety functions that operate in low demand mode of operation and those for safety functions that operate in high demand or continuous mode of operation. An example of the former would be an emergency stop facility on a machine. An example of the latter would be a temperature sensor monitoring a safety critical process. The quantitative values of the SILs are shown in Tables 17.1 and 17.2.

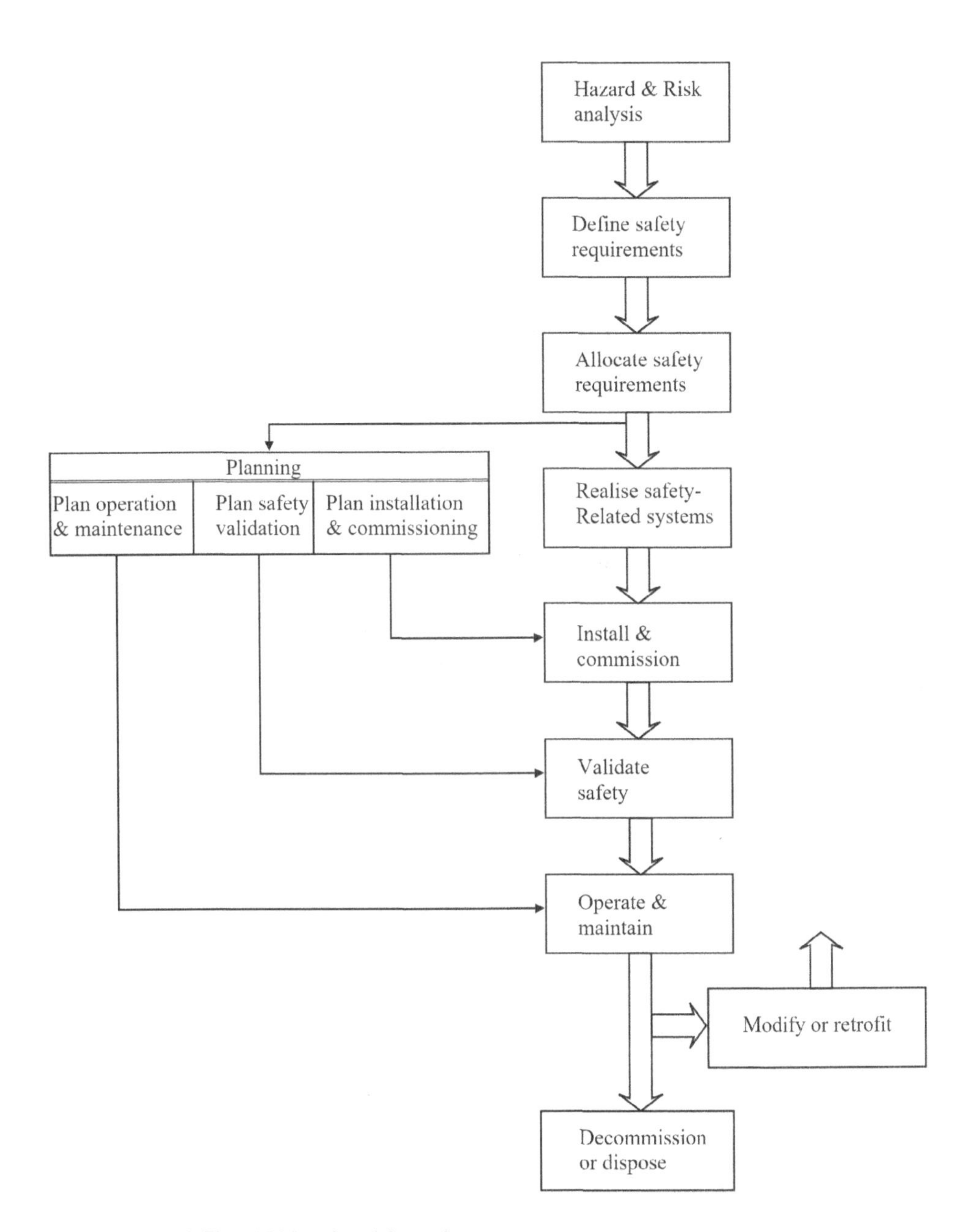

Figure 17.25 BS EN 61508 safety life cycle

Although the numerical values of the SILs can be tied into a quantitative risk analysis, the standard does allow the appropriate SIL for a safety function to be determined by qualitative means using risk graph techniques. The standard provides guidance on how the risk graph technique should be implemented.

Once a safety function has an assigned SIL, the designer needs to decide upon the architecture, diagnostic coverage, proof test interval and other criteria that need to be adopted

Table 17.1 Safety integrity levels – low demand mode

Safety integrity level	*Target failure measure (Average probability of failure to perform its design function on demand)*
1	$\geq 10^{-2}$ to $< 10^{-1}$
2	$\geq 10^{-3}$ to $< 10^{-2}$
3	$\geq 10^{-4}$ to $< 10^{-3}$
4	$\geq 10^{-5}$ to $< 10^{-4}$

Table 17.2 Safety integrity levels – high demand/continuous mode

Safety integrity level	*Target failure measure (Probability of dangerous failure per hour)*
1	$\geq 10^{-6}$ to $< 10^{-5}$
2	$\geq 10^{-7}$ to $< 10^{-6}$
3	$\geq 10^{-8}$ to $< 10^{-7}$
4	$\geq 10^{-9}$ to $< 10^{-8}$

Table 17.3 Equivalence between performance level and safety integrity level

Designated architecture category	*Performance level BS EN ISO 13849*	*Safety integrity level BS EN 62061*
Category B	PL a	None
Category 1	PL b	SIL 1
Category 2	PL c	SIL 1
Category 3	PL d	SIL 2
Category 4	PL d	SIL 3

to ensure that the system achieves the specified SIL. The standard provides comprehensive guidance on all these techniques.

In the context of programmable systems, the standard recognises the problem of systematic errors occurring in the hardware and software of complex systems and the difficulties associated with estimating the probability of these faults occurring. To overcome this problem, the standard provides guidance on the measures and techniques that may be adopted during the design stage to avoid the introduction of systematic errors. The specific measures and techniques are beyond the scope of this book.

BS EN 62061:2005+A2:2015 Safety of machinery. Functional safety of safety-related electrical, electronic and programmable electronic control systems

BS EN 62061 is the machinery sector implementation of BS EN 61508, covering the whole life cycle as per BS EN 61508. Rather than using the performance level measure of safety integrity used in BS EN ISO 13849, it uses the same SIL designation as BS EN 61508, with the required SIL for any one safety function being determined by risk assessment, although SIL4 is not used in machinery applications. The output of the risk assessment should be a safety integrity requirements specification covering the safety-related parts of the control system, as well as a functional requirements specification for the system.

In designing a system to meet the required SIL, designers need to take a systematic approach and consider factors such as the probability of dangerous failure per hour, hardware fault tolerance, safe failure fraction, proof test interval, diagnostic coverage and test interval and susceptibility to common cause failures. The similarity to the factors addressed in BS EN ISO 13849 can be seen. Indeed, there is a generally accepted, albeit simplistic, equivalence between BS EN ISO 13849's PLs and BS EN 62061's SILs, as shown in Table 17.3.

The standard suggests that systems are divided into subsystems which are amenable to analysis, with the overall system safety integrity level for each safety function being derived from that analysis. The hardware safety integrity level that can be claimed for a subsystem is limited by the reliability, hardware fault tolerance and safe failure fraction characteristics, leading to an architectural constraint quoted in terms of SIL Claim Limit, or SIL CL. This is an important feature of the standard.

The question arises about which standard designers of machinery safety-related control systems should use, given their similarities. It is unfortunate that the standards bodies have produced two standards that cover the same issue, a fact that undoubtedly creates confusion. However, the answer lies in the fact that BS EN 62061 addresses the safety requirements for complex programmable safety-related systems in a fashion that BS EN ISO 13849 does not; essentially, it calls for compliance with BS EN 61508, which is a complicated and onerous process. Therefore, designers of systems that incorporate complex electronic and programmable electronic components in the safety-related control system should use BS EN 62016, whereas those designing non-complex systems are best-advised to use the simpler BS EN ISO 13849.

18

Maintenance

Introduction

An electrical installation which has been properly designed, installed and verified to the requirements of relevant standards, such as BS 7671 in the case of low voltage systems, is likely to be safe and serviceable for some time depending on the nature of its use and misuse and the environmental conditions in which it is installed. Similarly, electrical equipment that has been designed and built to a relevant standard is likely to be safe at the time of its supply.

The result is that some management teams, particularly of smaller enterprises, who are preoccupied with running their businesses tend to ignore the need for maintenance and may well do nothing about it until the consequence of this neglect appears in the form of a breakdown, accident or fire. Those who have thought about it often dismiss maintenance as an unproductive cost with no return on the investment, spending as little as possible, doing nothing or next to nothing until there is an incident. In taking this all-too-common approach, they lose sight both of their legal obligation to maintain systems so as to prevent danger and the economic advantages of avoiding incidents; some careful consideration of these two aspects may induce them to inaugurate a planned maintenance system, a topic that is the subject of this chapter.

Another form of maintenance is corrective maintenance, meaning fault finding and repair work that needs to be done after a breakdown or failure has occurred. This type of work is not planned and is reactive in nature, often carried out under pressure to reinstate equipment to service. This type of work is not covered in this chapter but the importance of implementing safe working practices during it cannot be overstated.

Statutory requirements

In all locations where people are at work, excepting servants working in domestic premises (excluded by virtue of S51 of the HSW Act), maintenance of the electrical installations and systems is a statutory requirement covered by Regulation 4(2) of the Electricity at Work Regulations, although maintenance is required only in those circumstances where not to do it may lead to danger. This mirrors the general provision of Section 2(a) of the Health and Safety at Work Act that requires the provision and maintenance of plant that is, so far as is reasonably practicable, safe and without risks to health.

In addition to this, Regulation 5 of the Provision and Use of Work Equipment Regulations requires work equipment to be maintained in an efficient state, in efficient working order and in good repair – this requirement for maintenance embraces electrical equipment and apparatus and electrotechnical control systems on machinery and plant. Moreover, irrespective of whether or not work is being done, the Electricity Safety Quality and Continuity Regulations Supply, Regulation 3, requires electricity generators, distributors and meter operators to ensure that their equipment is so constructed, installed, protected (both electrically and mechanically), used and *maintained* as to prevent danger, interference with or interruption of supply, so far as is reasonably practicable.

This set of regulations requires that some form of maintenance is carried out to prevent danger, so as to reduce risks so far as is reasonably practicable. However, the law does not prescribe the form that the maintenance should take, simply requiring that maintenance is carried out so as to prevent danger. In the context of electrical systems energised at hazardous potentials, this means that some form of inspection, test and, if necessary, repair has to be done so as to anticipate dangerous faults and to ensure that they do not occur, so far as is reasonably practicable; this is preventive maintenance. It is evident that neglecting an installation or equipment until a dangerous incident occurs and then taking remedial action does not constitute preventive maintenance and does not comply with the law.

Definition and forms of maintenance

A planned preventive maintenance system comprises a procedure for the inspection, test and repair of the electrical installation and connected apparatus to avoid breakdowns and dangerous faults. It is most commonly regarded as a preventive surveillance system whereby apparatus is inspected and if necessary tested and repaired before it is likely to break down or develop a dangerous fault. The relative contributions of visual examinations and testing have already been considered elsewhere in this book.

In reality, there are a number of different forms that maintenance can take. Perhaps the most common is routine preventive maintenance of the type prescribed by BS 7671, where periodic inspections and tests are carried out to try to detect any failures or degradation that may lead to danger before that danger is realised. Whereas this is an effective technique, it does have the disadvantage that it requires intervention into the equipment on a periodic basis regardless of the actual condition of the equipment or the likelihood that a fault has occurred or the likelihood that the characteristics of the equipment may have degraded. This means that equipment that may be in perfectly good condition has to be dismantled, with the possibility that the intervention itself will cause failures or degradation. In the light of this, there have been attempts over the years to devise other maintenance strategies that take into account the actual condition of the equipment and/or its reliability.

One important development in this area has been that of 'condition-based maintenance' (CBM). The objective of CBM is to use non-invasive techniques to detect the current state of systems both to decide whether or not intervention is needed and accurately to predict their remaining useful lives. This enables engineers to perform maintenance only when needed to prevent operational deficiencies or failures, essentially eliminating costly periodic preventive maintenance and reducing the likelihood of plant failures. CBM uses a variety of sensor systems to detect and diagnose emerging equipment problems and to predict how long the equipment can effectively serve its operational purpose. The output of the procedures is maintenance personnel being alerted to developing problems, enabling maintenance activities to be scheduled and performed, as needed, before operational effectiveness or safety are compromised. The technique is applicable across a broad range of technologies but, in the electrical field, techniques such as partial discharge testing, thermographic imaging and oil sampling can be used.

- Partial discharge testing makes use of the fact that in high voltage systems any voids in insulation will usually fill with gas. The gas will ionise, and discharge activity caused by the high electric field strengths will take place. This discharge activity can be detected by radio frequency monitors, with the extent of the detected activity giving an indication both of the presence of voids and the likelihood of an insulation failure. This means that corrective action can be taken before a catastrophic insulation failure occurs. The technique is used on high voltage distribution boards to monitor the likes of compound-filled busbar chambers and cable boxes and cable joints.
- Thermographic imaging uses the fact that high resistance points in conductors, such as may occur at weak and loose connections, or locations where insulation has been damaged and there is leakage current between conductors at different voltages, will generate heat. This heat can be detected using cameras that produce images in the infrared band, allowing 'hot spots' to be detected and corrective action to be taken before a catastrophic failure occurs. The technique is increasingly being used to monitor both high and low voltage distribution boards and other electrical apparatus such as motors, especially in circumstances where a need for continuous operation means that the equipment cannot be switched off for invasive testing. Typical examples of this are refrigeration and freezer units in 24-hour supermarkets, and IT and power distribution systems supporting 24-hour banking and IT operations. Care has to be taken in its implementation because covers frequently have to be removed to expose the equipment inside enclosures and cabinets. This may lead to the imaging work having to be treated as 'live work' in the context of Regulation 14 of the Electricity at Work Regulations, in which case precautions will need to be taken against the possibility of the person using the camera making direct contact with exposed live conductors. Helpful guidance on best practice during thermal imaging has been published by the Building Services Research and Information Association (BSRIA) in its publication *Safe thermal imaging of electrical systems (up to and including 1000 V a.c.)*.
- Many items of high voltage switchgear, some older low voltage switchgear, and transformers contain mineral oil that acts both as an insulator and as a coolant. Samples of this oil taken periodically and analysed in laboratories for particulates and dissolved gases can provide information on the condition both of the oil and the electrical equipment. Internal faults such as arcing activity, overheating, and partial discharges can be detected and remedial action taken before a disruptive fault occurs. Guidance on the monitoring of mineral insulating oils in electrical equipment is provided in BS EN 60422:2013 *Mineral insulating oils in electrical equipment. Supervision and maintenance guidance.*

These techniques, on their own, do not provide a full 'health check' of an electrical system. Thermographic imaging, for example, does not give an indication of the continuity of earth conductors or the value of earth loop impedance on protective systems. It cannot, therefore, give an indication of the condition of the systems installed for fault protection, including earthed equipotential bonding and automatic disconnection protection against indirect contact injuries. They should therefore be treated as being just one element in the overall armoury of maintenance techniques available.

Another important concept in preventive maintenance is that of reliability centred maintenance (RCM). RCM is a systematic process of preserving a system's function by selecting and applying a range of preventive maintenance techniques. It differs from most approaches to preventive maintenance by focusing on function rather than equipment. In general, the concept of RCM is applicable in large and complex systems such as chemical plants, oil refineries and power stations.

The RCM approach arose in the late 1960s and early 1970s when the increasing complexity of systems (and consequent increasing size and cost of the preventive maintenance task) forced a rethink of maintenance policies among manufacturers and operators of large passenger aircraft. Pioneering work on the subject was done by United Airlines in the United States in the 1970s to support the development and licensing of the Boeing 747. The principles which define and characterise RCM are

- a focus on the preservation of system function;
- the identification of specific failure modes to define loss of function or functional failure;
- the prioritisation of the importance of the failure modes, because not all functions or functional failures have equal consequence; and
- the identification of effective and applicable preventive maintenance tasks for the appropriate failure modes.

It will be appreciated from this that there is a diverse range of techniques and methods that can be applied to the preventive maintenance of electrical systems. In simple and non-complex systems, the normal approach will be to adopt the simple planned preventive schemes advocated by BS 7671 and similar standards. In more complex systems, the innovative techniques associated with schemes such as CBM and RCM should be considered, although specialist advice will frequently be needed if informed judgements are to be made.

The following sections largely assume that a simple planned preventive scheme is appropriate.

The planner

A competent person familiar with the installation and the connected equipment, and their usage, is best placed to devise the planned maintenance system. For larger establishments with a maintenance department, the senior electrical person should do it. For smaller premises with no in-house maintenance staff or person with the appropriate competencies, a consultant or similar entity should be sought. As an alternative, the electrical contractor appointed to do the maintenance work will usually advise on how the work should be planned and implemented.

The maintenance scheme

Circuit and apparatus identification

If not already available, a circuit diagram or schematic of the installation should be prepared, and the circuits and connected apparatus on it should be identified. The numbers and other identifying labels marked on circuits, switchgear, controlgear and apparatus should match those in the drawings and schematics. Fixed, transportable and portable apparatus, which is not usually shown on a circuit diagram, should also be marked or recorded in a separate inventory. Existing circuit diagrams should be checked against the installation and amended if necessary to reflect any alterations.

Register

Next, a register of the circuits and apparatus should be prepared. It will probably be most convenient if this takes the form of a loose-leaf ledger or computer database or spreadsheet. In larger premises it will be worth considering investing in one of the many specialist software packages available to support preventive maintenance schemes. This will facilitate adding new items, removing those that are redundant and adding additional records for any new items as required. For each item there should be a description which identifies it and its location and gives essential information about it, such as the name and address of the maker, cost, location of the manufacturer's installation, use and maintenance instructions, and a list and location of spare parts. There should also be a maintenance schedule showing the frequency of inspections and what inspection, testing, adjusting and renewal of parts should be done on each occasion.

Additional pages should contain a checklist for the maintainer to tick the items done and to provide space for remarks. Other pages should be provided to record defects and breakdowns and the rectifying action that has been taken. The register could also contain copies of Electrical Installation Certificates, Minor Work Certificates and Electrical Installation Condition Reports (EICR) where these are raised.

Risk assessments

Maintenance work – both planned preventive maintenance and corrective or repair maintenance – should be covered by risk assessments. These should consider the main hazards and risks, such as electrical injury when carrying out live fault finding or working on a system that has not be securely isolated, falls from height, exposure to asbestos and chemical substances, and slips and trips. The control measures must be clearly laid out and communicated to the maintenance workers.

A good practice is to attach the risk assessment to any job cards or work orders that are raised for the work, and link them to any safety documents that may have to be raised, such as Permits for Work or Sanctions for Test. They could also be referenced in the aforementioned maintenance register.

Small installations

For small installations, such as most domestic accommodation, small shops and offices with only a few circuits for a cooker, water heater, lighting and socket-outlets fed from a consumer unit, the scheme can be simplified. For instance, if a circuit diagram is not available, a circuit schedule may be used and the inspections and tests carried out as recommended by BS 7671 at intervals of 5 years and 10 years for domestic installations unless otherwise recommended by the person carrying out the inspection and test.

Electrical equipment, be it fixed, movable, transportable, portable or hand-held, should be routinely inspected, with Class I appliances and higher risk Class II appliances being tested at a suitable frequency.

The 'events diary'

The inspection dates should be entered in a diary or call-forward system or planned preventive maintenance software package which is consulted by the supervisor when planning the work schedule.

Frequency of inspections and tests

The average installation

Except for those installations where there is a mandatory requirement, such as petrol filling stations, the frequency of inspections and tests is a matter of judgement by the planner based on manufacturers' instructions, experience and published guidance.

For a low voltage fixed installation, i.e. excluding the connected equipment, the IET has published a list of recommended initial inspection intervals in Table 3.2 of Guidance Note No 3 *Inspection and testing*. Examples are as follows:

- Commercial premises, offices and shops: Every 5 years or change of occupancy
- Residential accommodation (HMOs, student residences etc): Every 5 years or change of occupancy
- Industrial premises, theatres and leisure complexes: Every 3 years
- Caravan parks, marinas, swimming pools and launderettes: Annually

The installation should not be left alone during the intervening periods and should be regularly inspected, without any dismantling or opening of enclosures, to detect any obvious signs of damage or deterioration that might require remedial action. Guidance Note No 3 suggests this should be done annually in most installations, and more frequently in higher risk establishments such as schools, hospitals and construction sites.

These intervals should not be regarded as being fixed and immutable – they should be considered at each inspection and test, and the person carrying out the inspection and test should recommend the next interval based on his/her assessment of the installation's physical condition and conditions of use. The aim should be to choose the interval between inspections of an item so that it is checked just before it is likely to become defective. There is no virtue in unnecessarily increasing the frequency of inspections; unnecessary dismantling may itself cause trouble by damaging parts such as gaskets.

The manufacturers' recommendations on maintenance intervals are likely to be conservative. The intervals given by them may sometimes be safely extended, but only after the guarantee period has expired to avoid invalidating it. A good example of this is high voltage switchgear owned and operated by distribution network operators; over the years the maintenance intervals have been progressively extended on the basis of experience to the point where much of the switchgear is now given deep strip-down maintenance once every 12 years or so, with regular operational proof tests and inspections being conducted in the intervening period.

The failure rate/time curve for most items is shaped like a bath (the so-called bath-tub curve), the failure rate being plotted vertically and the time horizontally. The higher failure rate soon after commissioning is due to assembly errors and premature failure of defective components. For the more important items, at least, it is prudent to increase the number of inspections during this time. After this teething period, the failure rate should fall to a minimum until the apparatus nears the end of its economically useful life when general wear and fatigue stresses cause the failure rate to increase again.

Equipment and apparatus

Electrical equipment and apparatus needs to be subjected to inspection and testing to ensure its continuing safety. The extent to which this needs to be done depends upon the type of equipment, the conditions in which it is used, and the risks presented by the equipment from not maintaining it. For example, an 11 kV oil filled circuit breaker presents considerably higher risks from not maintaining it than, say, a photocopier in an office; a 230 V hand-held grinder used on a construction site presents considerable greater risks from not maintaining it than a personal computer located in a construction site office; and a flameproof Ex 'd' lamp located in a Zone 2 hazardous area presents considerable greater risks from not maintaining it than a luminaire suspended from the ceiling in a shop.

In recent years there has been a tendency for employers to over-test their appliances while ignoring the need to inspect and test fixed equipment and the installation. This topic figured in Professor Ragnar Löfstedt's report *Reclaiming health and safety for all: An independent review of health and safety legislation* dated November 2011, in which he recommended that the HSE issue clarification on the requirements for maintaining equipment and appliances, especially in relation to low risk premises such as offices and shops. This arose because a small number of people complained during a government consultation exercise known as the 'Red Tape Challenge' that the law on portable appliance testing was too burdensome, although the text of their complaints indicated that in many cases the complainants did not understand the law and may have been unduly influenced by companies touting for equipment maintenance business.

Nonetheless, this led the HSE to produce guidance note INDG236 *Maintaining portable electric equipment in low-risk environments* in 2013, in which it advised that in some circumstances appliances did not need to be tested at all – Class II double-insulated appliances in offices being just one example – although they would still require a visual examination to detect any hazardous damage. INDG236 supplements the HSE's core guidance on the topic, HSG107 *Maintaining portable equipment.* This freely available guidance contains a table that advises on the frequency of user checks, regular visual inspections, and combined inspections and tests for Class I and Class II equipment in a range of business environments and is recommended reading for anyone managing an equipment maintenance programme.

Another source of authoritative advice on the form that inspection and testing of equipment should take, and the appropriate inspection and test intervals, is the IET's *Code of practice for in-service inspection and testing of electrical equipment.* The fourth edition of this was published in 2012. This publication quite correctly stresses the need for before-use and regular visual examinations of equipment because most faults, such as scuffed/abraded insulation and cracked plugs, can be found simply through inspecting the equipment. It also reinforces the HSE's INDG236 and HSG107 by advising that in some benign environments, such as offices and hotels, there may well be no requirement to test items such as Class II equipment at all. This would not apply to equipment used in harsher environments, such as construction sites, for example, where regular combined inspections and tests are recommended, as frequently as monthly in the case of 230 V equipment.

Some electrical equipment associated with machinery such as lifting equipment (lifts, cranes and mobile elevating work platforms) and power presses, is subject to statutory inspections and thorough examinations under legislation such as the Lifting Operations and Lifting Equipment Regulations (LOLER) 1998 and the Provision and Use of Work Equipment Regulations 1998.

Special installations

In some cases, the maintenance has to be done at particular times. Continuous process plants are an example. They are run for prolonged periods and then shut down only at predetermined intervals for maintenance and modification/upgrade purposes. Electrical components, vital to the operation of the plant, usually merit inspection on each such occasion and any parts subject to wear deterioration should be replaced prematurely to ensure against failure during operation. Brushes, contactor contacts and the oil of oil-immersed switchgear and controlgear are examples.

Electrical equipment specified for use in potentially explosive atmospheres merits special consideration, particularly with regard to ensuring its continuing explosion protection properties and not creating an explosion risk during the maintenance work.

Construction sites are also a special case. The electrical installation is temporary and is altered from time to time to meet the changing needs of the construction programme. The installation is subject to adverse environmental conditions and rough handling. More frequent inspections are therefore needed to ensure safety. IET Guidance Note No. 3 recommends intervals not exceeding 3 months for the electrical installation, and the IET's code of practice for equipment maintenance recommends the same interval for inspecting and testing 110 V equipment.

Maintenance instructions

For most items, the maintenance engineer should have little difficulty listing the work to be done when the item is inspected. Companies that manufacture and supply electrical equipment and machinery containing electrical equipment should provide maintenance instructions, which should be referred to as necessary.

Amongst the literature available, BS 7671 and its supporting Guidance Note 3 describe the preventive maintenance of low voltage installations. The IET's code of practice and the HSE guidance material describe how to maintain low voltage electrical equipment and apparatus. BS 6423:2014 *Code of practice for maintenance of low-voltage switchgear and controlgear* and BS 6626:2010 *Maintenance of electrical switchgear and controlgear for voltages above 1 kV and up to and including 36 kV. Code of practice* give excellent advice on switchgear and controlgear maintenance. For the more complex apparatus the manufacturers' instructions should be obtained and followed.

Tools, special equipment, instruments and PPE

When compiling the register, the planner should have regard to the equipment needed to service particular items of apparatus, ensure that it is available and note it on the service schedule so that the maintainer will take it when going to inspect the relevant apparatus. This might include switchgear locking kits, safety padlocks and caution notices in circumstances where safe isolation procedures need to be implemented. For live testing, insulated tools should be provided together with insulating screens and mats and, if the risk assessment indicates, any PPE required to protect against injury such as electric shock and/or arc flashover burns. These matters should be addressed in the relevant risk assessments.

There must also be a range of instruments including, as appropriate, voltage detectors for proving dead, insulation testers, multimeters, loop impedance testers, RCD testers, current transformers, wattmeters and perhaps relay testing equipment where primary or secondary injection testing of protection relays on high voltage distribution systems is to be conducted. Expensive instruments which may be only occasionally required and which may require expert handling such as oscilloscopes are probably not worth buying, but can be hired when needed or can be provided by a specialist called in to deal with a problem that cannot be tackled in-house.

Operation of the maintenance scheme

Introducing the preventive maintenance programme

In the case of a new installation and equipment, it should not be too difficult to ensure that inspections and, where necessary, tests of each item are carried out on, or about, the due date because teething-trouble breakdowns and minor alterations to the installation should not last so long as to disrupt the preventive maintenance programme seriously.

It is usually much more difficult to introduce a preventive maintenance programme in premises where it has not previously existed and where the maintenance staff, if there are any, may be fully stretched coping with successive breakdowns which occur because of this neglect. One way out of this dilemma is to augment the staff temporarily so that some of them can concentrate on the inspections and tests. As more and more of the equipment is being properly serviced, the number of breakdowns should diminish until such time as the scheme is fully operational and the extra staff are no longer needed.

Disruptions

Extensions and alterations to the installation, breakdowns and staff shortages or inability to engage a suitable maintenance contractor all tend to disrupt the maintenance programme. If the inspections and tests are late, the breakdown rate may increase and further disrupt the programme and may induce a snowball effect, culminating eventually in the staff's time being solely taken up with breakdown repairs. It is essential to avoid such an eventuality, so when new work or alterations to the installation are required, they should only be carried out by the maintenance staff if it can be done without serious interference with the maintenance programme. If there is any doubt, the work could be placed with an electrical contractor.

There should be a prioritisation system to determine which inspections and tests must be done on time and which may be deferred. There should, however, be a safeguard to ensure the deferments are of limited duration to avoid an item's inspection being so late that it breaks down or deteriorates into a dangerous condition.

Recording and reporting

For the removal of doubt, the Electricity at Work Regulations do not require maintenance records to be kept, but the memorandum of guidance to the regulations (HSE publication HSR25) points out that records can aid demonstration of compliance and allow useful analysis of equipment condition. Moreover, retaining maintenance records, including test results, will allow the condition of the equipment and the effectiveness of maintenance policies to be monitored. Without effective monitoring, planners and managers cannot be certain that the requirement for maintenance has been complied with.

Electricians inspecting and testing low voltage installations should be encouraged to complete an EICR, the form prescribed by BS 7671, because it provides a well-codified and widely used method of recording the results. There is no legal duty to do this, but it does represent best practice. The form should have attached to it the schedules of inspection and test results completed during the inspection and test.

The inspector is required to declare on the EICR whether the installation is in a satisfactory or unsatisfactory condition and to note any observations and any defects found during the inspection and testing. The observations and defects should be prioritised according to the Classification Codes contained in Appendix 6 to BS 7671; there are three principal codes plus a further one, as shown in Table 18.1.

Table 18.1 EICR Classification Codes for the fixed installation

Classification Code	*Meaning*	*Examples*
Code C1	Danger present. There is a risk of injury and immediate remedial action is required.	• Live conductors exposed to touch • Reversed polarity
Code C2	Potentially dangerous. Not dangerous at the time of the inspection but may become dangerous if a fault or other event were to occur. Urgent remedial action is required.	• No earthing or bonding • Inadequately sized cables • Incorrectly rated overcurrent protection • No means of isolation • High earth fault loop impedance values
Code C3	Improvement recommended. There is no immediate or potential danger but improvements could considerably enhance the safety of the installation.	• No RCD protection for circuits in thin walls, or circuits supplying bathroom and shower rooms that have adequate supplementary bonding • Socket-outlet mounted at a position where it is susceptible to damage

(*Continued*)

Table 18.1 (Continued)

Classification Code	*Meaning*	*Examples*
Code F1	Further investigation is required because the inspection has revealed an apparent deficiency which, owing to the extent or limitations of the inspection, could not be fully identified and further investigation may reveal a code C1 or C2 item.	

The codification of observations is a matter for professional judgement on the part of the inspector, but any code C1 observations must be made safe when discovered or the matter urgently reported to the installation's owner or person in charge of it. Moreover, any code C1, C2 or F1 observations on the EICR should automatically result in the installation being marked on the form as 'unsatisfactory' to warn the owner about the need for remedial works and/or further investigation.

In addition to the foregoing, the inspector should complete the section of the EICR that recommends the interval to the next formal inspection and test. Again, this is a matter for professional judgement based on the findings of the inspection and test, together with the history of the installation, the environmental conditions, and the way in which it is used and operated.

A very good source for advice on the codification of installations is Electrical Safety First's Best Practice Guide 4 (issue 4) *Electrical installation condition reporting: Classification Codes for domestic and similar electrical installations*.

The inspection and testing of equipment and appliances that do not form part of the fixed installation should also be recorded, preferably including any insulation resistance and protective conductor continuity test results so that deterioration over time may be detected. Many manufacturers of test equipment offer software packages for recording test data.

Between inspections, some equipment will become defective and may become dangerous, so the users should be encouraged to look for and report such defects as soon as they are aware of them, preferably as part of a pre-use inspection. They should cease to use the equipment until the defect has been repaired. Portable apparatus with flexible cables and plugs are more likely to be damaged than fixed equipment. It is also more likely to be hazardous to the user, who should be reminded periodically of the danger and the need to report incipient defects before they become dangerous.

Training and competence

The success of any planned maintenance scheme is dependent on the skill of the staff and its conscientious application. Apart from the selection of competent people to carry out the work, in-house training is required to familiarise them with the installation and the planned maintenance scheme, and hopefully to inspire them with some enthusiasm for the task because, frankly, inspection and testing can be a tedious and repetitive activity. This might be done by explaining the importance of preventive maintenance, their role in it and its relevance to the safety of the staff and economic success of the business. A training course might include

- the safety rules and procedures in force in the premises, including safe isolation practices (including the use of Permits to Work and other safety documents where appropriate) and safe systems of work for live working;
- a brief history of the enterprise from its formation to the present time, and its current objectives;
- the circuit diagram(s) for the installation from the power source(s) to the connected equipment;
- the reasons for a planned maintenance scheme, including its economic justification;
- a detailed explanation of the scheme in force; and
- the task of the maintenance electrician, including

  (a) liaison with the production staff to arrange for the shutdown of the equipment for inspection and to obtain details of any defects they have observed,
  (b) the use of the checklist,
  (c) the use of the makers' instructions,
  (d) what to enter under 'Remarks' on the report sheet,
  (e) investigating the cause of failures,
  (f) dealing with insoluble problems by seeking help from the supervisor rather than guessing the answer, and
  (g) looking for trouble (i.e. using one's senses of sight, smell and hearing) when moving around in the premises, to detect possible defects. For example, be on the lookout for damage to flexible cords and cables, the odour of burnt insulation and the sound of a noisy motor bearing.

From time to time, the initial training should be supplemented to keep the staff informed about any problems that arise, new apparatus, changes in practice, and so forth. Such meetings should be of the discussion type to encourage full participation by all staff members.

There are qualifications available that provide evidence of competence in the inspection and testing of low voltage installations and the inspection and testing of equipment. For the former, City and Guilds qualification number 2394 is a level 3 qualification covering initial verification and certification of installations, and qualification number 2395 is a level 3 qualification covering the 'Principles, practices and legislation for the periodic inspection, testing and condition reporting of electrical installations'. These qualifications are usually obtained by qualified electricians. In the case of equipment maintenance, City and Guilds qualification number 2377 provides competence in the implementation of the IET's code of practice.

It has already be noted in Chapter 15 that there are specialist courses available under the COMPEX scheme for persons maintaining equipment installed in potentially explosive atmospheres.

Good housekeeping

Apparatus should be kept clean, and the legal requirement for it to be accessible must be observed. Dirt and obstructions can hide defects and hamper the task of the maintenance staff who should not have to move obstructions to obtain access to defective apparatus or spend time cleaning it, time which might be more usefully employed in repairing the defect.

Selection of equipment

The engineer responsible for the maintenance should also be consulted before new equipment is purchased so that advice can be obtained on its maintainability. Good quality apparatus, designed for easy maintenance, may prove to be more economical than cheaper apparatus of lesser quality which is more difficult to maintain. The engineer should also be familiar with the environmental conditions and can advise on any protective features required, e.g. whether the apparatus needs to be dust-tight, weatherproof, or suitable for use in a hot location.

Siting of new equipment

Again, the maintenance engineer should be consulted to ensure that the equipment is accessible and there is adequate space and, where necessary, lifting provisions to facilitate dismantling for maintenance. The siting should consider the proximity of other apparatus, the operation of which when too close may endanger the safety of the maintenance staff.

Specialist help

Some modern equipment is complex, and it may well prove to be too difficult for the ordinary electrician to find a fault and repair it. For example, if a photoelectric guard on a power press fails, the electrician can check the relevant limit switches and rectify a fault in them and can replace burnt-out lamps, but if the fault is in the electronic circuits, it will be necessary to call a specialist engineer, preferably from the maker. Unskilled tinkering with the electronic circuits or bypassing them to get the press back into production could have serious consequences for the press operator whose safety is dependent on the correct functioning of the guard. Where there is sufficient electronic apparatus to justify it, a specialist technician could be employed, but otherwise specialist help should be sought.

Not all electricians are capable of servicing HV equipment. Where it is confined to substation switchgear, for example, and perhaps a few large electric drives, it may be convenient to arrange for its maintenance by a distribution company rather than entrust it to less experienced in-house staff. It is crucial that HV work should only be carried out by competent personnel who have the appropriate skills, knowledge and experience, and who have been authorised to carry out the work.

Servicing and adjusting the relays of protective switchgear is another item that could be contracted out to a distribution company or other specialist contractor with the appropriate equipment and expertise. Major repairs, such as rewinding motors, are best carried out by electrical repair specialists. The periodic calibration of instruments is a matter for the makers or a metrology laboratory.

19

Competence

Introduction

Many accidents are caused by people making mistakes when working on electrical systems, with the mistakes often being directly attributable to weaknesses in the competences they needed to carry out their work activities safely. These accidents are commonly known as failures in the systems of work. In addition to these, system technical failures, be they in power systems or in control systems, are frequently the result of poor design features, again often resulting from lack of competence on the part of the engineers who specified, designed and produced the systems. It is clear, therefore, that personal competences play a particularly important role in the achievement of high standards of electrical and control system safety.

It is important that managers who are responsible for ensuring the competence of engineers and technicians understand the factors that determine whether or not an individual is competent to perform particular tasks. For many people in this position it is tempting simply to equate competence to training and qualifications, working on the basis that individuals only require appropriate technical qualifications and some task-specific training to enable them to carry out specified tasks safely.

Whereas training is undoubtedly an essential component of competence, it is not the only one. Furthermore, not all qualified people are competent, and a lack of qualifications does not automatically mean that a person is not competent. On the other hand, the holding of a particular qualification or trade grade may be one way of demonstrating competence. For example, a tradesman in the electrical contracting industry graded as an Approved Electrician is able legitimately to claim competence in the installation of electrical systems covered by BS 7671 by virtue of having completed a 4 year apprenticeship, gained a relevant Level 3 vocational qualification, passed the appropriate academic and practical tests and gained 2 years further relevant experience.

In the safety arena, for a person to be competent to undertake work safely he or she should have a mix of sufficient and suitable training, appropriate qualifications, experience, knowledge and other personal qualities. The 'other personal qualities' are largely associated with the individual's approach to work, with factors such as powers of concentration, sense of personal responsibility, diligence, integrity and maturity being important attributes. As an example of this, it is highly unlikely that a well qualified and experienced engineer could be competent to work on high voltage systems and make them safe for others to work on if he has a tendency to be a bit slapdash in his approach and to cut corners; such behaviour is a recipe for eventual disaster.

The issue of competence has been a concern to many organisations over the years and this has resulted in the development of competence-based qualifications that, increasingly, incorporate units on health and safety. For example, free-standing health and safety units have been incorporated into vocational qualification standards set by sector skills councils such as SummitSkills, the body that sets many of the standards for electrical vocational qualifications. This has led to the holding of appropriate National and Scottish Vocational Qualifications being one way of demonstrating basic skills and knowledge both in technical topics and in employment-specific health and safety issues.

In this chapter, the topic will be examined in the context of both electrical safety and the safety of electrotechnical control systems employing electrical, electronic and programmable electronic safety-related control systems. In addition, the increasingly common practice of using conformity assessment to demonstrate competence will be considered.

Legal aspects

There is no definition of competence in health and safety legislation, although the Management of Health and Safety at Work Regulations 1999, Regulation 7(5), says that "a person shall be regarded as competent . . . where he has sufficient training and experience or knowledge and other qualifications." There is no shortage of other legislation that makes reference to the need for people to have the necessary safety-related skills and knowledge for their work, and the HSE certainly puts considerable emphasis on competence issues in its guidance material and approved codes of practice.

Employers have a duty to ensure that those working for them have the necessary safety-related competences for the work being done. This is enshrined in the Health and Safety at Work Act, Section 2(c), which requires employers to provide such information, instruction, training and supervision as is necessary to ensure, so far as is reasonably practicable, the health and safety at work of their employees. This duty does not extend just towards their direct employees; employers must also ensure that people not in their employment whose work activity may put their employees at risk are given information and instruction. For example, an employer must make sure that contractors brought onto his premises to carry out electrical work are competent to do the work in a manner that does not put people on the premises at risk of injury. Of course, the contractor's managers have similar duties to their own employees.

These are the general legal requirements relating to the need to ensure that persons are competent. The Electricity at Work Regulations 1989 provide statutory requirements relating specifically to electricity by stipulating that persons must be competent to avoid electrical danger. It was noted in Chapter 6 that both Regulation 4(3) and Regulation 16 address the matter. Regulation 16 says:

> *No person shall be engaged in any work activity where technical knowledge or experience is necessary to prevent danger or, where appropriate, injury, unless he possesses such knowledge or experience, or is under such degree of supervision as may be appropriate having regard to the nature of the work.*

The Memorandum of Guidance to the regulations offers some help in interpreting these regulations by defining the scope of "technical knowledge or experience" as including

- adequate knowledge of electricity;
- adequate experience of electrical work;

- adequate understanding of the system to be worked on and practical experience of that class of system;
- understanding of the hazards which may arise during the work and the precautions which need to be taken; and
- ability to recognise at all times whether it is safe for work to continue.

These are very useful tests that can assist in determining the level of competence of individuals for conducting electrical work. Of course, the law also recognises that where persons do not have the appropriate competences they must either not carry out the work or be under sufficient supervision by persons who are competent so as to ensure that the work is carried out safely.

When considering the legal duties of those who design, manufacture and supply electrical equipment, including equipment that incorporates safety-related control systems, the main legal duty is the Health and Safety at Work Act, Section 6. This requires that articles for use at work, which includes electrical equipment, are designed and constructed so as to be safe and without risks to health when being set, used, cleaned or maintained by a person at work, so far as is reasonably practicable. Although this law does not specifically say so, it is clearly the responsibility of the suppliers to ensure that their design engineers have the appropriate competences to ensure that these legal obligations are satisfied. A similar comment applies to other 'supply side' legislation such as the Supply of Machinery (Safety) Regulations, the Electrical Equipment (Safety) Regulations and the Electromagnetic Compatibility Regulations.

Electrical competence

Electrical competence is no different than other competences in that it is gained from a mixture of qualifications (equating to knowledge), training, experience and personal qualities. In this regard, Regulation 16 of the Electricity at Work Regulations is at odds with common sense when it stipulates that technical knowledge *or* experience is required. Most engineers would consider that a mixture of technical knowledge *and* experience is necessary for a person to be able safely to carry out electrical work unsupervised.

Whether or not a person is competent to carry out electrical work safely will depend on the individual having the appropriate mix of formal qualifications, training and experience in relation to the complexities of the systems being worked on, the nature of the work, and the degree of risk. For example, an electrician who is skilled at installing cable systems but who does not energise or test them may not have the necessary competences for carrying out the higher risk activities of live testing and fault finding on the systems once they have been energised, despite the fact that he may hold formal qualifications as an electrician. He or she will almost certainly need to undergo specialist training to gain the competences needed to ensure safety during the live working activity.

In similar vein, the task of isolating and proving dead a low voltage circuit to allow work to be done on it requires different skills and knowledge when compared with the higher risk task of isolating and earthing a high voltage circuit, forming part of a high voltage distribution network.

As another example, a newly graduated electrical engineer with a first class honours masters degree may have considerable academic knowledge of electrical engineering science but no, or very little, practical knowledge and experience in implementing safe working practices and procedures on electrical systems.

Given these different levels of risk, knowledge and experience, the employer has the responsibility to ensure that people exposed to electrical risks have the correct mix of technical and behavioural competences and experience to manage the risk to themselves, and to others affected by their work, down to an acceptable level.

A common and well-proven technique adopted by many companies employing personnel to carry out electrical work, particularly where there are high voltage distribution systems on their premises, is to draft electrical safety rules which, among other things, define a range of electrical competences and authorisations. This may include definitions covering, for example, Competent Person (Electrical), Authorised Person (Low Voltage), Authorised Person (High Voltage) and Senior Authorised Person. People who are required to carry out electrical work are then allocated to one or more of these groupings according to an assessment of their individual competences and task needs.

A Competent Person (Electrical) would be a person who the company has assessed to have the required competences for carrying out work safely on its electrical systems. These competences will usually stem in the first instance from the individual having a formal qualification such as a National Certificate (at ordinary or higher level) or National/Scottish Vocational Qualification in electrical installation work, a diploma or a degree in electrical engineering, a City and Guilds qualification in electrical maintenance and so on.

For example, an electrician who intends to work as a qualified supervisor in an electrical contracting company would typically hold a Level 3 N/SVQ vocational qualification in installing electrotechnical systems and equipment plus a Level 3 award in the periodic inspection, testing and certification of electrical installations, or equivalent qualifications, and be able to demonstrate 2 years of relevant experience. This type of qualification is often obtained after serving a modern apprenticeship with an electrical contracting company or other electrical engineering business.

In order to become competent, the person will need to be trained on the particular characteristics of, and risks associated with, the electrical systems on which he will be working, as well as on the systems of work implemented by the company. He will need to become familiar with the formal risk assessments that should have been carried out, and which should detail the measures to be taken to control the risks. He will also, in many cases, be required to work under the supervision of a more experienced person for a period of time, during which his competence will be assessed and confirmed before being classed as a Competent Person (Electrical).

An Authorised Person (Low Voltage) would be a Competent Person (Electrical) who has the required competences for carrying out work on the low voltage systems in the premises. The person will be authorised in writing, with a clear definition of the nature of the work he or she is authorised to carry out. This would include activities such as isolating low voltage circuits and equipment and, after a risk assessment has been completed, carrying out fault finding and testing work on systems that may be live.

On high voltage systems, the higher level of risk justifies a more structured and formal safety management system covering the competence of the persons involved and the procedures to be adopted, which should usually be set out in High Voltage Safety Rules. Competent Persons (Electrical) who are authorised to work on high voltage systems must receive special-to-type training in high voltage operational safety and on the high voltage equipment itself. The company should allocate responsibility to an electrical engineer, either from their staff or contracted in to do the job, to assess the competence of engineers aspiring to be Authorised Persons (High Voltage). The assessment will check their general skills and knowledge, as well as their understanding of the electrical systems on which they will be working. The extent of their authorisations should be defined in writing, paying attention to the extent to which they are authorised to make systems safe and to issue electrical safety documents such as Permits to Work and Sanctions for Test; the issuing of such safety documents is commonly restricted to Senior Authorised Persons.

A Senior Authorised Person is normally the person, or one of a small number of persons, who is in charge of the electrical system and responsible for ensuring the safety of the system and of those responsible for working on it. A Senior Authorised Person will normally be a highly trained and experienced individual with a particularly good knowledge and understanding of the electrical safety procedures at the plant.

It goes without saying that Authorised Persons, especially Senior Authorised Persons, are in positions of considerable responsibility. They make systems safe for others to work on them and they therefore need to be conscientious and thorough, and these traits are essential elements of their overall competence. Any mistakes can have serious consequences, especially where high voltage work is concerned. There has been a small number of instances in which Authorised Persons who issued inappropriate Electrical Permits to Work or failed to manage the electrical safety of the systems under their control have been prosecuted following accidents involving people working under the terms of the permits.

Employers should always keep in mind that competences can be lost. A common example of this occurs in high voltage work. An engineer may attend high voltage operational training and, and after receiving local training, become authorised to operate high voltage switchgear. However, he or she may only have to carry out, on average, one high voltage switching operation per year, or sometimes even less frequently. In these circumstances not much operational experience is gained in the work and, over time, the procedures may be forgotten and competence lost. One way of guarding against this is for the person to attend refresher training, typically every 5 years.

The problem of checking the competence of electrical contractors has already been alluded to, but techniques are available to ensure that enough is done to confirm that a selected contractor is competent. One option is to ensure that the companies being considered at the tendering stage are members of a recognised trade association such as the Electrical Contractors' Association (in England and Wales) or SELECT (in Scotland), or are members of the National Inspection Council for Electrical Installation Contracting (NICEIC). The NICEIC is a non-profit making body and a registered charity that maintains a roll of Approved Contractors that meet its enrolment rules and national technical safety standards, including BS 7671. Inspecting engineers employed by these bodies make annual visits to member companies to assess their technical capability and to inspect samples of their work, so membership gives some confidence in the capabilities of the companies and the individuals they employ.

There is a government-sponsored Registered Competent Person Electrical register that went live on 30th June 2014 and was officially launched in Parliament on 2nd July 2014. It is linked to a competence management scheme operated in the electrical contracting industry in England and Wales known as the Competent Person Scheme, which allows individuals and enterprises to self-certify that their work on domestic electrical installations complies with the Building Regulations; see later in this chapter under 'Conformity Assessment'.

In addition to membership of these bodies, checks could be made of the companies' electrical safety rules, where they exist, to confirm that they are relevant to the planned work. The companies' risk assessments and method statements should also be checked, to ensure that a sound risk-based approach to work planning and control is being adopted.

Evidence should be sought of the competence of the individual tradesmen who would be working on the premises, possibly in the case of installation and maintenance electricians by checking their Electrotechnical Certification Scheme (ECS) grading cards issued by the Joint Industry Board in England/Wales, or the Scottish Joint Industry Board. These boards represent employers and trade unions and have the remit of setting the standards for employment, welfare, grading and apprentice training in the electrical contracting industry.

The grading cards issued to electricians show their individual qualifications and training backgrounds. There are different coloured cards, with the following designations:

- Gold cards are issued to fully qualified, skilled craftsmen, i.e. electricians, approved electricians and technicians.
- Brown cards are issued to labourers.
- Red cards are issued to trainees, i.e. apprentices and adult trainees.
- Black cards are issued to contract managers.
- Platinum cards are issued to site managers.
- Yellow cards are issued to office staff and visitors.
- White cards are issued to ECS related disciplines.

Thus, electricians holding a gold ECS card will have the appropriate level of vocational qualifications and experience for securing compliance of low voltage installations with BS 7671 and will have been assessed once every 3 years on their knowledge and understanding of health and safety matters. An advantage of holding an ECS card is that the scheme is affiliated with the Construction Skills Certification Scheme (CSCS), so the card acts as a passport allowing the holder access to construction sites on which the CSCS is implemented.

The scope of the ECS is broader than just installation and maintenance electricians; it also encompasses trades in data communications, fire detection and alarm systems, and emergency and security systems.

In addition to checking the components listed previously, evidence can be sought of prior work done satisfactorily by the companies in the same field as the planned work, and details of any reportable accidents and enforcement action taken by regulatory bodies. These types of checks, if conducted conscientiously and thoroughly, would be a reasonable approach to the task of selecting competent electrical contractors.

Electrical competence schemes are not unique to electricians working in the electrical contracting industry. For example, there is a sector skills council known as Energy and Utility Skills (EU Skills) which sets the competence standards for persons working in the utility sector, such as employees of DNOs, generators and meter operators. It operates a skills register known as the EU Skills Register and issues skills cards to individuals registered with the scheme. EU Skills also works closely with the National Skills Academy for Power to ensure that universities and other training centres deliver education and training that is relevant to the industry.

Safety-related control systems

The topic of safety-related electrotechnical control systems is covered in Chapter 17, where it is noted that such systems can comprise a mixture of electromechanical, electronic and programmable electronic equipment and can have designs that range from being quite simple to being extremely complex. In many cases, the systems can have safety functions or protection functions in which the consequences of failure can be severe. Safety-related control system failures leading to an explosion in a petrochemical site, or a collision between aircraft being controlled by an air traffic control system, or the loss of the protection systems in a nuclear power station are examples in which the potential loss of life and environmental damage would generally be regarded as unacceptable.

This places much emphasis on the engineering of sufficient safety integrity into the safety-related systems to ensure that risks affected by those systems are reduced to a level that is as low as reasonably practicable (ALARP). There are many methods and techniques for achieving this. As noted in Chapter 17, BS EN 61508 *Functional safety of electrical/electronic/programmable electronic safety-related systems*, for example, advocates a risk-based approach to determining the required safety performance of safety-related systems, and does so in the context of an overall safety life cycle. The success in the implementation of such an approach demands a suitable level of competence in the engineers involved in the determination of the safety requirements and in the realisation of such systems.

There has not been much guidance available on the competences required for working in this area. This was a deficiency that needed to be remedied as the progressive introduction of programmable elements into systems that performed safety functions, such as PLCs used in machinery interlocking circuits, resulted in the systems becoming more complex and increasingly difficult to assess for safety.

One of the first publications covering the matter was a joint document produced in 1999 by the Institution of Electrical Engineers (IEE, now the IET) and the British Computer Society (BCS), *Competency guidelines for safety-related system practitioners*. The document identified the following twelve functional areas in which competence is required to support the specification, development, procurement, operation and maintenance of safety-related control systems:

- Corporate functional safety management
- Project safety assurance management
- Safety-related system maintenance
- Safety-related system procurement
- Independent safety assessment
- Safety hazard and risk analysis
- Safety requirements specification
- Safety validation
- Safety-related system architecture design
- Safety-related system hardware realisation
- Safety-related software realisation
- Human factors safety engineering

For each of these functions, the document provided an outline of the key responsibilities and described the functional, as well as task-related, competences. It did this for three levels of competence; in increasing levels of competence these are the supervised practitioner, the practitioner and the expert. The document also provided guidance on how a competence scheme could be operated in these functional areas, how competence could be assessed and how the outcome of the competence assessment could be recorded.

The document did not gain much traction because of its complexity and the difficulty faced by people reading it and trying to assimilate its content. However, in July 2016 the IET published a new guidance document titled *Code of Practice for Achieving Competence for Safety-Related Systems Practitioners.* According to the IET's advance publicity, its purpose is to help organisations where safety critical systems are in use, including energy, transport, defence, aerospace and healthcare:

- to create or develop a scheme for assessing the competence of people and teams undertaking safety critical functions;
- to demonstrate to clients that the organisation has the necessary competence to undertake particular activities and that a recognised competence measurement scheme has been used;
- to provide clear levels of expertise and competence required prior to recruiting engineers in safety critical roles, and subsequently for appraising/training those personnel;
- to help in implementing an overall competence management system (CMS) for an engineering division or organisation;
- to comply with regulatory requirements/relevant standards, showing duty of care and compliance to regulations and EU directives (specifically IEC 61508 and HSE requirements); and
- to provide evidence of best practice and high levels of competence to any industry regulator and to avoid potential litigation.

The HSE had a go at producing guidance on the topic in its 2007 publication *Managing competence for safety-related systems.* This proposed establishing a competence management system in four phases, each of which contains one or more management principles, as set out in Table 19.1.

Table 19.1 Structure of HSE guidance on managing competence

Phase	*Management principle*
One – Plan	1. Define purpose and scope according to risk
Two – Design	2. Establish competence criteria
	3. Decide processes and methods
Three – Operate	4. Select and recruit staff
	5. Assess competence
	6. Develop competence
	7. Assign responsibilities
	8. Monitor competence
	9. Deal with failure to perform competently
	10. Manage assessors' and managers' competence
	11. Manage supplier competence
	12. Manage information
	13. Manage change
Four – Audit and review	14. Audit
	15. Review

Despite the complexity of this topic, it is advisable to persevere and to put in place the recommended procedures and practices, or procedures and practices that will achieve the same goals in the management of personal competences. The benefits far outweigh the problems.

Conformity assessment

One consequence of the fact that competence has a direct impact on health and safety has been the development of conformity assessment schemes covering competence. Conformity assessment, in its generality, is an activity concerned with determining directly or indirectly that relevant requirements have been met and is most commonly used to determine if a product, system, process or a person's competence meets a defined specification.

In the context of this chapter, conformity assessment relates to the process of determining whether people working for businesses in the electrical or safety-related control systems sectors can perform to the required level of competence and, if they can, awarding certificates of competence in one form or another to the businesses.

Organisations that carry out this type of assessment work are known as certifying bodies, and there are mechanisms in place whereby these bodies can be accredited as meeting appropriate standards. The sole government-approved accreditation body in the UK is the United Kingdom Accreditation Service (UKAS), which is a non-profit-distributing private company limited by guarantee.

An example of an accredited certifying body in the electrical sector is the NICEIC, which has been assessed by UKAS as meeting the requirements of EN 45011 (replaced by BS EN ISO/IEC 17065:2012), the European standard for the certification of products, processes and services. This means that the NICEIC, as well as other accredited certifying bodies such as the National Association of Professional Inspectors and Testers (NAPIT), is able to issue accredited enrolment certificates to its approved contractors. This includes the assessment of the competence of contractors under a scheme known as the Electrotechnical Assessment Specification which enables contractors to register as 'Assessed Enterprises'.

Such enterprises are recognised as competent to self-certify their work as complying with Part P of the Building Regulations in England and Wales under the terms of the Competent Person Scheme (see earlier in this chapter and also Chapter 9). This defines two types of operatives who can self-certify their work on electrical installations in dwellings: full scope operatives competent to design, install, inspect and test low voltage electrical installations, and defined scope operatives such as plumbers who carry out limited electrical work on systems such as boilers and electric showers. These operatives need to demonstrate that they meet the relevant minimum technical competence requirements which have been agreed and approved by all the bodies involved in the scheme. Allied to this are defined roles such as Principal Duty Holders, Qualified Supervisors, and Responsible Persons.

The Electrotechnical Assessment Specification and Part P Competent Person Scheme are described in greater detail in an IET document dated July 2015, titled *Electrotechnical Assessment Specification for use by Certification and Registration Bodies.*

To add to the complexity of these national competence schemes, Scotland has a different scheme for certifying compliance with the Building (Scotland) Regulations 2004, known as the Certification of Construction (Electrical Installations to BS 7671) Scheme (see Chapter 9).

Turning to the field of safety-related control systems, a significant initiative in the UK has been the scheme known as the Conformity Assessment of Safety-related Systems, or CASS. The scheme is concerned with conformity assessment for systems based on BS EN 61508. It has a broad scope, and covers all those involved in the design, development, manufacture, implementation, support and application of complete safety-related systems and their components.

The CASS methodology is owned and managed by the CASS Scheme Ltd, a non-trading company limited by guarantee with a membership of cross-industry institutions. It is mainly concerned with allowing companies who carry out these functions to have their products and processes certified as conforming with the requirements of BS EN 61508. The competence of those who are designing, developing, supporting and applying safety-related systems is a significant factor in the assessment process.

As a final observation, it is clear that these conformity assessment schemes have their roots in trade issues rather than in health and safety, but there are undeniable benefits in the health and safety field. For example, certification and registration schemes in the building industries, including the electrical sector, are favoured as one means of countering the 'cowboys' who charge exorbitant fees for minor works which are frequently of dubious quality and/or are unsafe. Registering competent practitioners allows clients to select contractors with demonstrably adequate competences and business practices, but it also provides some assurance about their health and safety practices and about the safety of any electrical work they do.

Electrical safety in the rented property sector

Introduction

One of the characteristic features of Great Britain's property market over recent years has been the growth in the number of families and individuals occupying rented properties. Taking England as an example, the Department of Communities and Local Government's (DCLG) *English Housing Survey Headline Report 2013–14* reported that 19% (4.4 million) of households were renting privately, up from 18% in 2012–13 and 11% in 2003. The proportion of households renting social housing remained steady at 17% (3.9 million), where the term 'social housing' includes local authority and housing association homes providing accommodation at a subsidised rent, in contrast to the private rented sector where accommodation is offered at market rents. The remainder of households were in owner occupation.

The same DCLG report highlighted that, in 2013, 4.8 million dwellings (21%) in England failed to meet the Decent Homes Standard, meaning that they were unsafe, were not in a reasonable state of repair, did not have reasonably modern facilities and services or did not have efficient heating and effective insulation – or a combination thereof. While this represented a reduction of 2.9 million homes since 2006, when around a third (35%) of homes failed to meet the standard, it is still a significant number. Many rented houses failed to meet the Decent Homes Standard because they contained serious safety hazards measured through a risk assessment process known as the Housing Health and Safety Rating System (HHSRS). Although there are a number of different hazards assessed through this system, as described later, one of them is a defective or sub-standard electrical installation.

The situation in Scotland is not too different to that in England, although the country has its own legislation and standards. The Scottish Government's *Scottish Household Survey* for 2015 reported that, of the 2.43 million homes in the country, 61% were owner occupied and 14% were leased from a private landlord, family member, friend or employer. Twenty-three percent of households were living in social housing, comprising local authority (13%) and housing association (10%) properties. The number of people occupying private rented accommodation has risen by 9% in the last 16 years, which is the same percentage as the fall in the percentage of households occupying social housing. So the trend has been for the percentage of households in owner occupation to stay steady and for a progressive transfer of households from social housing to the private rental sector.

In Wales, according to the Welsh Local Government Association report *Private Rented Sector Improvement Project: End of Project Report* dated February 2014, social housing provided by councils and housing associations comprise 16% of all Welsh homes, and privately rented homes accounted for 13% of the tenures.

This chapter explains the legal duties landlords are under to ensure the electrical installations in their properties are safe. It also describes the standards that apply and discusses an inspection and test procedure for assessing electrical installations in rented properties.

Legislation, standards and guidance

Legislation dealing with landlord and tenant relationships is quite complicated and is different in each of the three countries in Great Britain. Until recently it did not contain any specific electrical safety requirements, but there have been important developments in this field, as explained in the sections that follow. There follows a brief overview of the most important and relevant legislation and how it relates to electrical safety.

England

Section 11 of the Landlord and Tenant Act 1985 states that, for short leases (i.e. a lease for fewer than 7 years), the landlord has an implied responsibility to keep in repair and proper working order, inter alia, the electrical installation and any electrical appliances and fittings supplied as part of the lease. There is no specific requirement for how this is to be done, particularly no mention of periodic inspection and testing. However, the draft Housing and Planning Bill currently going through Parliament, first published in October 2015, contains an amendment that requires landlords in the private rented sector to carry out electrical safety checks every 5 years, presumed to mean inspections and tests, and it is to be hoped that this will be spelled out in the final version of the bill and any supporting guidance material. This will be a significant improvement to the legislation from an electrical safety perspective which stems from concerted lobbying by Electrical Safety First and partner organisations.

Once the bill is enacted it will bring the legislation in the private rented sector into line with the Management of Houses in Multiple Occupation (HMO) (England) Regulations 2006, which require in Regulation 6(3) that the HMO manager must ensure that every fixed electrical installation is inspected and tested at intervals not exceeding 5 years by a person qualified to undertake such inspection and testing, and obtain a certificate specifying the results of the test from the person conducting that test. A house is said to be in multiple occupation, and therefore classified as an HMO, if at least three tenants live in it, forming more than one household, and toilet, bathroom or kitchen facilities are shared with other tenants. A landlord of an HMO that is at least three storeys high, has five or more unrelated people living in it, and has two or more separate households living in it must be licensed by the local authority (so-called mandatory licensing), with the licence being renewed at a maximum interval of 5 years. A licence condition is that electrical appliances supplied by the landlord are kept in a safe condition and that declarations of their safety must be supplied to the local council on demand.

The Housing Act 2004 introduced the current system for assessing housing conditions and enforcing housing standards. In particular it introduced the HHSRS risk assessment approach to assessing hazards and risks in dwellings. The HHSRS is implemented by local authority inspectors who carry out assessments of dwellings using a numerical system for deriving a hazard score for the overall risk from a hazard, with identified hazards being classed as either Category 1 (the highest risk) or Category 2. Assessments are made in response to applications for HMO licences, and a privately rented property may be assessed by a council inspector if a tenant has complained about the state of a property or if an inspector has cause to believe the property may be hazardous.

Of the twenty-nine health and safety hazards covered in an HHSRS assessment, one of them is electricity with regard to the electrical installation and apparatus and the possibility of electric shock and burn injuries and fire. An electrical installation considered to be in a dangerous condition, for example with exposed live conductors or evidence of overheating, would be classed at the higher risk end of the assessment spectrum, leading to the local authority potentially using its extensive enforcement powers against the landlord to ensure the property is returned to an acceptable and safe state of repair; these powers include, in extremis, issuing a demolition order, but there are less draconian forms of enforcement notice available.

If a rented property in the local authority or social housing sectors contains one or more serious categories of harm (HHSRS Category 1), it will not meet the Decent Homes Standard, which will lead to action potentially being taken by local authority enforcement officers. This standard does not cover just health and safety risks but also considers matters such as quality of fittings and thermal comfort.

In addition to the foregoing legislation, the Electrical Equipment (Safety) Regulations 1994 requires landlords to ensure that any electrical equipment that may form part of the lease must be safe, as described in Chapter 8. Furthermore, the Electricity at Work Regulations 1989 will apply to the work activities of anyone engaged by the landlord to work on a property's electrical system.

Scotland

In Scotland, private landlords must register with their local authority under the terms of the Antisocial Behaviour etc (Scotland) Act 2004. This is an attempt to remove the scourge of cowboy landlords who provide accommodation that is dangerous and/or unfit for human habitation, do not maintain their properties, and/or charge exorbitant rents.

The Housing (Scotland) Act 2006 covers the legal and contractual obligations of private landlords to ensure that a property meets a minimum physical standard known as the 'repairing standard'. One of the requirements for a rented property to meet the repairing standard is that the electrical installation must be in 'a reasonable state of repair and in proper working order'. This duty was expanded with effect from 1 December 2015 when Sections 13 and 19 of the act were amended to require landlords to carry out an electrical safety inspection before a tenancy starts, and during the tenancy at intervals of no more than 5 years from the date of the previous inspection. This duty is effectively the same as that being introduced in England through the proposed Housing and Planning Bill, so it is instructive to review the published guidance.

This new legal requirement on landlords is explained in Scottish Government statutory guidance available at https://www.prhpscotland.gov.uk/repairs-downloads-landlords. This document stipulates that

- The electrical installation is to be inspected and tested by a competent person and the results recorded on an Electrical Installation Condition Report (EICR) of the type prescribed by BS 7671 and described in Chapter 18 of this book.

- All moveable electrical appliances supplied by the landlord must be inspected and tested; the document uses the tautological phrase 'PAT tested' and does not differentiate between Class I appliances that will need to be tested and Class II appliances that may need simply to inspected without any form of testing. The results must be recorded in a 'PAT Report', a label must be attached to each appliance that has been tested, and remedial action must be taken for any appliances that fail.
- An electrical safety inspection required by the legislation will result in both an EICR and a 'PAT Report', both of which must be given to the tenant and retained by the landlord for a period of at least 6 years.
- It is good practice for landlords to carry out more frequent annual visual inspections of the installation and appliances to look for any signs of deterioration or damage that may lead to danger.

One of the interesting features of this statutory guidance is that it effectively requires electrical installations to be inspected and tested against the standards contained in the current version of BS 7671, which is a best practice standard that contains requirements above and beyond the statutory requirement of 'a reasonable state of repair and in proper working order' set out in the Housing (Scotland) Act. The statutory guidance advises that 'reasonable state of repair' is not defined in the legislation, but it broadly means that the condition of the equipment is what a reasonable person would expect taking the circumstances into account; equipment that is not safe for use would not be in a reasonable state of repair. It can confidently be predicted that this will create tensions between, on the one hand, electricians who when completing their EICR advise substantial and potentially expensive modifications to secure compliance with the current version of BS 7671 and, on the other hand, landlords who argue that their installations meet the less onerous statutory standard. In the final analysis, however, compliance with the current version of BS 7671 will achieve a safer electrical installation.

Tenants who are unable to get their landlords to take action to meet the repairing standard may bring any complaints about a property not meeting the repairing standard to the Private Rented Housing Panel (PRHP), who may force a landlord to take action where necessary.

Another important concept in Scottish law with an electrical safety reference is that of the 'tolerable standard' which relates to the quality of the country's housing stock, not just dwellings in the rented sector. It was first enacted in The Housing (Scotland) Act 1969, replacing the previous concept of houses classified as being "unfit for human habitation". Properties that are below the tolerable standard are not fit to be lived in. Originally, the standard made no reference to the electrical installation, but this changed when the Housing (Scotland) Act 2006 included the electrical installation in the following terms:

> *A house meets the tolerable standard if it, in the case of a house having a supply of electricity, complies with the relevant requirements in relation to the electrical installations for the purposes of that supply;*
>
> - *"the electrical installation" is the electrical wiring and associated components and fittings, but excludes equipment and appliances;*
> - *"the relevant requirements" are that the electrical installation is adequate and safe to use*

As in the case of the repairing standard, the Scottish Government has issued statutory guidance to explain the tolerable standard; it is titled *Implementing the Housing (Scotland) Act 2006, Parts 1 and 2 Advisory and Statutory Guidance for Local Authorities Volume 4 Tolerable Standard* and dated March 2009. It offers advice on the types of features of an electrical installation that might cause it to be below the tolerable standard, examples being the absence of a consumer unit, the use of VIR and lead-sheathed rubber-insulated cables, light switches etc within reach of a person using a bath or shower (not ceiling mounted cord-pull type), 230 V socket-outlets in bathrooms (excluding shaver sockets) within 3m of the bath or shower, and charring or scorches around socket-outlets, or sparks from light switches.

In similar vein to the statutory guidance on the repairing standard, it advises that, where there is doubt, the installation should be inspected and tested by a competent person and an EICR completed and given to the local authority inspector for consideration. A house with a Code 1 deficiency would be considered to be below the tolerable standard, and the inspector would be expected to take appropriate action with the owner of the property. Again, the challenge here is to determine which edition of BS 7671 to inspect and test the installation against. Given that the requirement is for the installation to be 'safe', an effectively maintained installation constructed to the 16th edition of the IET Wiring Regulations will be safe, but not as safe as one constructed to BS 7671:2008+A3. The resolution of this requires professional judgement on the part of both the local authority inspector and the competent person producing the EICR.

Turning to the topic of HMOs, mandatory licensing introduced in Scotland in 2000 requires landlords (private and social housing) to obtain a licence from the local authority for all properties containing three or more people from three or more families. The majority of licences are issued for properties occupied by students, including student halls of residence, and they must be renewed at a frequency no lower than once every 3 years. In order for a licence to be granted the local authority must be satisfied that the property meets the required physical standards, one of which concerns the electrical installation and apparatus. In essence, the requirement is that the installation and landlord-supplied apparatus are in a safe condition and that they are inspected and tested at least once every 3 years, or more frequently if the tenancy changes. There are also specific requirements for the minimum number of socket-outlets in rooms in the property.

As in the case of England and Wales, the Electrical Equipment (Safety) Regulations 1994 and the Electricity at Work Regulations 1989 apply in Scotland.

Wales

The legal duties of landlords in Wales have tended to mirror those in England, with legislation such as the Landlord and Tenant Act 1985 applying to both countries. However, the Welsh Government is empowered to enact its own housing legislation and, accordingly, has produced the Housing (Wales) Act 2014 as part of its move to have a distinctive legislative framework. Under the act, private landlords have to register with their local authority, and anyone managing a rented property has to be licensed; this process is in the course of being introduced during 2016.

A new act, the Renting Homes (Wales) Act 2016, was introduced in January 2016. This legislation requires landlords to ensure that their properties are fit for human habitation based on an assessment using the same HHSRS as used in England and that, inter alia, electrical installations are kept in repair and proper working order. There is no specific requirement for the landlord periodically to insect and test the installation.

There is a Welsh Housing Quality Standard that stipulates that 'electrical lighting and power installations must be checked and certified safe by an appropriately qualified person at least every 10 years as a minimum'; this applies to landlords of social housing.

Assessing an electrical installation

Introduction

There is often a demand for reports on the condition of tenants' electrical installations in rented accommodation. This arises because some landlords fail to maintain the accommodation in a safe and serviceable condition and ignore their tenants' complaints. In circumstances such as this, a tenant may approach a local authority, the PRHP in Scotland, or a solicitor for assistance in resolving the matter, and this will frequently result in a building surveyor or an environmental health consultant being asked to inspect the accommodation and produce a report. If the tenant indicates that he or she has had electrical problems or suspects that the electrical installation is faulty, an electrical engineer will often be asked to inspect and test the installation and submit a report.

The following sections describe a procedure that can be adopted by electrical engineers who are asked to carry out this type of work.

Procedure

Papers

The engineer charged with investigating the complaint should obtain as much detail of it as possible. A copy of any statements made by the tenant and other potential witnesses and a copy of any reports concerning the complaint, especially any from a building surveyor, are good starting points. The latter can be useful, as, apart from any electrical comments, it will probably reveal wetness problems from leaking roofs or pipes, condensation, faulty damp courses, flooding, and so forth, which may have an adverse effect on the electrical installation.

Visit

A visit should be made during daylight hours, as the insulation and continuity testing involves switching off at the consumer unit, so there will be no lighting. It is advisable for the tenant to be there to provide first-hand information on the length of tenure and the electrical complaints.

If the complaint leads to litigation, it is improbable that the judge and counsel will visit the premises, but they are most likely to rely on the report when arguing the case. Therefore, an outline plan showing the layout and location of the electrical points, supported by photographs of any defects, is helpful and may enable them to understand it better, particularly as it is probable that they will have little electrical knowledge. Figure 20.1 is an example of a typical installation layout sketch.

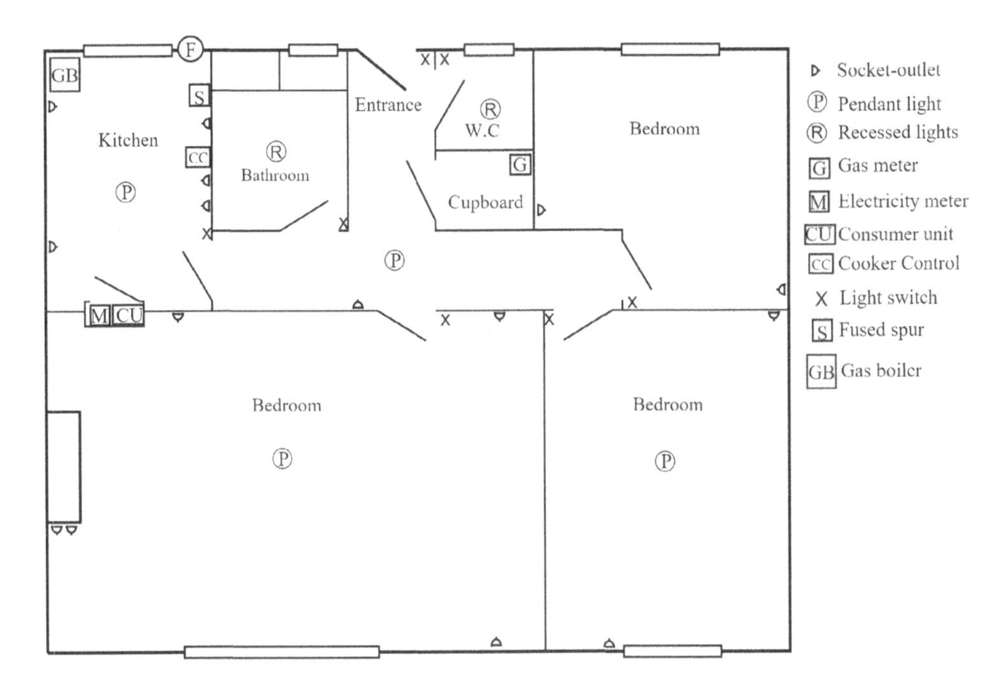

Figure 20.1 Sketch of a typical small domestic electrical installation

When going around the premises with the occupier to prepare the layout sketch, any defective items and infringements of the standards should be noted. Socket-outlets' polarity should be checked with a neon indicator (socket tester), which will also show whether the socket-outlet has an earth connection. Also, be sure to check the main equipotential bonding of the water service, gas installation, and other service metal pipes and also of the central heating, air conditioning systems, exposed metallic structural parts of the building, and the lightning protection installation, if any.

The intake

A normal modern intake into a house comprises a CNE concentric service cable terminating in a plastic sealing box. The DNO provides an earthing terminal connected to the cable sheath, which is the CNE conductor, and a cut-out consisting usually of a 60 A or 100 A BS 1361 fuse. The meter tails are generally PVC-insulated and sheathed conductors. The Electricity Safety Quality and Continuity Regulations (ESQCR) require them to be marked for polarity. Quite often the meter tail wiring is untidy and not secured, so as to relieve its terminations of strain.

The earthing conductor should have a minimum cross-sectional area of 10 mm^2 if the service cable does not exceed 35 mm^2. This size also applies to the main equipotential bonding conductors.

Older installations may have an armoured paper insulated and lead-sheathed service cable providing an SNE supply. The consumer's earthing terminal should be connected to the lead sheath of this service cable. Look out for earthing to the incoming water service pipe, which is no longer permitted. Check that the size of the earthing conductor complies with BS 7671 Chapter 54 or, if the consumer has his or her own buried earth electrode, it is in accordance with standard's requirements.

In rural areas there are still some TT supply systems where consumers have to supply their own earth, in which case it is usually necessary to protect the installation by means of a current operated earth leakage circuit breaker (RCD) because of the difficulty of providing an earth of sufficiently low resistance to ensure that the protection operates within the prescribed time in the event of an earth fault. Note that voltage operated earth leakage circuit breakers are no longer acceptable.

Check the position of the meter. It should be installed in the consumer's premises unless it is more reasonable for it to be located elsewhere, in which case it should be accessible to the consumer. This is more likely to occur in blocks of flats and HMOs. In such premises the DNO may bring in only one service cable, and the Building Network Operator then feeds each tenant via risers, laterals and cables in conduit or sometimes in MIMS cable. The latter is not a good choice because the insulation is hygroscopic, and if the end seals fail, the cable can break down. It may also fail if subject to HV transients, so such cables should be fitted with suppressors. If the conduit or sheath of the MIMS cable is used as the earthing conductor, check that the connections are satisfactory and of low resistance and any joints in the conduit provide adequate continuity.

As service cables tend to have highly rated overload and short circuit protection, a fault may cause arcing until it burns itself out, so there is a potential fire risk although such breakdowns are uncommon. Nevertheless, where such cables are inside the premises, it is worthwhile checking that they are so located or shielded that the arcing would not ignite anything else and start a fire. Where the supply is from an overhead line, there is a small potential fire risk from high voltage spikes caused by a lightning strike or near by strike on the line conductors. Again, DNOs do not always protect their service cables against lightning, so suppressors could be fitted where connected equipment is not protected against such transients and/or where there is a history of frequent thunderstorm activity.

The consumer's consumer unit

It is rare nowadays to find installations without a consumer unit, which is a small distribution board with a double pole isolating switch and a single phase busbar to feed the fuses or circuit breakers controlling the final circuits. There will be terminal blocks or bars for the neutrals and protective conductors. The earth terminal block or bar may be utilised as the consumer's earthing terminal. Some very old models may have wood frames and/or may be backless, although these are now rare. As BS 7671 requires connections to be made in non-flammable enclosures, the wooden framed type do not comply and the backless ones are acceptable only if mounted on non-flammable material, but it is good practice for any units like this to be changed for modern units. Note that amendment 3 to BS 7671:2008 requires that, with effect from January 2016, new consumer units are metal boxes or, if not, are mounted inside non-combustible enclosures.

If rewirable fuses to BS 3036 are used, check that the correct size of fuse wire has been used. For cartridge fuses, check that blown fuses have not been 'repaired', with a bit of fuse wire spanning the contacts or the cartridge replaced by a nail, hairpin or the like. There should be only one final circuit connected to each fuse carrier or circuit breaker.

The circuits should be identified on a diagram or schedule, and there will usually be separate circuits for lighting, socket-outlets, cooker, immersion heater and central heating; there may be a separate circuit for smoke and heat alarms, although it is allowable to supply these from a regularly used lighting circuit. Although not mandatory, it will usually be found desirable and convenient to have a separate socket-outlet ring main for the kitchen to cater for the substantial load of the appliances therein, a number of which may operate at the same time, and it is also convenient to have a separate socket-outlet ring main for the socket-outlets on each floor. Again, it is desirable to have several separate lighting circuits to avoid a fault on one plunging the whole house into darkness. For protection against electric shock, consumer units are often of the split variety, with an RCD providing earth leakage protection on the socket-outlet circuits; as described in Chapter 12, RCD protection is a requirement in the current edition of BS 7671 for socket-outlet circuits (apart from socket-outlets supplying essential services), for circuits supply bathrooms and shower rooms, and for circuits with cables buried at a depth less than 50 mm in walls.

Some pre-World War II installations may be found with an ironclad splitter switchfuse, usually with four or six rewirable fuses in porcelain carriers, both phase and neutral being fused. This is, of course, no longer permitted. It stems from the time when most small dwellings had few electrical appliances and electricity was mainly used for fixed lighting. Sometimes there were a few BS 546 2 A and 5 A socket-outlets to supply table and standard lamps and perhaps a vacuum cleaner. Subsequently, more appliances were introduced, and the BS 546 outlets were replaced by BS 1363 13 A socket-outlets by the occupier or a 'cowboy' electrician without altering the distribution system. So the socket-outlet wiring will probably be found to be of inadequate size for the potential load, and oversized fuse wire may be found in the fuse carriers. Alternatively, there may be a double pole switch or switchfuse controlling a small distribution board which may be wooden-cased with a glazed front. This is unacceptable because the wood is flammable.

Wiring

The ends of the outgoing circuits at the consumer unit will indicate the type of wiring employed. If these ends are exposed, they must be secured to relieve their terminations of strain, and cable sheaths should terminate inside the enclosure. Entry holes should be bushed. Look out for VIR lead-covered cables. There was a vogue for these in the 1920s and 1930s. The lead sheath was usually used as the protective conductor, and continuity was a problem because the clips used on the sheath for the purpose loosened, owing to the cold flow properties of lead, and corroded. Impurities in the lead and sulphur from the vulcanised rubber could cause perforations in the lead sheaths and, as it aged, the rubber perished. So any such wiring found will be unserviceable and should be discarded. VIR conductors were also used with tough rubber sheaths. With age the rubber hardens and cracks and, if disturbed, will part company with the conductor, thereby no longer being serviceable. VIR cables were also made with a tape binding over the rubber, and then a compounded textile braid was applied over the tape. These cables were installed in metal conduit or ducting and sometimes in wood capping and casing. This type of enclosure is no longer permissible because it is flammable.

VIR and TRS cables were superseded by PVC-insulated and sheathed cables in the 1950s. Any TRS or VIR cables now found will almost certainly have perished insulation, will no longer be serviceable and will need replacing. Screwed steel conduit may be utilised as the protective conductor, but light gauge conduit, too thin to be screwed or with close joints connected by pin grips or slip jointed, cannot be relied on for continuity and is no longer allowed. There were also clamped joint versions which, although not as good as screwed steel conduit, may provide adequate earth continuity, which is ascertainable by testing. Since the introduction of PVC insulation, most domestic wiring has been in two core and earth PVC-insulated and sheathed cables concealed in the structure and often without further protection. Check for the following requirements of BS 7671 for potential infringements:

- The bare protective conductor has to be sleeved, with green/yellow sleeving at terminations where it emerges from the sheath.
- Under floorboards where the cable traverses joists, the cable has to be at least 50 mm from the floorboards to avoid damage from fixing nails or screws.
- Mechanically unprotected cables concealed in walls or partitions at a depth of less than 50 mm have to be protected by a 30 mA RCD and be run within 150 mm of the top of the wall or partition or within 150 mm of the angle formed by two adjoining walls or partitions. Where the cable is connected to a point outside these zones, such as into an accessory, it has to be run in a line either vertically or horizontally. This can be checked with a metal detector.
- There should be no loose wiring. Cables should be secured except where normally inaccessible, such as in roof spaces.
- Any polystyrene thermal insulation found should not be in contact with the cables, as PVC is adversely affected by it.
- Cable sizes, and hence their current-carrying capacity, should be correct for the expected load current. Overcurrent protection should be rated lower than the current-carrying capacity.
- Where cables are run in thermally insulated spaces or in hot locations or grouped with other cables, ensure that the conductor sizes are adequate when derated.
- Ensure that the protective conductor is available at every point.
- Where cables pass through fire barriers, check that the opening is sealed.
- Ensure that segregation of the telecommunication, fire alarm, emergency lighting and extra-low voltage circuits from the rest of the installation circuits has been effected.
- Cables are not subjected to condensation dripping from metal sinks, cold water pipes and tanks or made wet from leaks.
- In damp locations and where condensation occurs, Class II accessories such as switches and socket-outlets are watertight or otherwise designed to prevent a film of moisture providing a current path from a 'live' part inside to the exterior where it may be touched.
- Ensure that there is adequate separation between cables and non-electrical services such as gas installation pipes and equipment so that any work on one service does not adversely affect the other.

Main equipotential bonding

The main equipotential bonding cable sizes are given in BS 7671. For most dwellings, having a service cable neutral conductor not exceeding 35 mm^2 and where the supply is

PME (TN-C-S, or CNE), the minimum size is 10 mm^2. Otherwise it should be not less than half the size of the earthing conductor, with a minimum of 6 mm^2. In practice, 6mm^2 is generally adequate for non-PME supplies. If separate conductors are not used for each point, then a conductor common to several points must be continuous so that it is not broken if detached at one point. Check that all connections are accessible for periodic testing, have the safety label and are not corroded. Connections under sinks are most likely to be affected by leaks or dripping condensation.

Rooms containing a bath or shower basin

For rooms containing a bath or shower basin, check for compliance with Section 701 of BS 7671. There must be no surface wiring in a metallic enclosure, and heaters with touchable elements must be out of reach of anyone using a bath or shower. The only socket-outlets allowed within 3 metres of the zone 1 boundary are for SELV socket-outlets and shaver sockets to BS EN 61558–2–5.

It is important to check that the supplementary equipotential bonding has been properly done, unless all the circuits are protected by a 30 mA RCD. This entails connecting together all simultaneously accessible exposed conductive and extraneous conductive parts by means of an insulated conductor. In effect this means, for example, connecting a metal bath to its hot and cold water and waste pipes. The pipework itself is not regarded as an equipotential conductor for this purpose. Figure 12.3 shows the connections required. As it is unlikely that mechanical protection will be provided for these bonding conductors, the minimum size will be 4 mm^2 in most cases. Although most bathrooms will have wooden floors, bathrooms in flats, basements or on ground floors may have solid floors which are conductive and earthed. The standards do not require such floors to have equipotentially bonded grids, except where electric heating is embedded in the floor. For safety, it is suggested that such floors should have an insulating cover when the bath is in use.

Socket-outlets

Check for the presence of 30 mA RCD protection on socket-outlet circuits and note any circuits that aren't so protected.

Sockets should not be mounted on skirting boards where they are vulnerable to damage from floor cleaners and the like, but should be at a sufficient height above the floor or worktop to enable the plug to be inserted without unduly bending the flexible cord; the modern standard in building standards is for a minimum height above floor level of 450 mm. The connections should be made in a non-flammable enclosure, which usually means a sunken steel box or stud-wall plastic box for flush socket-outlets or a plastic or steel one for the surface type. The sheaths of sheathed cables should terminate in the box, and for surface socket-outlets where the wiring is on the surface, the cables should be secured to relieve the terminations of strain. Box entry holes should be bushed and bare protective conductors sleeved. If conduit is used as the protective conductor, the earthing terminal of the socket-outlet has to be connected to the box earthing terminal by means of a separate conductor. Any exposed external socket-outlets should be weatherproof to IP55.

Switches

Although most switches will be Class II, the protective conductor should be available at the point. Again, the connections should be made within non-flammable enclosures, the sheaths of sheathed cables terminated inside the enclosures, bare protective conductors sleeved and, for surface wiring, the cables secured to relieve their terminations of strain. Check whether the incoming phase terminal has been used for looping instead of at the ceiling rose. This practice is not acceptable for steel conduit systems, as the currents will not be in balance and the magnetic effect will cause a circulating current in the conduit.

Flexible cables

Flexible cables are often used to supply extractor fans, immersion heaters, pumps and controls for the central heating and domestic hot water systems. They are subject to the same requirements as for fixed wiring and where accessible should be secured so as to relieve their terminations of strain. Check the condition of lighting pendants which may have deteriorated from age or been damaged by a hot lampholder.

Downlighters

Ceiling recessed downlighters have caused fires. Samples should be removed to check whether they are fire-rated or there is adequate ventilation space in the ceiling void to dissipate the lamp heat, particularly where there is thermal insulation. Check also that the cables are of adequate size where the luminaire does not have an integral transformer and the extra-low voltage lamps are fed from a separate transformer.

Smoke, heat and carbon monoxide alarms

The presence of smoke, heat and carbon monoxide alarms should be confirmed according the required standards.

The Smoke and Carbon Monoxide Alarm (England) Regulations 2015 require private landlords in England during any period beginning on or after 1st October 2015 to ensure that there is a smoke alarm on each storey of the premises on which there is a room used wholly or partly as living accommodation and a carbon monoxide alarm in any room of the premises which is used wholly or partly as living accommodation and contains a solid fuel (e.g. wood or coal) burning combustion appliance. The landlord must also ensure that each prescribed alarm is in proper working order on the day the tenancy begins if it is a new tenancy. There is no prescription on whether the alarms should be mains or battery-powered.

In Scotland, the repairing standard requires that a house should have satisfactory provision for detecting fires and for giving warning in the event of fire or suspected fire. The government's statutory guidance advises that there should be at least

- one functioning smoke alarm in the room which is frequently used by the occupants for general daytime living purposes;
- one functioning smoke alarm in every circulation space, such as hallways and landings;
- one heat alarm in every kitchen; and
- all alarms should be interlinked.

The alarms should be mains powered with battery back-up.

In order to comply with the repairing standard, carbon monoxide detectors should be installed where there is

- a fixed combustion appliance (excluding cookers) in the dwelling;
- a fixed combustion appliance in an interconnected space (e.g. an integral garage); or
- a combustion appliance necessarily located in a bathroom (advice would be to locate it elsewhere) – the CO detector should be sited outside the room as close to the appliance as possible but allowing for the effect humid air might have on the detector when the bathroom door is open.

In Wales, the housing quality standard requires that there are mains powered battery-backed smoke detectors on each floor. There is no specific mention of carbon monoxide detectors, but clearly landlords would be well advised to have them fitted.

Testing

It is not necessary to carry out all the tests prescribed by BS 7671, unless the limited tests that are carried out indicate significant deficiencies with the installation. Sufficient tests are required, however, to prove the safety or otherwise of the installation at the time of the inspection. The method of testing is described in Chapter 16. The recommended tests are as follows:

(i) Continuity. The continuity of the protective conductors in sample circuits should be checked with an ohmmeter. It is not necessary to test the phase and neutral conductors as any break in them will be self-evident, but all the conductors in the socket-outlet ring circuits should be checked as any break could lead to overloading of the live conductors and a failure of the protective conductor under fault conditions. After testing the protective conductors, it is worth applying the test again to the protected item after reconnecting the conductor. A higher reading indicates a defective joint.

(ii) Insulation. The tests described in Chapter 16 can be curtailed and reduced as follows:

  (a) Unplug portable appliances and isolate fixed equipment, as it is only the wiring installation that is being tested. Open the main switch. Connect the phase busbar to the neutral terminal block. Disconnect or short out any suppressors, electronic apparatus such as a programmer, or lighting dimmer switches, but close other lighting switches. Then apply a 500 V d.c. insulation test between the phase and neutral conductors and the earth terminal block. If the reading is low, remove the temporary connection between the phase busbar and neutral terminal block, open all lighting switches and test between the phase busbar and earth terminal block to determine whether the fault is on the phase or neutral conductors. If on the phase, test each circuit by removing the fuses or opening the circuit breakers in turn. If on the neutral conductors, each circuit will need to be tested by disconnecting them in turn from the neutral terminal block.

  (b) It is not usually necessary to test between the phase and neutral, but if it is done, electronic equipment will need to be disconnected or shorted out and one pole of the bell transformer, indicator lamps and capacitors disconnected and lighting switches opened. A modern installation usually has an insulation value of over 20 megohms. A reading of only 0.5 megohms is required, but low readings of the order of 2 megohms should be investigated.

(iii) Earthing. The connection to earth has to be of sufficiently low resistance for the protective devices to operate within the prescribed time in the event of an earth fault. To test this, an earth loop impedance test should be done. In most cases, a test at the socket-outlet furthest from the consumer unit is sufficient. On TN systems where the supply company provide an earth, readings in the order of 0.3 Ω are usual and sufficiently low for the purpose. On TT systems, where consumers provide their own earth, higher readings are likely, and overcurrent protection for earth faults may not be adequate for fault protection, in which case earth leakage protection is required by means of an RCD. The device can be tested for operation by pressing the test button, but for operating time, test with the appropriate instrument.

Tenant's complaints

Apart from the obvious defects, such as damaged or unserviceable points, sometimes there are intermittent faults which are more of a problem, such as the occasional rupturing of a particular fuse or the tripping of one of the circuit breakers. Questioning the occupier about the circumstances of such outages will often provide answers which point to the possible cause. For example, an upstairs landing light when switched on during or after rain may have caused the relevant fuse to rupture. It was found that there was a roof leak, and on opening the ceiling rose, it was evident that polluted water had entered and caused a flashover between the phase and neutral and/or the protective conductor. After drying out, the circuit functioned normally.

A similar problem occurred in another house, but in this case there was no roof leak, but the upstairs lighting circuit's circuit breaker tripped occasionally when the lighting switch was operated. Again, the ceiling rose showed evidence of flashovers, and it was found that the point was underneath a cold water tank in the loft. Condensation sometimes occurred on the exterior tank walls and base and dripped through the cable hole in the ceiling plaster into the ceiling rose.

In another case, the circuit breaker controlling the downstairs socket-outlet ring main tripped after the bath on the first floor had been used. There was a defective joint in the waste pipe under the bath, and polluted water from it ran down a hollow partition wall nearby and entered a socket-outlet box recessed in the partition, causing tracking across the insulation between the terminals at the back of the socket-outlet and again triggering a flashover when wet.

Outages also occur when there is nothing switched on, during the silent hours for example. These are usually due to defective insulation vulnerable to a flashover triggered by HV or other transients which attain a higher potential when there is no load impedance to attenuate them.

Complaints about tripping out also occur, where the whole installation is controlled by an RCD. There is always some leakage from cooker, kettle and immersion heater elements, which together with leakage from radio interference suppressors can often be enough to cause RCD tripping, which is one reason why installation designers sometimes restrict RCDs to socket-outlet circuits. There are often complaints about short lives of tungsten filament lamps. Where there is any substance in such complaints, it is usually due to an over-high supply voltage. If this exceeds the statutory limits, the remedy is to get the DNO to put it right.

Warning

As a professional person, the engineer carrying out the checks has a duty to warn the tenant of any substantial dangers present and what he or she should do to be safe until the landlord rectifies them. Some dangers, such as exposed live conductors accessible to children, need immediate temporary measures such as making the circuit dead, applying insulation or a barrier. Other items, such as the failure to apply equipotential bonding, are less of a threat and do not demand such immediate attention. Where the engineer considers that the dangers require it, and the landlord will not resolve the matter immediately, he should advise the tenants' solicitor to seek a court injunction for immediate remedial attention.

The report

The report should state the qualifications and experience of the engineer either at the start or in an appendix, as they will be called for in court if the case goes to trial. Then there should be a brief outline of the circumstances requiring the engineer's services. There should be a section on the details of the installation and the defects found, and another for the results of the tests. Then comes a schedule of the BS 7671 infringements, quoting the number of each relevant regulation; attaching a completed EICR form is a good practice because it includes all the information that would be expected, including the comprehensive schedules of inspection and test results. References should also be made to the applicable standard, such as the Decent Homes Standard and HHSRS in England and the Repairing and Tolerable Standards in Scotland.

In the conclusions, the risks of electric shock and/or burn or fire to which the tenant has been exposed by the installation defects found due to the landlord's negligence should be listed. The final section consists of the work needed to remedy the defects. In cases where the installation is obsolete and where it would be unsatisfactory and uneconomical to repair, the engineer should say so and recommend a rewire.

Submitting the report

A solicitor to whom the report is submitted will usually require at least three copies of the report, one of which he will normally send to the landlord to request him to remedy the defects or face a court action. If the report has been properly prepared and it is evident to the landlord that the repairs are necessary, he will usually have them done. If there is any doubt about the remedial work, the solicitor will probably ask the report's author to revisit and check it. In the event of the landlord's refusal to do the repairs, resulting in court action, the engineer may be required to attend the court to give evidence as an expert witness. In this role, his or her responsibility is to the court, not to the client, and he or she needs to assist the court by providing an unbiased opinion.

Index

Note: The numbers in **bold** refer to tables and the numbers in *italics* refer to figures.